IGNEOUS ROCKS: A CLASSIFICATION AND GLOSSARY OF TERMS

Decades of field and microscope studies and more recent quantitative geo-chemical analyses have resulted in a vast, and sometimes overwhelming, array of nomenclature and terminology associated with igneous rocks. Under the auspices of the International Union of Geological Sciences (IUGS), a group of petrologists from around the world has laboured for more than 30 years to collate these terms, gain international agreement on their usage, and reassess the methods by which we categorize and name igneous rocks.

This book presents the results of their work and gives a complete classifi-cation of igneous rocks based on all the recommendations of the IUGS Sub-commission on the Systematics of Igneous Rocks. Revised from the 1st edition (1989), it shows how igneous rocks can be distinguished in the sequence of pyroclastic rocks, carbonatites, melilite-bearing rocks, kalsilite-bearing rocks, kimberlites, lamproites, leucite-bearing rocks, lamprophyres and charnockites. It also demonstrates how the more common plutonic and volcanic rocks that remain can then be categorized using the familiar and widely accepted modal QAPF and chemical TAS classification systems. The glossary of igneous terms has been fully updated since the 1st edition and now includes 1637 entries, of which 316 are recommended by the Subcommission, 312 are regarded as local terms, and 413 are now considered obsolete.

Incorporating a comprehensive list of source references for all the terms included in the glossary, this book will be an indispensable reference guide for all geologists studying igneous rocks, either in the field or the laboratory. It presents a standardized and widely accepted naming scheme that will allow geologists to interpret terminology found in the primary literature and provide formal names for rock samples based on petrographic analyses.

Work on this book started as long ago as 1958 when Albert Streckeisen was asked to collaborate in revising Paul Niggli's well-known book *Tabellen zur Petrographie und zum Gesteinbestimmen* (*Tables for Petrography and Rock Determination*). It was at this point that Streckeisen noted significant problems with all 12 of the classification systems used to identify and name igneous rocks at that time. Rather than propose a 16th system, he chose instead to write a review article outlining the problems inherent in classifying igneous rocks and invited petrologists from around the world to send their comments. In 1970 this lead to the formation of the Subcommission of the Systematics of Igneous Rocks, under the IUGS Commission on Petrology, who published their conclusions in the 1st edition of this book in 1989. The work of this international body has continued to this day, lead by Bruno Zanettin and later by Mike Le Bas. This fully revised 2nd edition has been compiled and edited by Roger Le Maitre, with significant help from a panel of co-contributors.

IGNEOUS ROCKS

A Classification and Glossary of Terms

Recommendations of the
International Union of Geological Sciences
Subcommission on the Systematics of Igneous Rocks

R.W. LE MAITRE (EDITOR), A. STRECKEISEN, B. ZANETTIN,
M.J. LE BAS, B. BONIN, P. BATEMAN, G. BELLIENI, A. DUDEK,
S. EFREMOVA, J. KELLER, J. LAMEYRE, P.A. SABINE,
R. SCHMID, H. SØRENSEN, A.R. WOOLLEY

PUBLISHED BY THE PRESS SYNDICATE OF THE UNIVERSITY OF CAMBRIDGE
The Pitt Building, Trumpington Street, Cambridge, United Kingdom

CAMBRIDGE UNIVERSITY PRESS
The Edinburgh Building, Cambridge CB2 2RU, UK
40 West 20th Street, New York NY 10011–4211, USA
477 Williamstown Road, Port Melbourne, VIC 3207, Australia
Ruiz de Alarcón 13, 28014 Madrid, Spain
Dock House, The Waterfront, Cape Town 8001, South Africa

http://www.cambridge.org

First published 2002
Reprinted 2003
First paperback edition 2004

Typeface Times 11/14 pt   *System* LATEX 2$_\varepsilon$ [TB]

A catalogue record for this book is available from the British Library

Library of Congress Cataloguing in Publication data

Igneous rocks : IUGS classification and glossary : recommendations of the International
Union of Geological Sciences, Subcommission on the Systematics of Igneous Rocks /
R.W. Le Maitre (editor) ... [et al.]. – [2nd ed.].
    p. cm.
Includes bibliographical references and index.
ISBN 0 521 66215 X hardback
1. Rocks, Igneous – Classification. I. Le Maitre, R.W. (Roger Walter) II. International Union
of Geological Sciences. Subcommission on the Systematics of Igneous Rocks.
QE461 .I446 2002
552′.1–dc21   2001052450

ISBN 0 521 66215 X hardback
ISBN 0 521 61948 3 paperback

'The URL (active at the time of going to press) from which to download IUGSTAS is :
http://www.cambridge.org/052166215X'

Contents

Figures

Tables

Albert Streckeisen

8 November 1901 – 29 September 1998

Albert Streckeisen was born on 8 November 1901 in Basel, Switzerland, into an old Basel family. His father Dr Adolf Streckeisen was a Professor in Medicine. Later he studied geology, mineralogy and petrology in Basel, Zürich and Berne under famous teachers like the Professors Buxdorf, Reinhard and Paul Niggli.

In 1927, under the supervision of Prof. Reinhard, he presented his doctoral thesis dealing with the geology and petrology of the Flüela group in the Grisons of Eastern Switzerland.

In the same year, at the age of 26, he took up the position of ordinary Professor in Mineralogy and Petrology at the Polytechnic of Bucharest in Romania. He also became a member of the Romanian Geological Service and was very active in the mapping programme in the Carpathians. In addition to his interests in alpine petrography and structural analysis he became interested the petrography of the interesting and unique nepheline syenite massif of Ditro in Transylvania, on which he published eight papers. This is almost certainly where his interest in the petrographic classification of igneous rocks started.

In the 1930s Albert Streckeisen returned to Switzerland, as to remain professor in Bucharest he would have been forced to change his nationality. He then decided to become a school teacher and taught Natural Sciences in Swiss high schools until his retirement in Berne in 1939. This also enabled him to become an honorary professorial associate at the University of Berne (1942) and to take part in the scientific and teaching life of the Earth Sciences at Berne, where he was nominated extraordinary professor.

Albert Streckeisen – Albert to his many friends in the Commission and the world over – started his work on the classification and systematics of igneous rocks at an age of over 60. This kept him scientifically busy for more than 35 years.

Photographed in Venice 1979

The IUGS asked him to create and lead the then Commission on the Systematics of Magmatic Rocks, that became the IUGS Subcommission on the Systematics of Igneous Rocks when similar groups for Metamorphic and Sedimentary Rocks were formed. This commission, of which Albert Streckeisen was founder and *spiritus rector,* will certainly remain as the "Streckeisen Commission" in the same way and spirit that the QAPF classification will remain the "Streckeisen double triangle".

It is certainly due to his concilient, but determined, firm personality and authority that agreement in his Subcommission on "general recommendations" was achieved. As a determined petrographic observer Albert Streckeisen's heart was with a quantitative modal approach – what could be observed and quantified under the microscope. However, when the explosive development of geochemical analysis provided large chemical data sets for igneous rocks, Albert directed the work of the Subcommission towards a chemical classification of volcanic rocks as expressed in the now generally accepted and adopted TAS diagram. For his devotion and energy, for his achievements in the systematics of igneous rocks he was honoured by the Deutsche Mineralogische Gesellschaft with the Abraham-Gottlob-Werner medal in 1984.

Albert Streckeisen died in Berne aged 97 in October 1998, during the early stages of the preparation of this second edition, in which he made a considerable contribution.

All members of the Subcommission and igneous petrologists worldwide owe Albert Streckeisen an enormous debt of gratitude for his generosity of spirit, his leadership and inspiration, and for his encyclopaedic knowledge of igneous petrology which enabled so much to be achieved.

SELECTED PUBLICATIONS

1964. Zur Klassifikation der Eruptivgesteine. *Neues Jahrbuch für Mineralogie. Stuttgart. Monatshefte.* p.195–222.

1965. Die Klassifikation der Eruptivgesteine. *Geologische Rundschau. Internationale Zeitschrift für Geologie.* Vol.55, p.478–491.

1967. Classification and nomenclature of igneous rocks. Final report of an inquiry. *Neues Jahrbuch für Mineralogie. Stuttgart.* *Abhandlungen.* Vol.107, p.144–240.

1973. Plutonic rocks. Classification and nomenclature recommended by the IUGS Subcommission on the Systematics of Igneous Rocks. *Geotimes.* Vol.18, No.10, p.26–30.

1974. Classification and nomenclature of plutonic Rocks. Recommendations of the IUGS Subcommission on the Systematics of Igneous Rocks. *Geologische Rundschau. Internationale Zeitschrift für Geologie. Stuttgart.* Vol.63, p.773–785.

1976. To each plutonic rock its proper name. *Earth Science Reviews.* Vol.12, p.1–33.

1978. IUGS Subcommission on the Systematics of Igneous Rocks. Classification and nomenclature of volcanic rocks, lamprophyres, carbonatites and melilitic rocks. Recommendations and suggestions. *Neues Jahrbuch für Mineralogie. Stuttgart. Abhandlungen.* Vol.134, p.1–14.

1979. Classification and nomenclature of volcanic rocks, lamprophyres, carbonatites, and melilitic rocks. Recommendations and suggestions of the IUGS Subcommission on the Systematics of Igneous Rocks. *Geology.* Vol.7, p.331–335.

1980. Classification and nomenclature of volcanic rocks, lamprophyres, carbonatites and melilitic rocks. IUGS Subcommission on the Systematics of Igneous Rocks. Recommendations and suggestions. *Geologische Rundschau. Internationale Zeitschrift für Geologie.* Vol.69, p.194–207.

1986. (with Le Bas, Le Maitre & Zanettin) A chemical classification of volcanic rocks based on the total alkali – silica diagram. *Journal of Petrology. Oxford.* Vol.27, p.745-750.

Foreword to 1st edition

In the early summer of 1958 Ernst Niggli asked Theo Hügi and me if we would be willing to collaborate in revising Paul Niggli's well-known book *Tabellen zur Petrographie und zum Gesteinsbestimmen* which had been used as a text for decades at the Federal Polytechnical Institute of Zürich. We agreed and I was placed in charge of the classification and nomenclature of igneous rocks. Quite soon I felt that the scheme used in the Niggli Tables needed careful revision but, as maybe 12 other classification schemes had already been published, Eduard Wenk warned that we should not propose an ominous 13th one; instead he proposed that it would be better to outline the inherent problems of igneous rock classification in an international review article and should present a provisional proposal, asking for comments and replies. This was dangerous advice!

However, the article was written (Streckeisen, 1964), and the consequence was an avalanche of replies, mostly consenting, and many of them with useful suggestions. It thus became clear that the topic was of international interest and that we had to continue. A short report (Streckeisen, 1965) summarized the results of the inquiry. Subsequent discussions with colleagues from various countries led to a detailed proposal (Streckeisen, 1967), which was widely distributed. This was accompanied by a letter from Professor T.F.W. Barth, President of the International Union of Geological Sciences (IUGS), who emphasized the interest in the undertaking, and asked for comments.

The IUGS Commission on Petrology then established a Working Group on Rock Nomenclature, which made arrangements to discuss the nomenclature of magmatic rocks at the International Congress in Prague; the discussion was fixed for 21 August 1968. For this meeting a large amount of documentation was provided; it contained an Account of the previous work, a Report of the Petrographic Committee of the USSR, a Report of the Geological Survey of Canada, and comments from colleagues throughout the world. But political events prevented the intended discussion.

At this stage, Professor T.F.W. Barth, as President of the IUGS, suggested the formation of an International Commission. The Subcommission on the Systematics of Igneous Rocks was formed, under the IUGS Commission on Petrology, to deliberate the various problems of igneous rock nomenclature and to present definite recommendations to the IUGS.

The Subcommission began its work in March 1970. This was done by way of correspondence with subsequent meetings for discussions and to make decisions. Tom Barth suggested beginning with plutonic rocks, as this was easier; his advice was followed.

It was agreed that plutonic rocks should be classified and named according to their modal mineral contents and that the QAPF double triangle should serve for their presentation. A difficulty arose in discussing the nomenclature of granites; the most frequent granites were named quartz-monzonite in America and adamellite in England. With energetic intervention, A.K. and M.K. Wells (Contribution No. 12) advocated that a logical classification would demand that quartz-monzonite in relation to monzonite must have the same status as quartz-syenite to syenite and quartz-diorite to diorite. On this critical point, Paul Bateman made an inquiry concerning this topic among leading American geologists (Contribution

No.21) with the result that 75% of the respondents declared themselves willing to accept quartz-monzonite as a term for field 8*, i.e. for a straight relation to monzonite.

In this relatively short period of time, the Subcommission had, therefore, discussed the problems of plutonic rocks, so that in spring 1972 at the Preliminary Meeting in Berne it made recommendations, which were discussed and, with some modifications, accepted by the Ordinary Meeting in Montreal, in August 1972. A special working group had been set up to discuss the charnockitic rocks and presented its recommendations in 1974.

Work then started on the problems of the classification of volcanic and pyroclastic rocks. The latter was dealt with by a special working group set up under the chairmanship of Rolf Schmid. After much discussion and some lengthy questionnaires this group published a descriptive nomenclature and classification of pyroclastic rocks (Schmid, 1981).

The first problem to be addressed for volcanic rocks was whether their classification and nomenclature should be based on mineralogy or chemistry. Strong arguments were put forward for both solutions. However, crucial points were the fine-grained nature and presence of glass that characterize many volcanic rocks, which means that modal contents can be extremely difficult to obtain. Similarly, the calculation of modes from chemical analyses was considered to be too troublesome or not sufficiently reliable. After long debate the Subcommission decided on the following important principles:

(1) if modes are available, volcanic rocks should be classified and named according to their position in the QAPF diagram

(2) if modes are not available, chemical parameters should be used as a basis for a chemical classification which, however, should be made comparable with the mineralogical QAPF classification.

Various methods of chemical classification were considered and tested by a set of combined modal and chemical analyses. Finally, the Subcommission agreed to use the total alkali – silica (TAS) diagram that Roger Le Maitre had elaborated and correlated with the mineralogical QAPF diagram (Le Maitre, 1984) by using the CLAIR and PETROS databases. After making minor modifications, the TAS diagram was accepted by the Subcommission (Le Bas *et al.*, 1986).

Later on, Mike Le Bas started work on distinguishing the various types of volcanic nephelinitic rocks using normative parameters; similar work is also underway to distinguish the various volcanic leucitic rocks. An effort to classify high-Mg volcanic rocks (picrite, meimechite, komatiite) united Russian and western colleagues at the closing meeting in Copenhagen in 1988.

The intention to compile a glossary of igneous rock names, which should contain recommendations for terms to be abandoned, and definitions of terms to be retained, had already been expressed at the beginning of the undertaking (Streckeisen, 1973, p.27). A first approach was made in October 1977 by a questionnaire which contained a large number of igneous rock terms. Colleagues were asked whether, in their opinion, the terms were of common usage, rarely used today, or almost never been used. More than 200 detailed replies were received, and almost all heartily advocated the publication of a glossary, hoping it would be as comprehensive as possible. At the final stage of our undertaking, Roger Le Maitre has taken over the heavy burden of compiling the glossary, for which he will be thanked by the entire community of geologists.

The work of the Subcommission began with the Congress of Prague and will end with that of Washington, a space of 20 years. During this time, 49 circulars, containing 145 contributions and comments amounting to some 2000 pages, were sent to members and interested colleagues — a huge amount of knowledge, stimulation, ideas and suggestions. Unfortunately only part of this mass of knowledge was able to be incorporated in the final documents. However, all the documents will be deposited in the British Museum (Natural History) in London [Ed.: now called the Natural History Museum], so that they will be available for use in the future.

Within this period of time, a large number of geologists have been collaborating as colleagues, whether as members of the Subcommission or of working groups, as contributors of reports and comments, as guests of meetings, and in other ways: in short, a family of colleagues from many different countries and continents, united in a common aim.

On behalf of the Subcommission, I thank all those colleagues who have helped by giving advice, suggestions, criticisms and objections, and I am grateful for the continual collaboration we have enjoyed.

Albert Streckeisen
Berne, Switzerland

November 1988

Chairman's Preface

This 2nd edition contains the same essentials of the QAPF and TAS classification systems as the 1st edition, but with a few corrections and updates. Bigger changes have been made in the area of the alkaline and related rocks. In the ten years between the 1st and 2nd editions, several Working Groups have been working hard on the kimberlitic, lamproitic, leucitic, melilitic, kalsilitic, lamprophyric and picritic rocks with varied success. The Subcommission thanks each member of the Working Groups for their assiduous and constructive contributions to the revised classifications. Being too numerous to identify individually here, they are named in Appendix A.

The lamproites and kimberlites continue to defy precise classification, but we do now have improved characterizations rather than the definitions normally required to give limits between one rock type and the next. The work of the Subcommission continues, not only to resolve these problematic areas but also to tackle new issues as they arise. With the publication of the 2nd edition, I shall retire from the chair and am pleased to pass the reins over to the capable hands of Prof. Bernard Bonin of Université Paris-Sud.

Particular tribute is due to the late Albert Streckeisen who died during the early stages of preparation of this 2nd edition. His long association with igneous rock nomenclature began in earnest in 1964 when he published a review article in which he evaluated the dozen igneous rock classifications current at that time. That stirred much international interest and produced many enquiries, the result of which was that in 1965 he wrote his "Die Klassifikation der Eruptivgesteine". It established the QAPF as the primary means of classification.

Further discussions led to his 1967 paper written in English which he considered would be the "final report of an enquiry". This plenary study aroused the interest of IUGS and was not the final report he had anticipated. Instead, it led to arrangements being made for a discussion meeting at the 1968 International Congress in Prague, but the Russian invasion prevented that taking place. In its place, IUGS created the Commission of Petrology and its Subcommission on the Systematics of Igneous Rocks, with Streckeisen as the Chairman of both.

The Subcommission began with the plutonic rocks and gave a progress report to the 1972 International Congress in Montreal. This resulted in several papers published in 1973–74, all without the author's name. The two most significant ones were a simplified version in *Geotimes* in 1973 and a fuller account in *Geologische Rundschau* for 1974. This was followed by the definitive 1976 paper "To each plutonic rock its proper name". Such was the demand that he rapidly ran out of reprints.

Recommendations on volcanic rocks swiftly followed in 1978, 1979 and 1980, which were longer and shorter versions of the same recommendations, but in journals reaching different readers.

Now the entire geological community was receiving recommendations on how to name igneous rocks. More followed from Streckeisen's tireless efforts with the Subcommission: pyroclastic rocks, charnockitic rocks, alkaline and other rocks, all classified in numerous papers some written by him as sole author, others with co-authors. By means of patient listening, discussing and careful proposals, he was able to produce consensus, and he was acclaimed "The father of igneous nomenclature and classification". He never owned a computer but produced innumerable spreadsheets of data all laboriously handwritten (and referenced) and then plotted on graph paper, which would be circulated to all members of the Subcommission for discussion. His industry and assiduity were profound. His command of most European languages served him well in finding the best terminology that would stand the test of maintaining meaning during translation. This, he told me, had first been put to the test in 1937 when he was personal assistant to Paul Niggli at Berne University, and under that tutelage had been commissioned to produce a French translation of a lecture that Niggli had given in Paris on petrochemistry. It took, he said, several weeks plus a visit to Paris to attain a satisfactory text.

After 18 years at the helm, he felt in 1980 that he should introduce new leadership to the Subcommission and Bruno Zanettin took over. I followed in 1984. Streckeisen remained a powerful influence on the workings of the Subcommission, offering valuable advice and criticism on the construction of the first edition of this book. He strongly supported the creation of the TAS classification for volcanic rocks (1986). Although no longer Chairman, he continued contributing to the discussions until 1997 when ill-health slowed him down and the stream of authoritative letters ceased. I particularly recall his vigour and valuable advice at an *ad hoc* meeting at EUG95 in Strasbourg, and gladly acknowledge my debt to him for his tutorship since 1972 in the business of naming igneous rocks.

Besides writing up several Swiss geological map sheets for the Survey, he began in the 1980s writing a book *Systematik der Eruptivgesteine* to be published by Springer. He completed some chapters which would "discuss the problem of classification and no-

menclature, and present not only the rules but also the reasons by which we were guided in elaborating our proposals." Having been given the opportunity by Streckeisen to see his critique of the CIPW and other normative analyses and of other classification schemes such as the R1–R2 scheme of De La Roche, it is regrettable that this potentially valuable book never reached publication.

Sincere thanks are also owed to Roger Le Maitre for his skilful and painstaking editing of this 2nd edition. Without his commitment to produce on his Mac all the text, figures and tables ready for print by Cambridge University Press, this book would be vastly more expensive. He has served science well, for which we are all most grateful. I am particularly grateful to Alan Woolley for his exemplary secretaryship during my 17 years as Chairman. His unfailing cheerful outlook, good advice, efficiency and all-round helpfulness were his hallmarks. He has also been instrumental in getting all the papers, reports and circulars put into the archives of the Natural History Museum, London where they may be consulted. A full set has also been deposited by Henning Sørensen in Geologisk Central Institut in Copenhagen. The keen co-operation of Wang Bixiang in producing a Chinese translation of the 1st edition, published in Beijing in 1991, and of Slava Efremova for the Russian edition published in 1997 is also gratefully acknowledged.

I would also like to pay tribute to Jean Lameyre who died in 1992 and who contributed so considerably to the 1st edition.

Mike Le Bas
Chairman, IUGS Subcommission on
the Systematics of Igneous Rocks,
School of Ocean and Earth Science,
University of Southampton, UK

August 2001

Editor's Preface

As the member of the Subcommission once again given the responsibility for compiling and producing this publication, my task in editing the 2nd edition, which has taken well over a year, has been somewhat easier than editing the 1st edition.

This is due to several facts. Firstly, I was not directly involved in any of the working groups; secondly, only minor editing had to be done to the Glossary; and thirdly, improvements in computer technology – in particular e-mail which, with over 450 transmissions, enabled me to obtain quick responses to my editorial queries with colleagues around the world. However, the occasional phone call to speak to another human being also made life more bearable and speeded things up.

This edition has been much easier to produce than the 1st edition, which was produced from photo-ready copy. That involved printing the entire book on a Laserwriter and sending large parcels of paper to the publishers. To produce this edition all I have had to do is to generate PDF (Portable Document Format) files which I have then sent to the publishers by e-mail. The book was then printed directly from the PDF files by the printer – a much simpler task!

The software used to do this included Adobe PageMaker®, for editing the entire text and producing the PDF files; Adobe Illustrator® for producing all the figures and tables; and FileMaker Pro® for maintaining the relational databases of rock descriptions, references, journal names and contributors.

In addition, FileMaker Pro® was scripted to export the information in rich text format (RTF) so that when imported into Adobe PageMaker® the text was italicized, bolded, capitalized etc. in all the right places. The glossary, bibliography and appendices were all produced in this manner. This, of course, saved an enormous amount of editing time and minimized the possibility of errors.

However, since the 1st edition the amount of information has increased considerably, with the main changes, additions and deletions being outlined in the Introduction (see p.1–2). As a result the number of pages has increased from 193 to 236.

In the Glossary an extra 51 rock terms have been added to bring the total number to 1637, of which 316 are recommended by the Subcommission, 413 are regarded as obsolete and 312 are regarded as local terms.

Of the 316 recommended rock names and terms 179 are strictly speaking IUGS root names; 103 are subdivisons of these root names, including 33 specific names for the various "foid" root names, e.g. nepheline syenite; and 34 are rock terms.

The Bibliography has 18 new references bringing the total number of references to 809, and an extra 37 people have contributed to the classification in various ways, bringing the total number of contributors to 456 – from 52 different countries.

To take account of the extra data in the Glossary and the list of references, the statistics given in Chapters 3 and 4 have been completely recalculated. Unfortunately, during this process I discovered that, in some cases, the number of references used in the 1st edition had included some that should not have been present. I apologise for these errors and, after much checking, am now sure that the present numbers are correct.

Without the help of my colleagues this task would not have been possible. My thanks to all those who have helped in the preparation and proof reading of this edition, in particular: Mike Le Bas and Alan Woolley for much guidance, helpful comments and suggestions; Giuliano Bellieni, Bernard Bonin, Arnost Dudek, Jörg Keller, Peter Sabine, Henning Sørensen and Bruno Zanettin for meticulous proofreading, many helpful suggestions and locality checking; in addition Jörg Keller for checking some of the older German references and helping to update the pyroclastic classification; George J. Willauer of Connecticut College for checking on an early American reference; Louise Simpson of the Earth Sciences Information Centre (located via the internet), Natural Resources Canada, for help with an early Annual Report; and Mrs Z.J.X. Frenkiel of the Natural History Museum, London, for help with the Russian references.

I have also been able to include a new Appendix C, with the approval of the Subcommission, giving details of a C++ package called IUGSTAS for determining the TAS name of an analysis. Although this is code has been used for a considerable time by myself and many of my colleagues it has never been generally available until now. As IUGSTAS was developed on a Power Macintosh, I would like to thank John Semmens for making sure that the code also ran on a PC under Windows and for writing the small amount of machine specific code required to allow the user to abort execution at any time – a feature not available with standard C++.

I would also like to thank Susan Francis (my CUP editor) for being extremely helpful in promptly dealing with my many queries with what is not a normal run-of-the-mill book; and Anna Hodson (my copy editor) for patiently explaining the idiosyncrasies of the CUP style (most of which were adopted) and for meticulously correcting my punctuation and grammar.

Finally, I would like to sincerely thank my wife, Vee, for once more putting up with me in editorial mode.

Roger Le Maitre
"Lochiel"
Ross 7209
Tasmania
Australia

August 2001

1 Introduction

This book is the result of over three decades of deliberation by The International Union of Geological Sciences (IUGS) Subcommission on the Systematics of Igneous Rocks.

The Subcommission was originally set up after the International Geological Congress meeting in Prague in 1968 as the result of an earlier investigation into the problems of igneous rock classification that had been undertaken by Professor Albert Streckeisen from 1958 to 1967 (Streckeisen, 1967). He was appointed the first Chairman of the Subcommission, a position he held from 1969 to 1980 and was followed by Bruno Zanettin (1980–1984, Italy), Mike Le Bas (1984–2001, UK) and Bernard Bonin (2001–, France). The secretaries of the Subcommission have been V. Trommsdorff (1970–75, Switzerland), Rolf Schmid (1975–80, Switzerland), Giuliano Bellieni (1980–84, Italy) and Alan Woolley (1984–2001, UK).

During this time the Subcommission has held official meetings in Bern (1972), Montreal (1972), Grenoble (1975), Sydney (1976), Prague (1977), Padova (1979), Paris (1980), Cambridge (1981), Granada (1983), Moscow (1984), London (1985), Freiburg im Breisgau (1986), Copenhagen (1988), Washington D.C. (1989), Southampton (1996) and Prague (1999).

For these meetings the secretaries distributed 52 circulars to the members of the Subcommission containing a total of 164 contributions from petrologists throughout the world. All of these contributions have now been deposited in the Department of Mineralogy at the Natural History Museum in London and in the Library of the Geological Museum, University of Copenhagen.

Records of the Subcommission also indicate that 456 people from 52 countries participated in the formulation of the recommendations in various ways. Of these, 52 were official members of the Subcommission representing their countries at various times; 201 were members of various working groups that were periodically set up to deal with specific problems; 176 corresponded with the Subcommission; and 27 attended meetings as guests. These people are listed in **Appendix A**.

All the recommendations of the Subcommission were published as individual papers as soon as they were agreed upon. However, it was decided, at the 1986 meeting in Freiburg im Breisgau, to present all the results under one cover to make access easier, even though parts of the classification were still unresolved. This resulted in the first edition of this book being published (Le Maitre *et al.*, 1989).

Although the concept of a glossary was mentioned by the Subcommission in 1976, it was not until late 1986 that the work on creating it was started in earnest. The original idea for the glossary was that it should only include those names that were recommended for use by the Subcommission. However, it soon became obvious that for it to be really useful it should be as complete as possible.

1.1 CHANGES TO THE 1ST EDITION

During the last decade a considerable amount of work has been undertaken by the Subcommission to resolve those loose ends left after the publication of the 1st edition. In particular, the Subcommission has had two very active groups working on the problems of the "high-Mg" rocks and on the classification of the lamproites, lamprophyres and kimberlites. Also discussed at length were the classification of the melilite-, kalsilite- and leucite-bearing rocks and the chemical distinction between basanites and nephelinites.

All of these recommendations were approved at the 1999 meeting in Prague, which meant that the Subcommission was in a position to publish a much more comprehensive classification than that presented in the 1st edition.

Hence this book, which is in effect the second edition, although it does have a different title and publisher from the 1st edition.

Apart from minor rewriting and corrections, the main changes to this edition are as follows:

(1) following the Contents is a **List of Figures** on p.vi and a **List of Tables** on p.vii

(2) a change in the **hierarchy of classification** (section 2.1.3, p.6)

(3) a rewrite of the **pyroclastic classification** (section 2.2, p.7) to bring it into line with the latest volcanological terminology

(4) a complete rewrite of the **melilite-bearing rocks** (section 2.4, p.11)

(5) a new section on the **kalsilite-bearing rocks** (section 2.5, p.12)

(6) the replacement of the section on "lamprophyric rocks", which is no longer approved by the Subcommission, with three new individual sections, i.e. **kimberlites** (section 2.6, p.13), **lamproites** (section 2.7, p.16) and **lamprophyres** (section 2.9, p.19). Certain melilite-bearing rocks that were previously included in the lamprophyre classification are now classified under melilite-bearing rocks

(7) a new section on the **leucite-bearing rocks** (section 2.8, p.18)

(8) the section on detecting certain rock types, such as **"high-Mg" rocks**, before using the TAS classification has been rewritten and had **nephelinites** and **melanephelinites** added to it (section 2.12.2, p.34)

(9) the sections dealing with TAS fields **U1** and **F** have been rewritten (section 2.12.2, p.38–39)

(10) the section on **basalts** in TAS (section 2.12.2, p.36) has been expanded

(11) all the **Figures** have been redrawn and the **Tables** redrafted, hopefully for the better

(12) all figures, tables and sections of the book referred to in the glossary are now accompanied by a **page number**

(13) the **statistics** given in Chapters 3 and 4 have been completely **recalculated** in accordance with the extra entries in the glossary. Unfortunately during this process it was discovered that, in some cases, the number of references used in the 1st edition had included some that should not have been present. This has now been corrected

(14) the **glossary** now contains an extra 51 terms giving a total of 1637, of which 316 or 19% have been recommended and defined by the Subcommission and are given in bold capitals in the glossary in Chapter 3. These names are also listed in **Appendix B** at the end of the book for easy reference. The glossary rock **descriptions** have been changed in accordance with recommendations made by the International Mineralogical Association. However, with the **amphiboles** and **pyroxenes** the old names have been retained for historical and other reasons as explained in section 3.1.2 (p.44)

(15) the **bibliography** now contains a total of 809 references, an increase of 18 over the previous edition. The names of terms in square brackets for which the reference is not the prime source are now given in **italics**

(16) the **List of Circulars** (Appendix A in the 1st edition) has been omitted

(17) a new **Appendix C** giving details of a C++ software package **IUGSTAS** to determine the TAS name of an analysis has been added

2 Classification and nomenclature

This chapter is a summary of all the published recommendations of the IUGS Subcommission on the Systematics of Igneous Rocks together with some other decisions agreed to since the last Subcommission meeting in Prague in 1999.

2.1 PRINCIPLES

Throughout its deliberations on the problems of classification the Subcommission has been guided by the following principles, most of which have been detailed by Streckeisen (1973, 1976) and Le Bas & Streckeisen (1991).

(1) For the purposes of classification and nomenclature the term "*igneous rock*" is taken to mean "Massige Gesteine" in the sense of Rosenbusch, which in English can be translated as "igneous or igneous-looking". Igneous rocks may have crystallized from magmas or may have been formed by cumulate, deuteric, metasomatic or metamorphic processes. Arguments as to whether charnockites are igneous or metamorphic rocks are, therefore, irrelevant in this context.

(2) The **primary classification** of igneous rocks should be based on their **mineral content** or **mode**. If a mineral mode is impossible to determine, because of the presence of glass, or because of the fine-grained nature of the rock, then other criteria may be used, e.g. chemical composition, as in the TAS classification.

(3) The term **plutonic rock** is taken to mean an igneous rock with a phaneritic texture, i.e. a relatively coarse-grained (> 3 mm) rock in which the individual crystals can be distinguished with the naked eye and which

is presumed to have formed by slow cooling. Many rocks that occur in orogenic belts have suffered some metamorphic overprinting, so that it is left to the discretion of the user to decide whether to use an igneous or metamorphic term to describe the rock (e.g. whether to use gneissose granite or granitic gneiss).

(4) The term **volcanic rock** is taken to mean an igneous rock with an aphanitic texture, i.e. a relatively fine-grained (< 1 mm) rock in which most of the individual crystals cannot be distinguished with the naked eye and which is presumed to have formed by relatively fast cooling. Such rocks often contain glass.

(5) Rocks should be named according to what they are, and not according to what they might have been. Any manipulation of the raw data used for classification should be justified by the user.

(6) Any useful classification should correspond with natural relationships.

(7) The classification should follow as closely as possible the historical tradition so that well-established terms, e.g. granite, basalt, andesite, are not redefined in a drastically new sense.

(8) The classification should be simple and easy to use.

(9) All official recommendations should be published in English, and any translation or transliteration problems should be solved by members in their individual countries. However, publications by individual Subcommission members, in languages other than English, were encouraged in order to spread the recommendations to as wide an audience as possible.

2.1.1 PARAMETERS USED

The primary modal classifications of plutonic rocks and volcanic rocks are based on the relative proportions of the following mineral groups for which **volume modal data** must be determined:

Q = quartz, tridymite, cristobalite

A = alkali feldspar, including orthoclase, microcline, perthite, anorthoclase, sanidine, and albitic plagioclase (An_0 to An_5)

P = plagioclase (An_5 to An_{100}) and scapolite

F = feldspathoids or **foids** including nepheline, leucite, kalsilite, analcime, sodalite, nosean, haüyne, cancrinite and pseudoleucite.

M = mafic and related minerals, e.g. mica, amphibole, pyroxene, olivine, opaque minerals, accessory minerals (e.g. zircon, apatite, titanite), epidote, allanite, garnet, melilite, monticellite, primary carbonate.

Groups Q, A, P and F comprise the **felsic** minerals, while the minerals of group M are considered to be **mafic** minerals, from the point of view of the modal classifications.

The sum of $Q + A + P + F + M$ must, of course, be 100%. Notice, however, that there can never be more than four non-zero values, as the minerals in groups Q and F are mutually exclusive, i.e. if Q is present, F must be absent, and vice versa.

Where modal data are not available, several parts of the classification utilize chemical data. In these cases all oxide and normative values are in **weight %,** unless otherwise stated. All normative values are based on the rules of the CIPW norm calculation (see p.233).

2.1.2 NOMENCLATURE

During the work of the Subcommission it was quickly realized that the classification schemes would rarely go beyond the stage of assigning a general **root name** to a rock. As such root names are often not specific enough, especially for specialist use, the Subcommission encourages the use of additional qualifiers which may be added to any root name.

These additional qualifiers may be mineral names (e.g. biotite granite), textural terms (e.g. porphyritic granite), chemical terms (e.g. Sr-rich granite), genetic terms (e.g. anatectic granite), tectonic terms (e.g. post-orogenic granite) or any other terms that the user thinks are useful or appropriate. For general guidance on the use of qualifiers the Subcommission makes the following points.

(1) The addition of qualifiers to a root name must not conflict with the definition of the root name. That means that a biotite granite, porphyritic granite, Sr-rich granite, and post-orogenic granite must still be granites in the sense of the classification. Quartz-free granite, however, would not be permissible because the rock could not be classified as a granite, if it contained no quartz.

(2) The user should define what is meant by the qualifiers used if they are not self-explanatory. This applies particularly to geochemical terms, such as Sr-rich or Mg-poor, when often no indications are given of the threshold values above or below which the term is applicable.

(3) If more than one mineral qualifier is used the mineral names should be given in order of increasing abundance (Streckeisen, 1973, p.30; 1976, p.22), e.g. a hornblende-biotite granodiorite should contain more biotite than hornblende. Notice that this is the opposite of the convention often adopted by metamorphic petrologists.

(4) The use of the suffix **-bearing,** as applied to mineral names, has not been consistently defined, as it is used with different threshold values. For example, in the

QAPF classification, 5% Q in Q + A + P is used as the upper limit of the term quartz-bearing, while 10% F in A + P + F is used as the upper limit of the term foid-bearing. The value of 10% is also used for plagioclase-bearing ultramafic rocks (Fig. 2.6, p.25), but for glass-bearing rocks 20% is the upper limit (Table 2.1, p.5).

(5) For volcanic rocks containing glass, the amount of glass should be indicated by using the prefixes shown in Table 2.1 (from Streckeisen, 1978, 1979). For rocks with more than 80% glass special names such as **obsidian**, **pitchstone** etc. are used. Furthermore, for volcanic rocks, which have been named according to their chemistry using the TAS diagram, the presence of glass can be indicated by using the prefix **hyalo-** with the root name, e.g. hyalo-rhyolite, hyalo-andesite etc. For some rocks special names have been given, e.g. **limburgite** = hyalo-nepheline basanite

(6) the prefix **micro-** should be used to indicate that a plutonic rock is finer-grained than usual, rather than giving the rock a special name. The only exceptions to this are the long-established terms **dolerite** and **diabase** (= microgabbro) which may still be used. These two terms are regarded as being synonymous. The use of diabase for Palaeozoic or Precambrian basalts or for altered basalts of any geological age should be avoided.

(7) The prefix **meta-** should be used to indicate that an igneous rock has been metamorphosed, e.g. meta-andesite, meta-basalt etc., but only when the igneous texture is still preserved and the original rock type can be deduced.

(8) Volcanic rocks for which a complete mineral mode cannot be determined, and have not yet been analysed, may be named provisionally following the terminology of Niggli (1931, p.357), by using their visible minerals (usually phenocrysts) to assign a name which is preceded by the prefix **pheno-** (Streckeisen, 1978, p.7; 1979, p.333). Thus a rock containing phenocrysts of sodic plagioclase in a cryptocrystalline matrix may be provisionally called pheno-andesite. Alternatively the provisional "field" classifications could be used (Fig. 2.19, p.39).

(9) The colour index M' is defined (Streckeisen, 1973, p.30; 1976, p.23) as M minus any muscovite, apatite, primary carbonates etc., as muscovite, apatite, and primary carbonates are considered to be colourless minerals for the purpose of the colour index. This enables the terms leucocratic, mesocratic, melanocratic etc. to be defined in terms of the ranges of colour index shown in Table 2.2. Note that these terms are applicable only to rocks and must not be used to describe minerals.

Table 2.1. *Prefixes for use with rocks containing glass*

% glass	Prefix
0 – 20	glass-bearing
20 – 50	glass-rich
50 – 80	glassy

Table 2.2. *Colour index terms*

Colour index term	Range of M'
hololeucocratic	0 – 10
leucocratic	10 – 35
mesocratic	35 – 65
melanocratic	65 – 90
holomelanocratic	90 – 100

2.1.3 USING THE CLASSIFICATION

One of the problems of classifying igneous rocks is that they cannot all be classified sensibly by using only one system. For example, the modal parameters required to adequately define a felsic rock, composed of quartz and feldspars, are very different from those required to define an ultramafic rock, consisting of olivine and pyroxenes. Similarly, lamprophyres have usually been classified as a separate group of rocks. Also modal classifications cannot be applied to rocks which contain glass or are too fine-grained to have their modes determined, so that other criteria, such as chemistry, have to be used in these examples.

As a result several classifications have to be presented, each of which is applicable to a certain group of rocks, e.g. pyroclastic rocks, lamprophyres, plutonic rocks. This, however, means that one has to decide which of the classifications is appropriate for the rock in question. To do this in a consistent manner, so that different petrologists will arrive at the same answer, a hierarchy of classification had to be agreed upon. The basic principle involved in this was that the "special" rock types (e.g. lamprophyres, pyroclastic rocks) must be dealt with first so that anything that was not regarded as a "special" rock type would be classified in either the plutonic or volcanic classifications which, after all, contain the vast majority of igneous rocks. The sequence that should be followed is as follows:

(1) if the rock is considered to be of **pyroclastic origin** go to section 2.2 "Pyroclastic Rocks and Tephra" on p.7

(2) if the rock contains > 50% of **modal carbonate** go to section 2.3 "Carbonatites" on p.10

(3) if the rock contains > 10% of **modal melilite** go to section 2.4 "Melilite-bearing Rocks" on p.11

(4) if the rock contains **modal kalsilite** go to section 2.5 "Kalsilite-bearing Rocks" on p.12

(5) check to see if the rock is a **kimberlite** as described in section 2.6 on p.13

(6) check to see if the rock is a **lamproite** as described in section 2.7 on p.16

(7) if the rock contains **modal leucite** go to section 2.8 "Leucite-bearing Rocks" on p.18

(8) check to see if the rock is a **lamprophyre** as described in section 2.9 on p.19. Note that certain melilite-bearing rocks that were previously included in the lamprophyre classification should now be classified as melilite-bearing rocks

(9) check to see if the rock is a **charnockite** as described in section 2.10 on p.20

(10) if the rock is **plutonic**, as defined in section 2.1, go to section 2.11 "Plutonic rocks" on p.21

(11) if the rock is **volcanic**, as defined in section 2.1, go to section 2.12 "Volcanic rocks" on p.30

(12) if you get to this point, either the rock is not igneous or you have made a mistake.

2.2 PYROCLASTIC ROCKS AND TEPHRA

This classification has been slightly modified from that given in the 1st edition.

It should be used only if the rock is considered to have had a pyroclastic origin, i.e. was formed by fragmentation as a result of explosive volcanic eruptions or processes. It specifically excludes rocks formed by the autobrecciation of lava flows, because the lava flow itself is the direct result of volcanic action, not its brecciation.

The nomenclature and classification is purely descriptive and thus can easily be applied by non-specialists. By defining the term "pyroclast" in a broad sense (see section 2.2.1), the classification can be applied to air fall, flow and surge deposits as well as to lahars, subsurface and vent deposits (e.g. intrusion and extrusion breccias, tuff dykes, diatremes).

When indicating the grain size of a single pyroclast or the middle grain size of an assemblage of pyroclasts the general terms "mean diameter" and "average pyroclast size" are used, without defining them explicitly, as grain size can be expressed in several ways. It is up to the user of this nomenclature to specify the method by which grain size was measured in those examples where it seems necessary to do so.

2.2.1 PYROCLASTS

Pyroclasts are defined as fragments generated by disruption as a direct result of volcanic action.

The fragments may be individual crystals, or crystal, glass or rock fragments. Their shapes acquired during disruption or during subsequent transport to the primary deposit must not have been altered by later redepositional processes. If the fragments have been altered they are called "reworked pyroclasts", or "epiclasts" if their pyroclastic origin is uncertain.

The various types of pyroclasts are mainly distinguished by their size (see Table 2.3, p.9):

Bombs — pyroclasts the mean diameter of which exceeds 64 mm and whose shape or surface (e.g. bread-crust surface) indicates that they were in a wholly or partly molten condition during their formation and subsequent transport.

Blocks — pyroclasts the mean diameter of which exceeds 64 mm and whose angular to subangular shape indicates that they were solid during their formation.

Lapilli — pyroclasts of any shape with a mean diameter of 64 mm to 2 mm

Ash grains — pyroclasts with a mean diameter of less than 2 mm They may be further divided into **coarse ash grains** (2 mm to 1/16 mm) and **fine ash (or dust) grains** (less than 1/16 mm).

2.2.2 PYROCLASTIC DEPOSITS

Pyroclastic deposits are defined as an assemblage of pyroclasts which may be unconsolidated or consolidated. They must contain more than 75% by volume of pyroclasts, the remaining materials generally being of epiclastic, organic, chemical sedimentary or authigenic origin. When they are predominantly consolidated they may be called **pyroclastic rocks** and when predominantly unconsolidated they may be called **tephra**. Table 2.3 shows the nomenclature for tephra and well-sorted pyroclastic rocks.

However, the majority of pyroclastic rocks are polymodal and may be classified according to the proportions of their pyroclasts as shown in Fig. 2.1 as follows:

Agglomerate — a pyroclastic rock in which bombs > 75%.

Pyroclastic breccia — a pyroclastic rock in which blocks > 75%.

Tuff breccia — a pyroclastic rock in which bombs and/or blocks range in amount from 25% to 75%.

Lapilli tuff — a pyroclastic rock in which bombs and/or blocks < 25%, and both lapilli and ash < 75%.

Lapillistone — a pyroclastic rock in which lapilli > 75%.

Tuff or **ash tuff** — a pyroclastic rock in which ash > 75%. These may be further divided into **coarse (ash) tuff** (2 mm to 1/16 mm) and **fine (ash) tuff** (less than 1/16 mm). The fine ash tuff may also be called **dust tuff**. Tuffs and ashes may be further qualified by their fragmental composition, i.e a **lithic tuff** would contain a predominance of rock fragments, a **vitric tuff** a predominance of pumice and glass fragments, and a **crystal tuff** a predominance of crystal fragments.

Any of these terms for pyroclastic deposits may also be further qualified by the use of any other suitable prefix, e.g. air-fall tuff, flow tuff, basaltic lapilli tuff, lacustrine tuff, rhyolitic ash, vent agglomerate etc. The terms may also be replaced by purely genetic terms, such as hyaloclastite or base-surge deposit, whenever it seems appropriate to do so.

2.2.3 MIXED PYROCLASTIC–EPICLASTIC DEPOSITS

For rocks which contain both pyroclastic and normal clastic (epiclastic) material the Subcommission suggests that the general term **tuffites** can be used within the limits given in Table 2.4. Tuffites may be further divided according to their average grain size by the addition of the term "tuffaceous" to the normal sedimentary term, e.g. tuffaceous sandstone.

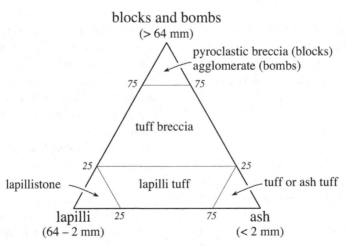

Fig. 2.1. Classification of polymodal pyroclastic rocks based on the proportions of blocks/bombs, lapilli and ash (after Fisher, 1966).

Table 2.3. *Classification and nomenclature of pyroclasts and well-sorted pyroclastic rocks based on clast size*

Average Clast size in mm	Pyroclast	Pyroclastic deposit	
		Mainly unconsolidated: tephra	Mainly consolidated: pyroclastic rock
	bomb, block	agglomerate bed of blocks or bomb, block tephra	agglomerate pyroclastic breccia
64	lapillus	layer, bed of lapilli or lapilli tephra	lapillistone
2	coarse ash grain	coarse ash	coarse (ash) tuff
1/16	fine ash grain (dust grain)	fine ash (dust)	fine (ash) tuff (dust tuff)

Source: After Schmid (1981, Table 1).

Table 2.4. *Terms to be used for mixed pyroclastic–epiclastic rocks*

Average clast size in mm	Pyroclastic		Tuffites (mixed pyroclastic –epiclastic)	Epiclastic (volcanic and/or non-volcanic)
64	agglomerate, pyroclastic breccia		tuffaceous conglomerate, tuffaceous breccia	conglomerate, breccia
2	lapillistone			
1/16	(ash) tuff	coarse	tuffaceous sandstone	sandstone
1/256		fine	tuffaceous siltstone	siltstone
			tuffaceous mudstone, shale	mudstone, shale
Amount of pyroclastic material	100% to 75%		75% to 25%	25% to 0%

Source: After Schmid (1981, Table 2).

2.3 CARBONATITES

This classification should be used only if the rock contains more than 50% modal carbonates (Streckeisen, 1978, 1979). Carbonatites may be either plutonic or volcanic in origin. Mineralogically the following classes of carbonatites may be distinguished:

Calcite-carbonatite — where the main carbonate is calcite. If the rock is coarse-grained it may be called **sövite**; if medium- to fine-grained, **alvikite.**

Dolomite-carbonatite — where the main carbonate is dolomite. This may also be called **beforsite.**

Ferrocarbonatite — where the main carbonate is iron-rich.

Natrocarbonatite — essentially composed of sodium, potassium, and calcium carbonates. At present this unusual rock type is found only at Oldoinyo Lengai volcano in Tanzania.

Qualifications, such as dolomite-bearing, may be used to emphasize the presence of a minor constituent (less than 10%). Similarly, igneous rocks containing less than 10% of carbonate may be called calcite-bearing ijolite, dolomite-bearing peridotite etc., while those with between 10% and 50% carbonate minerals may be called calcitic ijolite or carbonatitic ijolite etc.

If the carbonatite is too fine-grained for an accurate mode to be determined, or if the carbonates are complex Ca–Mg–Fe solid solutions, then the chemical classification shown in Fig. 2.2 can be used for carbonatites with $SiO_2 < 20\%$.

However, if $SiO_2 > 20\%$ the rock is a **silicocarbonatite**. For a more detailed chemical classification of **calciocarbonatites, magnesiocarbonatites** and **ferrocarbonatites** refer to Gittins & Harmer (1997) and Le Bas (1999).

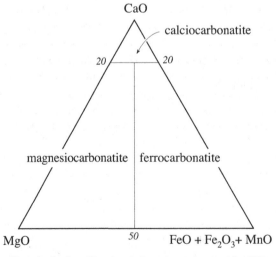

Fig. 2.2. Chemical classification of carbonatites with $SiO_2 < 20\%$ using wt % oxides (Woolley & Kempe, 1989). Carbonatites in which $SiO_2 > 20\%$ are silicocarbonatites.

2.4 MELILITE-BEARING ROCKS

This classification is used for rocks which contain > 10% modal melilite and, if feldspathoids are present, melilite > feldspathoid. The general term for plutonic melilite-bearing rocks is **melilitolite**, and for volcanic melilite-bearing rocks it is **melilitite**. For rocks with > 10% melilite and/or containing **kalsilite** go to section 2.5 "Kalsilite-bearing Rock" on p.12.

2.4.1 MELILITOLITES

The plutonic melilitic rocks, melilitolites, are classified according to their mineral content. Those with melilite < feldspathoid and with feldspathoid > 10% are classified by QAPF as melilite foidolites. However, the majority of melilitolites have M > 90 and may be classified according to their mineral content, e.g. pyroxene melilitolite.

In a recent paper on the classification of melilitic rocks, Dunworth & Bell (1998) suggested that melilitolites with melilite > 65% be termed "ultramelilitolites".

Besides melilite, other principal mineral components include perovskite, olivine, haüyne,

nepheline and pyroxene. If these minerals comprise > 10% of the rock and melilite is < 65% then the following names may be used:
1) if perovskite > 10% then it is an afrikandite
2) if olivine > 10% then it is a kugdite
3) if haüyne > 10% and melitite > haüyne then it is an okaite
4) if nepheline > 10% and melitite > nepheline then it is a turjaite
5) if pyroxene > 10% then it is an uncompahgrite.

If a third mineral is present in amounts greater than 10% then it can be applied as a modifier, e.g. magnetite-pyroxene melilitolite.

2.4.2 MELILITITES

If a mode can be determined, the appropriate name can be obtained from Fig. 2.3. However, if it falls in the foidite field of QAPF (Fig. 2.11, p.31) the name melilite should precede the appropriate foidite name, e.g melilite nephelinite, if the predominant foid is nepheline.

If the mode cannot be determined and a chemical analysis is available, then the TAS classification should be used (see description of field F on p.38).

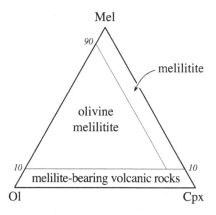

Fig. 2.3. Modal classification of volcanic rocks containing melilite (after Streckeisen, 1978, Fig. 5) based on the values of melilite (Mel), olivine (Ol) and clinopyroxene (Cpx).

2.5 KALSILITE-BEARING ROCKS

The principal minerals of the kalsilite-bearing rocks include clinopyroxene, kalsilite, leucite, melilite, olivine and phlogopite, as shown in Table 2.5. Rocks with kalsilite but no leucite or melilite may be called **kalsilitite**. If the rock is plutonic the term pyroxenite may be more appropriately employed.

The rock types mafurite and katungite, together with the closely associated leucite-bearing rock ugandite (which is excluded from Table 2.5, as it does not contain kalsilite and is more logically classified as an olivine leucitite), are the principal constitutents of the kamafugitic series of Sahama (1974).

From the point of view of the IUGS classifi-cation system, the presence of essential melilite and/or leucite indicates that either the classification of melilite-bearing or leucite-bearing rocks should be applied. However, the presence of kalsilite and leucite is considered petrogenetically so distinctive and important that the accepted term **kamafugite** should be retained for this consanguineous series of rocks. Table 2.6 indicates their nomenclature as a function of mineral assemblage.

Plutonic kalsilite-bearing rocks of the Aldan and North Baikal petrological provinces of Russia, which are not kamafugitic, may be distinguished by the prefix "kalsilite". Thus, the rock type synnyrite becomes kalsilite syenite, and yakutite becomes kalsilite-biotite pyroxenite.

Table 2.5. *Mineral assemblages of kalsilite-bearing volcanic rocks*

Rock	Phlogopite	Clinopyroxene	Leucite	Kalsilite	Melilite	Olivine	Glass
Mafurite	–	×	–	×	–	×	×
Katungite	–	–	×	×	×	×	×
Venanzite	×	×	×	×	×	×	–
Coppaelite	×	×	–	×	×	–	–

× = present; – = absent.
Source: Mitchell & Bergman (1991, Table 2.3).

Table 2.6. *Nomenclature of the kamafugitic rock series*

Historical name	Recommended name
Mafurite	Olivine-pyroxene kalsilitite
Katungite	Kalsilite-leucite-olivine melilitite
Venanzite	Kalsilite-phlogopite-olivine-leucite melilitite
Coppaelite	Kalsilite-phlogopite melilitite
Ugandite	Pyroxene-olivine leucitite

2.6 KIMBERLITES

Kimberlites are currently divided into Group I and Group II (Smith *et al.*, 1985; Skinner, 1989). The **Group I kimberlites** corresponds with archetypal rocks from Kimberley, South Africa, which were formerly termed "basaltic kimberlites" by Wagner (1914). The **Group II kimberlites**, on the other hand, correspond to the micaceous or lamprophyric kimberlites of Wagner (1914).

Petrologists actively studying kimberlites have concluded that there are significant petrological differences between the two groups, although opinion is divided as to the extent of the revisions required to their nomenclature. Some wish to retain the *status quo* (Skinner, 1989), whereas others (e.g. Mitchell, 1986; Mitchell & Bergman, 1991; Mitchell, 1994) believe that the terminology should be completely revised (see below). However, the Subcommission agreed that, because of the mineralogical complexity of the rocks, a single succinct definition cannot be used to describe both rock types, but that characterizations can be given (Woolley *et al.*, 1996).

Following a concept originally developed by Dawson (1980), the rocks may be recognized as containing a characteristic mineral assemblage. The following characterization of Group I kimberlites is after Mitchell (1995) which is based essentially on that of Mitchell (1986, 1994) and evolved from earlier "definitions" given by Clement *et al.* (1984) and Mitchell (1979).

2.6.1 GROUP I KIMBERLITES

Group I Kimberlites are a group of volatile-rich (dominantly CO_2) potassic ultrabasic rocks commonly exhibiting a distinctive inequigranular texture resulting from the presence of macrocrysts (a general term for large crystals, typically 0.5–10 mm diameter) and, in some cases, megacrysts (larger crystals, typically 1–20 cm) set in a fine-grained matrix. The macrocryst–megacryst assemblage, at least some of which are xenocrystic, includes anhedral crystals of olivine, magnesian ilmenite, pyrope, diopside (sometimes subcalcic), phlogopite, enstatite and Ti-poor chromite. Olivine macrocrysts are a characteristic and dominant constituent in all but fractionated kimberlites.

The matrix contains a second generation of primary euhedral-to-subhedral olivine which occurs together with one or more of the following primary minerals: monticellite, phlogopite, perovskite, spinel (magnesian ulvospinel-Mg-chromite-ulvospinel-magnetite solid solutions), apatite, carbonate and serpentine. Many kimberlites contain late-stage poikilitic micas belonging to the barian phlogopite–kinoshitalite series. Nickeliferous sulphides and rutile are common accessory minerals. The replacement of earlier-formed olivine, phlogopite, monticellite and apatite by deuteric serpentine and calcite is common.

Evolved members of Group I may be poor in, or devoid of, macrocrysts, and composed essentially of second-generation olivine, calcite, serpentine and magnetite, together with minor phlogopite, apatite and perovskite.

It is evident that kimberlites are complex hybrid rocks in which the problem of distinguishing the primary constituents from the entrained xenocrysts precludes simple definition. The above characterization attempts to recognize that the composition and mineralogy of kimberlites are not entirely derived from a parent magma, and the non-genetic terms macrocryst and megacryst are used to describe minerals of cryptogenic, i.e. unknown origin.

Macrocrysts include forsteritic olivine, Cr-pyrope, almandine-pyrope, Cr-diopside, magnesian ilmenite and phlogopite crystals, that are now generally believed to originate by the disaggregation of mantle-derived lherzolite, harzburgite, eclogite and metasomatized peridotite xenoliths. Most diamonds, which are excluded from the above "definition", belong to this suite of minerals but are much less common.

Megacrysts are dominated by magnesian ilmenite, Ti-pyrope, diopside, olivine and enstatite that have relatively Cr-poor compositions ($< 2\%$ Cr_2O_3). The origin of the megacrysts is still being debated (e.g. Mitchell, 1986), and some petrologists believe that they may be cognate.

Both of these suites of minerals are included in the characterization because of their common presence in kimberlites. It can be debated whether reference to these characteristic constituents should be removed from the "definition" of kimberlite. Strictly, minerals which are known to be xenocrysts should not be included in a petrological definition, as they have not crystallized from the parental magma.

Smaller grains of both the macrocryst and megacryst suite minerals also occur but may be easily distinguished on the basis of their compositions. In this respect, it is important to distinguish pseudoprimary groundmass diopside from macrocrystic or megacrystic clinopyroxene. Group I kimberlites do not usually contain the former except as a product of crystallization induced by the assimilation of siliceous xenoliths (Scott Smith *et al.*, 1983). The primary nature of groundmass serpophitic serpentine was originally recognized by Mitchell & Putnis (1988).

2.6.2 GROUP II KIMBERLITES

Recent studies (Smith *et al.*, 1985; Skinner, 1989; Mitchell, 1994, 1995; Tainton & Browning, 1991) have demonstrated that Group I and Group II kimberlites are mineralogically different and petrogenetically separate rock-types.

A definition of Group II kimberlites has not yet been agreed as they have been insufficiently studied. Mitchell (1986, 1994, 1995) has suggested that these rocks are not kimberlitic at all, and should be termed "orangeite", in recognition of their distinct character and unique occurrence in the Orange Free State of South Africa. Wagner (1928) previously suggested that the rocks which he initially termed micaceous kimberlite (Wagner, 1914) be renamed "orangite" (*sic*). The following characterization of the rocks currently described as Group II kimberlites or micaceous kimberlites follows that of Mitchell (1995).

Group II kimberlites (or orangeites) belong to a clan of ultrapotassic, peralkaline volatile-rich (dominantly H_2O) rocks, characterized by phlogopite macrocrysts and microphenocrysts, together with groundmass micas which vary in composition from "tetraferriphlogopite" to phlogopite. Rounded macrocrysts of olivine and euhedral primary crystals of olivine are common, but are not invariably major constituents.

Characteristic primary phases in the groundmass include: diopside, commonly zoned to, and mantled by, titanian aegirine; spinels ranging in composition from Mg-bearing chromite to Ti-bearing magnetite; Sr- and REE-rich perovskite; Sr-rich apatite; REE-rich phosphates (monazite, daqingshanite); potassian barian titanates belonging to the hollandite group; potassium triskaidecatitanates ($K_2Ti_{13}O_{27}$); Nb-bearing rutile and Mn-bearing ilmenite. These are set in a mesostasis that may contain calcite, dolomite, ancylite and other rare-earth carbonates, witherite, norsethite and serpentine.

Evolved members of the group contain groundmass sanidine and potassium richterite. Zirconium silicates (wadeite, zircon, kimzeyitic garnet, Ca-Zr-silicate) may occur as late-stage groundmass minerals. Barite is a common deuteric secondary mineral.

Note that these rocks have a greater mineralogical affinity to lamproites than to Group I kimberlites. However, there are significant differences in the compositions and overall assemblage of minerals, as detailed above, to permit their discrimination from lamproites (Mitchell 1994, 1995).

2.7 LAMPROITES

The lamproite classification system described by Mitchell & Bergman (1991) is recommended and involves both mineralogical and geochemical criteria.

2.7.1 MINERALOGICAL CRITERIA

Lamproites normally occur as dykes or small extrusions. Mineralogically they are characterized by the presence of widely varying amounts (5 – 90 vol %) of the following primary phases:

(1) titanian, Al-poor phenocrystic phlogopite (TiO_2 2% – 10%; Al_2O_3 5% – 12%)
(2) groundmass poikilitic titanian "tetraferriphlogopite" (TiO_2 5% – 10%)
(3) titanian potassium richterite (TiO_2 3% – 5%; K_2O 4% – 6%)
(4) forsteritic olivine
(5) Al-poor, Na-poor diopside (Al_2O_3 < 1%; Na_2O < 1%)
(6) non-stoichiometric iron-rich leucite (Fe_2O_3 1% – 4%)
(7) Fe-rich sanidine (typically Fe_2O_3 1% – 5%).

The presence of all the above phases is not required in order to classify a rock as a lamproite. Any one mineral may be dominant and this, together with the two or three other major minerals present, suffices to determine the petrographic name.

Minor and common accessory phases include priderite, wadeite, apatite, perovskite, magnesiochromite, titanian magnesiochromite and magnesian titaniferous magnetite with less commonly, but characteristically, jeppeite, armalcolite, shcherbakovite, ilmenite and enstatite.

The presence of the following minerals **precludes** a rock from being classified as a lamproite: primary plagioclase, melilite, monticellite, kalsilite, nepheline, Na-rich alkali feldspar, sodalite, nosean, haüyne, melanite, schorlomite or kimzeyite.

2.7.2 CHEMICAL CRITERIA

Lamproites conform to the following chemical characteristics:

(1) molar K_2O / Na_2O > 3, i.e. they are **ultrapotassic**
(2) molar K_2O / Al_2O_3 > 0.8 and often > 1
(3) molar $(K_2O + Na_2O)$ / Al_2O_3 typically > 1, i.e they are **peralkaline**
(4) typically FeO and CaO are both < 10%, TiO_2 1% – 7% , Ba > 2000 ppm (commonly > 5000 ppm), Sr > 1000 ppm, Zr > 500 ppm and La > 200 ppm.

2.7.3 NOMENCLATURE

The subdivision of the lamproites should follow the scheme of Mitchell & Bergman (1991), in which the historical terminology is discarded in favour of compound names based on the predominance of phlogopite, richterite, olivine, diopside, sanidine and leucite, as given in Table 2.7. It should be noted that the term "madupitic" in Table 2.7 indicates that the rock contains poikilitic groundmass phlogopite, as opposed to phlogopite lamproite in which phlogopite occurs as phenocrysts.

The complex compositional and mineralogical criteria required to define lamproites result from the diverse conditions involved in their genesis, compared with those of rocks that can be readily classified using the IUGS system. The main petrogenetic factors contributing to the complexity of composition and mineralogy of lamproites are the variable nature of their metasomatized source regions in the mantle, depth and extent of partial melting, coupled with their common extensive differentiation.

Table 2.7. *Nomenclature of lamproites*

Historical name	Recommended name
Wyomingite	Diopside-leucite-phlogopite lamproite
Orendite	Diopside-sanidine-phlogopite lamproite
Madupite	Diopside madupitic[a] lamproite
Cedricite	Diopside-leucite lamproite
Mamilite	Leucite-richterite lamproite
Wolgidite	Diopside-leucite-richterite madupitic[a] lamproite
Fitzroyite	Leucite-phlogopite lamproite
Verite	Hyalo-olivine-diopside-phlogopite lamproite
Jumillite	Olivine-diopside-richterite madupitic[a] lamproite
Fortunite	Hyalo-enstatite-phlogopite lamproite
Cancalite	Enstatite-sanidine-phlogopite lamproite

[a] Madupitic = containing poikilitic groundmass phlogopite.

2.8 LEUCITE-BEARING ROCKS

The leucite-bearing rocks, after the elimination of the lamproites and kamafugites, should be named according to the volcanic QAPF diagram (Fig. 2.11, p.31) with the prefix "leucite" or "leucite-bearing" as appropriate. Rocks containing little or no feldspar, i.e. falling into field 15 (foidite), are **leucitite**, which is divided into three subfields (shown in Fig. 2.12, p.32):

(1) QAPF subfield 15a, **phonolitic leucitite** in which foids are 60–90% of the light-coloured constituents and alkali feldspar > plagioclase.

(2) QAPF subfield 15b, **tephritic leucitite** in which foids are 60–90% of the light-coloured constituents and plagioclase > alkali feldspar

(3) QAPF subfield 15c, **leucitite** *sensu stricto* in which foids are 90–100% of the light-coloured constituents and leucite is practically the sole feldspathoid.

The essential mineralogy of the principal leucite-bearing rocks is given in Table 2.8.

No unambiguous chemical criteria have been found to distinguish this group of rocks. On TAS (Fig. 2.14, p.35), leucitites extend significantly beyond the foidite field into adjacent fields (see Le Bas *et al.*, 1992, Fig. 23). They are better distinguished from lamproites by other compositional parameters, although even here some overlap occurs. The chemical characteristics of the potassic rocks and attempts at distinguishing lamproites from certain leucite-bearing rocks, using a variety of criteria, are explored by Foley *et al.* (1987) and Mitchell & Bergman (1991).

Table 2.8. *Mineralogy of the principal groups of leucite-bearing volcanic rocks* [a]

Rock	Clinopyroxene	Leucite	Plagioclase	Sanidine [b]	Olivine
Leucitite	×	×	–	–	> 10%
Tephritic leucitite	×	×	plagioclase > sanidine		×
Phonolitic leucitite	×	×	plagioclase < sanidine		×
Leucite tephrite	×	×	×	–	< 10%
Leucite basanite	×	×	×	–	> 10%
Leucite phonolite	×	×	–	×	–

× = present; – = absent.

[a] These rocks may also contain some nepheline.

[b] Includes products of its exsolution.

2.9 LAMPROPHYRES

Lamprophyres are a diverse group of rocks that chemically cannot be separated easily from other normal igneous rocks. Traditionally they have been distinguished on the following characteristics:

(1) they normally occur as dykes and are not simply textural varieties of common plutonic or volcanic rocks

(2) they are porphyritic, mesocratic to melanocratic ($M' = 35-90$) but rarely holomelanocratic ($M' > 90$)

(3) feldspars and/or feldspathoids, when present, are restricted to the groundmass

(4) they usually contain essential biotite (or Fe-phlogopite) and/or amphibole and sometimes clinopyroxene

(5) hydrothermal alteration of olivine, pyroxene, biotite, and plagioclase, when present, is common

(6) calcite, zeolites, and other hydrothermal minerals may appear as primary phases

(7) they tend to have contents of K_2O and/or Na_2O, H_2O, CO_2, S, P_2O_5 and Ba that are relatively high compared with other rocks of similar composition.

The Subcommission no longer endorses the terms "lamprophyric rocks", or "lamprophyre clan", as used by Le Maitre *et al.* (1989) and Rock (1991) to encompass lamprophyres, lamproites and kimberlites, because lamproites and kimberlites are best considered independently of lamprophyres.

The recommended mineralogical classification of these rocks is given Table 2.9.

Table 2.9. *Classification and nomenclature of lamprophyres based on their mineralogy*

Light-coloured constituents		Predominant mafic minerals		
feldspar	foid	biotite > hornblende, ±diopsidic augite, (±olivine)	hornblende, diopsidic augite, ±olivine	brown amphibole, Ti-augite, olivine, biotite
or > pl	–	minette	vogesite	–
pl > or	–	kersantite	spessartite	–
or > pl	feld > foid	–	–	sannaite
pl > or	feld > foid	–	–	camptonite
–	glass or foid	–	–	monchiquite

or = alkali feldspar; pl = plagioclase; feld = feldspar; foid = feldspathoid.
Source: Modified from Streckeisen (1978, p.11).
Note: Alnöite and polzenite are no longer in the lamprophyre classification and rocks of this type should now be named according to the melilite-bearing rock classification (p.11).

2.10 CHARNOCKITIC ROCKS

This classification should be used only if the rock is considered to belong to the charnockitic suite of rocks, which is characterized by the presence of orthopyroxene (or fayalite plus quartz) and, in many of the rocks, perthite, mesoperthite or antiperthite (Streckeisen, 1974, 1976). They are often associated with norites and anorthosites and are closely linked with Precambrian terranes.

Although many show signs of metamorphic overprinting, such as deformation and recrystallization, they conform to the group of "igneous and igneous-looking rocks" and have, therefore, been included in the classification scheme.

The classification is based on the QAP triangle, i.e. the upper half of the QAPF double triangle (Fig. 2.4, p.22). The general names for the various fields are given in Table 2.10, together with a number of special names that may be applied to certain fields.

However, as one of the characteristics of charnockites is the presence of various types of perthite, this raises the common problem of how to distribute the perthites between A and P. The Subcommission has, therefore, recommended that in charnockitic rocks the perthitic feldspars should be distributed between A and P in the following way:

Perthite — assign to A as the major component is alkali feldspar.

Mesoperthite — assign equally between A and P as the amounts of the alkali feldspar and plagioclase (usually oligoclase or andesine) components are roughly the same.

Antiperthite — assign to P as the major component is andesine with minor albite as the alkali feldspar phase.

To distinguish those charnockitic rocks that contain mesoperthite it is suggested that the prefix m-, being short for mesoperthite, could be used, e.g. **m-charnockite**.

Table 2.10. *Nomenclature of charnockitic rocks*

QAPF field	General name	Special name
2	orthopyroxene alkali feldspar granite	alkali feldspar charnockite
3	orthopyroxene granite	charnockite
4	orthopyroxene granodiorite	opdalite or charno-enderbite
5	orthopyroxene tonalite	enderbite
6	orthopyroxene alkali feldspar syenite	–
7	orthopyroxene syenite	–
8	orthopyroxene monzonite	mangerite
9	monzonorite (orthopyroxene monzodiorite)	jotunite
10	norite (orthopyroxene diorite), anorthosite (M < 10)	–

Source: Modified from Streckeisen (1974, p.355).

2.11 PLUTONIC ROCKS

This classification should be used only if the rock is considered to be plutonic, i.e. it is assumed to have formed by slow cooling and has a relatively coarse-grained (> 3 mm) texture in which the individual crystals can easily be seen with the naked eye.

There is, of course, a gradation between plutonic rocks and volcanic rocks and the Subcommission suggests that, if there is any uncertainty as to which classification to use, the plutonic root name should be given and prefixed with the term "micro". For example, microsyenite could be used for a rock that was considered to have formed at considerable depth even if many of the individual crystals could not be seen with the naked eye.

The classification is based on modal parameters and is divided into three parts:

(1) if M is less than 90% the rock is classified according to its felsic minerals, using the now familiar QAPF diagram (Fig. 2.4), often simply referred to as the QAPF classification or the QAPF double triangle (section 2.11.1)

(2) if M is greater or equal to 90%, it is an **ultramafic rock** and is classified according to its mafic minerals, as shown in section 2.11.2, p.28

(3) if a mineral mode is not yet available, the "field" classification of section 2.11.3, p.29, may be used provisionally.

2.11.1 PLUTONIC QAPF CLASSIFICATION (M < 90%)

The modal classification of plutonic rocks is based on the QAPF diagram and was the first to be completed and recommended by the IUGS Subcommission (Streckeisen, 1973, 1976). The diagram is based on the fundamental work of many earlier petrologists, which is fully summarized by Streckeisen (1967).

The root names for the classification are given in Fig. 2.4 and the field numbers in Fig. 2.5.

To use the classification the modal amounts of **Q, A, P, and F must be known** and **recalculated so that their sum is 100%.**

For example, a rock with Q = 10%, A = 30%, P = 20%, and M = 40% would give recalculated values of Q, A, and P as follows:

$$Q = 100 \times 10 / 60 = 16.7$$
$$A = 100 \times 30 / 60 = 50.0$$
$$P = 100 \times 20 / 60 = 33.3$$

Although at this stage the rock can be plotted directly into the triangular diagram, if all that is required is to name the rock it is easier to determine the plagioclase ratio where:

plagioclase ratio $= 100 \times P / (A + P)$

as the non-horizontal divisions in the QAPF diagram are lines of constant plagioclase ratio. The field into which the rock falls can then easily be determined by inspection.

In the above example rock the plagioclase ratio is 40 so that it can be seen by inspection that the rock falls into QAPF field 8* (Fig. 2.5) and should, therefore, be called a quartz monzonite (Fig. 2.4).

Similarly, a rock with A = 50%, P = 5%, F = 30%, and M = 15% would recalculate as follows:

$$A = 100 \times 50 / 85 = 58.8$$
$$P = 100 \times 5 / 85 = 5.9$$
$$F = 100 \times 30 / 85 = 35.3$$
$$\text{Plagioclase ratio} = 9$$

This rock falls into QAPF field 11 and should, therefore, be called a foid syenite. Furthermore, if the major foid in the rock is nepheline, it should be called a nepheline syenite.

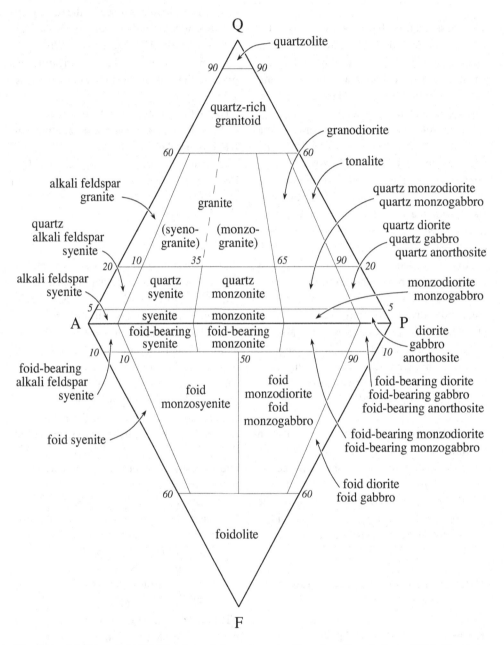

Fig. 2.4. QAPF modal classification of plutonic rocks (based on Streckeisen, 1976, Fig. 1a).
The corners of the double triangle are Q = quartz, A = alkali feldspar, P = plagioclase and F
= feldspathoid. This diagram must not be used for rocks in which the mafic mineral content,
M, is greater than 90%.

The location of the numerical QAPF fields are shown in Fig. 2.5.

Field 2 (alkali feldspar granite) — rocks in this field have been called alkali granite by many authors. The Subcommission, however, recommends that the term **peralkaline granite** be used instead for those rocks that contain sodic amphiboles and/or sodic pyroxenes. The term **alaskite** may be used for a light-coloured (M < 10%) alkali feldspar granite.

Field 3 (granite) — the term granite has been used in many senses; in most English and American textbooks it has been restricted to

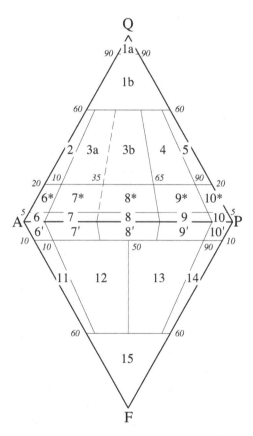

Fig. 2.5. QAPF field numbers (Streckeisen, 1976, Fig. 1a). The fields 6* to 10* are slightly silica oversaturated variants of fields 6 to 10, respectively, while 6' to 10' are slightly silica undersaturated variants.

subfield 3a, whereas subfield 3b has contained terms such as adamellite and quartz monzonite. In the European literature, however, granite has been used to cover both subfields, a view adopted by the Subcommission. The Subcommission has also recommended that the term adamellite should no longer be used, as it has been given several meanings, and does not even occur in the Adamello Massif as commonly defined (Streckeisen, 1976). Although the term quartz monzonite has also been used with several meanings, the Subcommission decided to retain the term in its original sense, i.e. for rocks in field 8*.

Field 4 (granodiorite) — the most widespread rocks in this field are granodiorites, commonly containing oligoclase, more rarely andesine. It seems advisable to add the condition that the average composition of the plagioclase should be $An_0 - An_{50}$ in order to distinguish the common granodiorites from the rare granogabbro in which the plagioclase is $An_{50} - An_{100}$.

Field 5 (tonalite) — the root name tonalite should be used whether hornblende is present or not. **Trondhjemite** and **plagiogranite** (as used by Russian petrologists) may be used for a light-coloured (M < 10%) tonalite.

Field 6′ (alkali feldpar nepheline syenite) — the general term **agpaite** may be used for peralkaline varieties characterized by complex Zr and Ti minerals, such as eudialyte, rather than simple minerals such as zircon and ilmenite.

Field 8 (monzonite) — many so-called "syenites" fall into this field.

Field 9 (monzodiorite, monzogabbro) — the two root names in this field are separated according to the average composition of their plagioclase – monzodiorite (plagioclase $An_0 - An_{50}$), monzogabbro (plagioclase $An_{50} - An_{100}$). The terms **syenodiorite** and **syenogabbro** may be used as comprehensive names for rocks between syenite and diorite/gabbro, i.e. for monzonites (field 8) and monzodiorite/monzogabbro, respectively.

Field 10 (diorite, gabbro, anorthosite) — the three root names in this field are separated according to the colour index and the average composition of their plagioclase – anorthosite (M < 10%), diorite (M > 10%, plagioclase An_0 – An_{50}), gabbro (M > 10%, plagioclase An_{50} – An_{100}). Gabbros may be further subdivided, as shown below. Either of the two synonymous terms **dolerite** or **diabase** may be used for medium-grained gabbros rather than the term microgabbro, if required.

Gabbroic rocks — the gabbros (*sensu lato*) of QAPF field 10, may be further subdivided according to the relative abundances of their orthopyroxene, clinopyroxene, olivine and hornblende as shown in Fig. 2.6. Some of the special terms used are:

Gabbro (*sensu stricto*) = plagioclase and clinopyroxene

Norite = plagioclase and orthopyroxene

Troctolite = plagioclase and olivine

Gabbronorite = plagioclase with almost equal amounts of clinopyroxene and orthopyroxene

Orthopyroxene gabbro = plagioclase and clinopyroxene with minor amounts of orthopyroxene

Clinopyroxene norite = plagioclase and orthopyroxene with minor amounts of clinopyroxene

Hornblende gabbro = plagioclase and hornblende with pyroxene < 5%.

Field 11 (foid syenite) — although foid syenite is the root name, the most abundant foid present should be used in the name, e.g. **nepheline syenite, sodalite syenite**.

Field 12 (foid monzosyenite) — the root name foid monzosyenite may be replaced by the synonym **foid plagisyenite**. Wherever possible, replace the term foid with the name of the most abundant feldspathoid. **Miaskite**, which contains oligoclase, may also be used.

Field 13 (foid monzodiorite, foid monzogabbro) — the two root names in this field are separated according to the average composition of their plagioclase, i.e. foid monzodiorite (plagioclase An_0 – An_{50}), foid monzogabbro (plagioclase An_{50} – An_{100}). Wherever possible, replace the term foid with the name of the most abundant feldspathoid. The term **essexite** may be applied to nepheline monzodiorite or nepheline monzogabbro.

Field 14 (foid diorite, foid gabbro) — again the two root names in this field are separated according to the average composition of their plagioclase, i.e. foid diorite (plagioclase An_0 – An_{50}), foid gabbro (plagioclase An_{50} – An_{100}). Wherever possible, replace the term foid with the name of the most abundant feldspathoid. Two special terms may continue to be used, **theralite** for nepheline gabbro and **teschenite** for **analcime gabbro**.

Field 15 (foidolite) — this field contains rocks in which the light-coloured minerals are almost entirely foids and is given the root name **foidolite** to distinguish it from the volcanic equivalent which is called foidite. As these rocks are rather rare the field has not been subdivided. Again note that the most abundant foid should appear in the name, e.g. **nephelinolite** (urtite, ijolite, melteigite).

Leuco- and Mela- variants — for rocks in the QAPF classification the Subcommission suggests (Streckeisen, 1973, p.30; 1976, p.24) that the prefixes **leuco-** and **mela-** may be used to designate the more felsic (lower colour index) and mafic (higher colour index) types within each rock group, when compared with the "normal" types in that group. As the threshold values of M' varies from rock group to rock group, the limits are shown diagrammatically in Fig. 2.7 and Fig. 2.8 for the rock groups to which the terms may be applied. The prefixes should precede the root name, e.g. biotite leucogranite, biotite melasyenite.

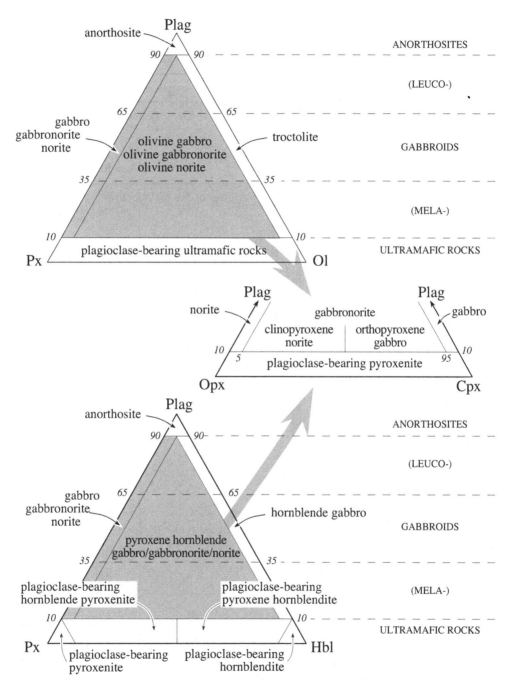

Fig. 2.6. Modal classification of gabbroic rocks based on the proportions of plagioclase (Plag), pyroxene (Px), olivine (Ol), orthopyroxene (Opx), clinopyroxene (Cpx), and hornblende (Hbl) (after Streckeisen, 1976, Fig. 3). Rocks falling in the shaded areas of either triangular diagram may be further subdivided according to the diagram pointed to by the arrows.

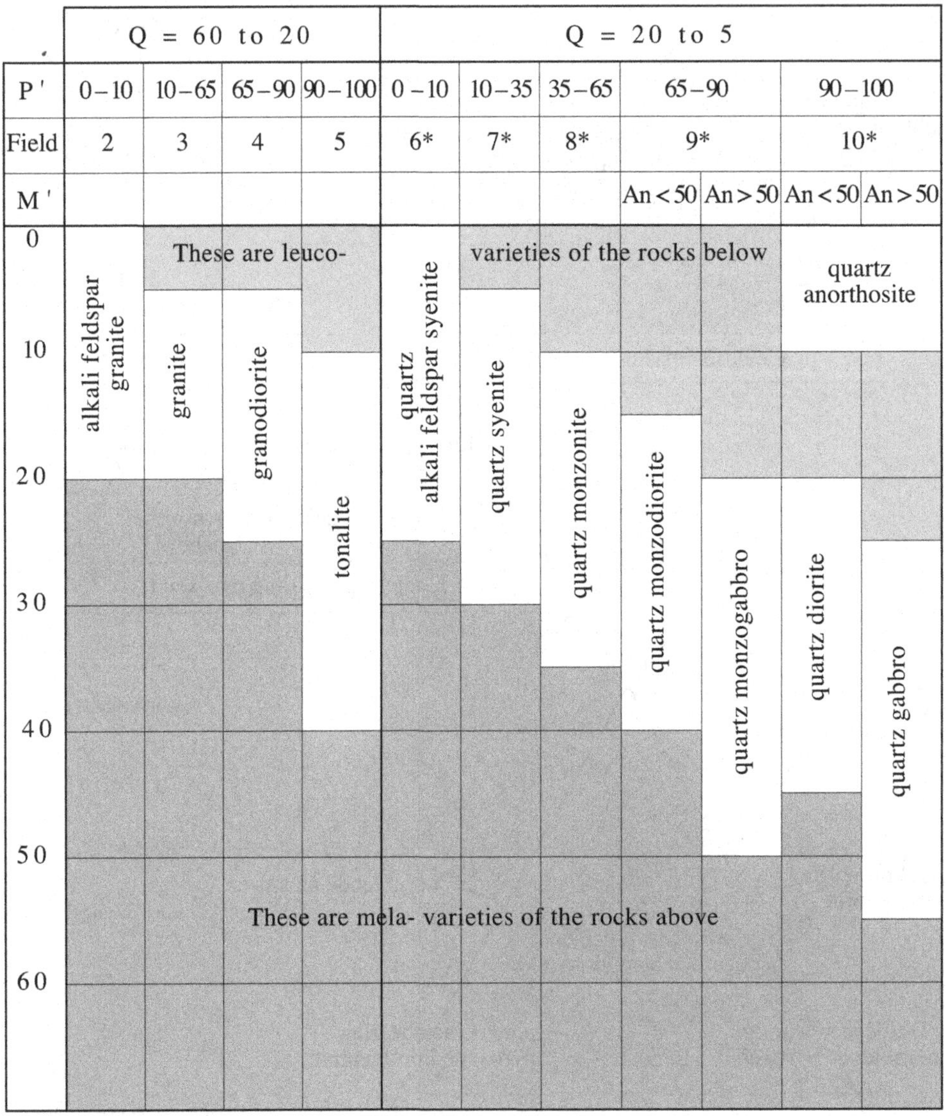

P '	Q = 60 to 20				Q = 20 to 5				
P '	0–10	10–65	65–90	90–100	0–10	10–35	35–65	65–90	90–100
Field	2	3	4	5	6*	7*	8*	9*	10*
M '							An<50 | An>50	An<50 | An>50

These are leuco- varieties of the rocks below

alkali feldspar granite — granite — granodiorite — tonalite — quartz alkali feldspar syenite — quartz syenite — quartz monzonite — quartz monzodiorite — quartz monzogabbro — quartz diorite — quartz gabbro — quartz anorthosite

(M' scale: 0, 10, 20, 30, 40, 50, 60)

These are mela- varieties of the rocks above

Fig. 2.7. Use of the terms mela- and leuco- with QAPF plutonic rocks with Q > 5% (after Streckeisen, 1976, Fig. 5). Abbreviations: P' = 100 * P / (A + P); M' = colour index; An = anorthite content of plagioclase.

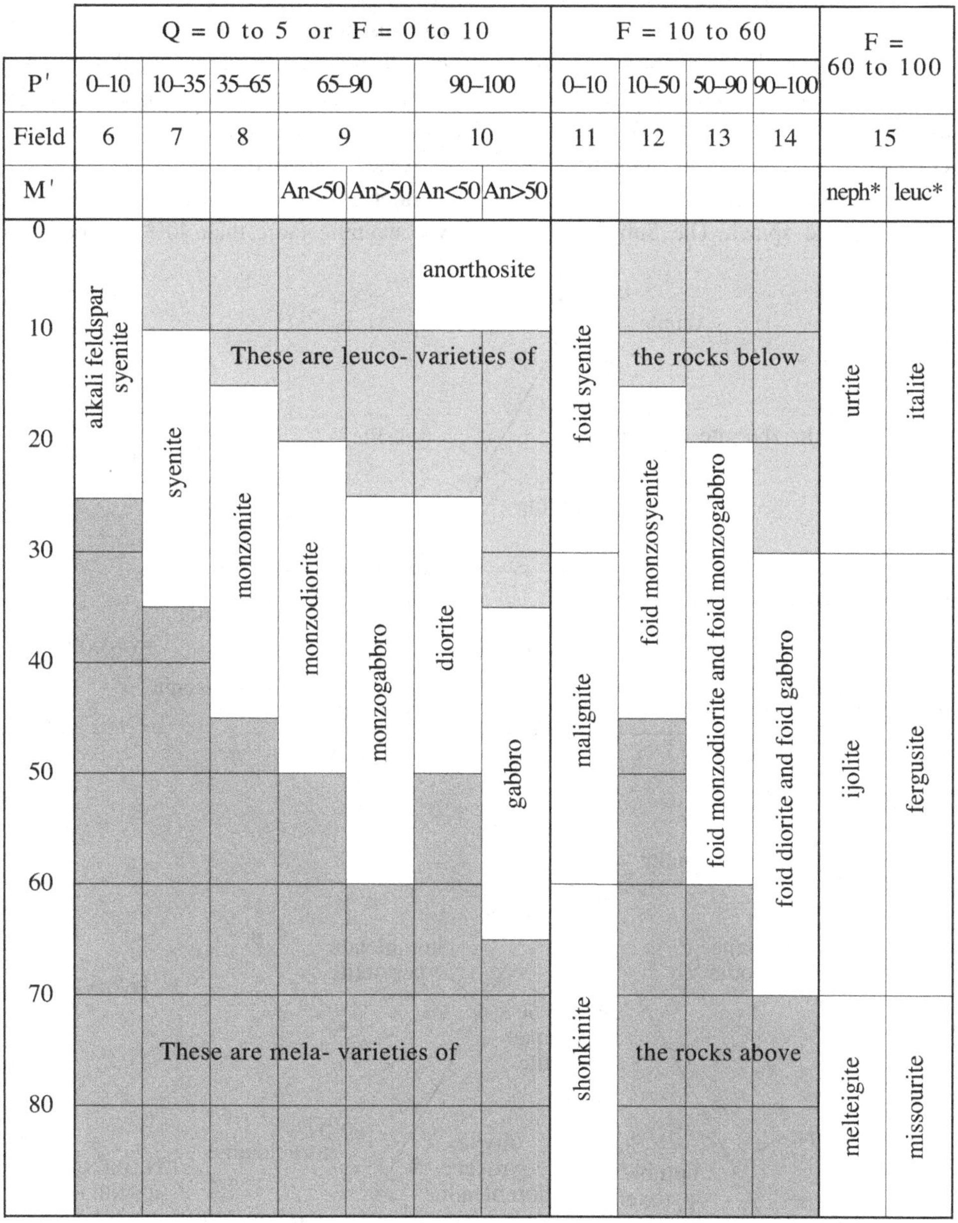

	Q = 0 to 5 or F = 0 to 10					F = 10 to 60				F = 60 to 100	
P'	0–10	10–35	35–65	65–90	90–100	0–10	10–50	50–90	90–100		
Field	6	7	8	9	10	11	12	13	14	15	
M'				An<50 / An>50	An<50 / An>50					neph*	leuc*

Rock names (by field):

- alkali feldspar syenite
- syenite
- monzonite
- monzodiorite
- monzogabbro
- diorite
- gabbro
- anorthosite
- foid syenite
- foid monzosyenite
- foid monzodiorite and foid monzogabbro
- foid diorite and foid gabbro
- malignite
- shonkinite
- urtite / italite
- ijolite / fergusite
- melteigite / missourite

These are leuco- varieties of ... the rocks below

These are mela- varieties of ... the rocks above

M' scale: 0, 10, 20, 30, 40, 50, 60, 70, 80

Fig. 2.8. Use of the terms mela- and leuco- with QAPF plutonic rocks with Q < 5% or F > 0% (after Streckeisen, 1976, Fig. 5). Abbreviations: P' = 100 * P / (A + P); M' = colour index; An = anorthite content of plagioclase; neph* = nepheline is the predominant foid; leuc* = leucite is the predominant foid. Note: some special names are applicable in certain parts of the diagram.

2.11.2 ULTRAMAFIC ROCKS (M > 90%)

The ultramafic rocks are classified according to their content of mafic minerals, which consist essentially of olivine, orthopyroxene, clinopyroxene, hornblende, sometimes with biotite, and various but usually small amounts of garnet and spinel. The Subcommission (Streckeisen, 1973, 1976) recommended two diagrams, both of which are shown in Fig. 2.9. One is for rocks consisting essentially of olivine, orthopyroxene, and clinopyroxene, and the other for rocks containing hornblende, pyroxenes, and olivine.

Peridotites are distinguished from **pyroxenites** by containing more than 40% olivine. This

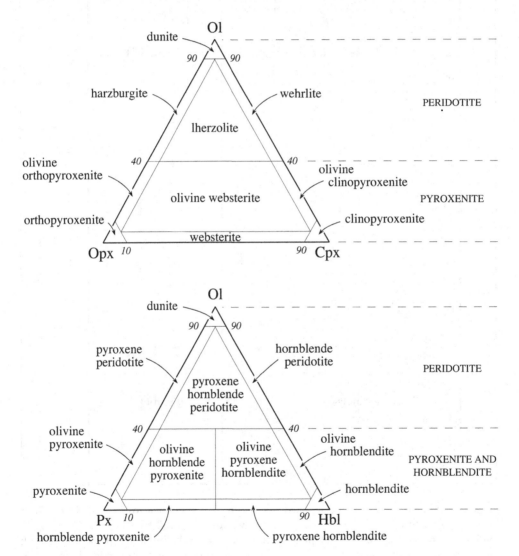

Fig. 2.9. Modal classification of ultramafic rocks based on the proportions of olivine (Ol), orthopyroxene (Opx), clinopyroxene (Cpx), pyroxene (Px) and hornblende (Hbl) (after Streckeisen, 1973, Figs. 2a and 2b).

value, rather than 50%, was chosen because many lherzolites contain up to 60% pyroxene. The peridotites are basically subdivided into **dunite** (or **olivinite** if the spinel mineral is magnetite), **harzburgite**, **lherzolite** and **wehrlite**.

The **pyroxenites** are further subdivided into **orthopyroxenite**, **websterite** and **clinopyroxenite**.

Ultramafic rocks containing garnet or spinel should be qualified in the following manner. If garnet or spinel is less than 5% use garnet-bearing peridotite, chromite-bearing dunite etc. If garnet or spinel is greater than 5% use garnet peridotite, chromite dunite etc.

2.11.3 PROVISIONAL "FIELD" CLASSIFICATION

The "field" classification of plutonic rocks should be used only as a provisional measure when an accurate mineral mode is not yet available. When available, the plutonic QAPF diagram should be used.

The classification is based on a simplified version of the plutonic QAPF diagram (Streckeisen, 1976) and is shown in Fig. 2.10. If the suffix "-oid" is felt to be linguistically awkward then the alternative adjectival form "-ic rock" may be used, i.e. use syenitic rock in place of syenitoid.

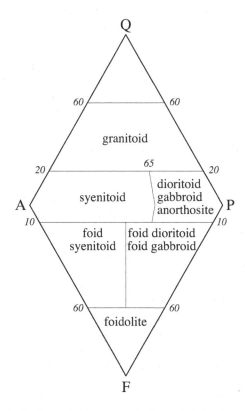

Fig. 2.10. Preliminary QAPF classification of plutonic rocks for field use (after Streckeisen, 1976, Fig. 6).

2.12 VOLCANIC ROCKS

This classification should be used only if the rock is considered to be volcanic, i.e. it is assumed to have been associated with volcanism and has a relatively fine-grained texture in which most of the individual crystals cannot be seen with the naked eye.

The classification of volcanic rocks is divided into three parts:

(1) if a mineral mode can be determined, use the QAPF classification (Fig. 2.11) of section 2.12.1

(2) if a mineral mode cannot be determined and a chemical analysis is available, use the TAS classification of section 2.12.2, p.33

(3) if neither a mineral mode nor chemical analysis is yet available, the "field" classification of section 2.12.3, p.39, may be used provisionally.

2.12.1 Volcanic QAPF classification (M < 90 %)

This classification should be used only if the rock is considered to be volcanic and if a mineral mode can be determined (Streckeisen, 1978 and 1979). The root names for the classification are given in Fig. 2.11.

The numbers of the QAPF fields are the same as those for the plutonic rock classification (see Fig. 2.5, p.23) except that field 15 has been divided into three subfields (Fig. 2.12, p.32).

Field 2 (alkali feldspar rhyolite) — the root name corresponds with alkali feldspar granite. The term **peralkaline rhyolite**, in preference to alkali rhyolite, can be used when the rock contains alkali pyroxene and/or amphibole. The name rhyolite may be replaced by the synonym **liparite**.

Fields 3a and 3b (rhyolite) — in an analogous manner to the granites, this root name covers both fields 3a and 3b. **Liparite** may be used as a synonym. The term **rhyodacite**, which has been used ambiguously for rocks of fields 3b and 4, can be used for transitional rocks between rhyolite and dacite without attributing it to a distinct field.

Fields 4 and 5 (dacite) — rocks in both these fields are covered by the root name dacite in the broad sense. Volcanic rocks of field 5, to which terms such as "plagidacite" and "quartz andesite" have been applied, are frequently also described as dacite, which is the recommended name.

Fields 6 (alkali feldspar trachyte), 7 (trachyte), 8 (latite) — rocks with these root names, which contain no modal foids but do contain nepheline in the norm, may be qualified with "*ne*-normative" to indicate that they would fall in subfields 6′–8′, respectively. **Peralkaline trachyte**, rather than alkali trachyte, should be used for trachytes containing sodic pyroxene and/or sodic amphibole.

Fields 9 and 10 (basalt, andesite) — these two fields contain the large majority of volcanic rocks. Basalt and andesite are tentatively separated using colour index, at a limit of 40 wt % or 35 vol %, and 52% SiO_2 as shown in Table 2.11. A plagioclase composition (at a limit of An_{50}) is less suitable for the distinction between basalt and andesite, because many andesites commonly contain "phenocrysts" of

Table 2.11. *Classification of QAPF fields 9 and 10 volcanic rocks into basalt and andesite, using colour index and wt % SiO_2*

Colour index		SiO_2 wt %	
vol %	wt %	< 52	> 52
> 35	> 40	basalt	mela-andesite
< 35	< 40	leuco-basalt	andesite

Source: Streckeisen (1978, Fig. 2).

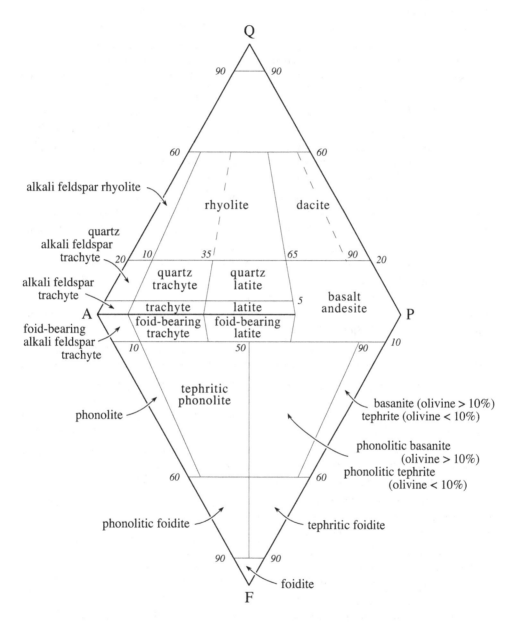

Fig. 2.11. QAPF modal classification of volcanic rocks (based on Streckeisen, 1978, Fig. 1). The corners of the double triangle are Q = quartz, A = alkali feldspar, P = plagioclase and F = feldspathoid. This diagram must not be used for rocks in which the mafic mineral content, M, is greater than 90%.

labradorite or bytownite. Although this may seem rather unsatisfactory, it is unlikely that many of these rocks will be classified using the QAPF diagram, as the modes of most basalts and andesites are difficult to determine accurately so that the TAS classification will have to be used.

Field 11 (phonolite) — the root name phonolite is used in the sense of Rosenbusch for rocks consisting essentially of alkali feldspar, any feldspathoid and mafic minerals. The nature of the predominant foid should be added to the root name, e.g. leucite phonolite, analcime phonolite, leucite-nepheline phonolite (with nepheline > leucite) etc. Phonolites containing nepheline and/or haüyne as the main foids are commonly described simply as "phonolite". Phonolites that contain sodic pyroxene and/or sodic amphibole may be called **peralkaline phonolite**.

Field 12 (tephritic phonolite) — these rocks are rather rare. Although it was originally suggested that the term tephriphonolite is a synonym (Streckeisen, 1978), it is probably better to reserve this term for the root name of TAS field U3, to indicate that the name has been given chemically and may not be identical to those of QAPF field 12.

Field 13 (phonolitic basanite, phonolitic tephrite) — these two root names are separated on the amount of olivine in the CIPW norm. If normative olivine is greater than 10% the rock is called a phonolitic basanite; if less than 10% it is a phonolitic tephrite. Although it was originally suggested that the term phonotephrite was a synonym of phonolitic tephrite (Streckeisen, 1978), it is probably better to reserve this term for the root name of TAS field U2, to indicate that the name has been given chemically and may not be identical to those of QAPF field 13. There is no conflict if the term phonobasanite is used as a synonym for phonolitic basanite, as the term is not used in TAS.

Field 14 (basanite, tephrite) — these two root names are separated on the amount of olivine in the CIPW norm. If normative olivine is greater than 10% the rock is called a basanite; if less than 10% it is a tephrite. The nature of the dominant foid should be indicated in the name, e.g. nepheline basanite, leucite tephrite etc.

Field 15 (foidite *sensu lato*) — the general root name of this field is **foidite**, but as these rocks occur relatively frequently the field has been subdivided into three: fields 15a , 15b and 15c as shown in Fig. 2.12.

Field 15a (phonolitic foidite) — wherever possible replace the term foidite with a more specific term, such as phonolitic nephelinite. Alternatively, the term alkali feldspar foidite could be used as the root name, which would give specific terms such as sanidine nephelinite.

Field 15b (tephritic foidite, basanitic foidite) — these two root names are separated according to their olivine content, as in field 14. wherever possible replace the term foidite with a more specific term, such as tephritic leucitite, basanitic nephelinite.

Field 15c (foidite *sensu stricto*) — the root name is **foidite** and should be distinguished by the name of the predominant foid, e.g. nephelinite, leucitite, analcimite.

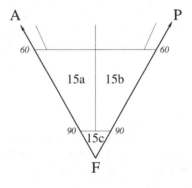

Fig. 2.12. Subdivision of volcanic QAPF field 15 into fields 15a, 15b and 15c.

2.12.2 THE TAS CLASSIFICATION

The TAS (Total Alkali – Silica) classification should be used only if:

(1) the rock is considered to be volcanic
(2) a mineral mode cannot be determined, owing either to the presence of glass or to the fine-grained nature of the rock
(3) a chemical analysis of the rock is available.

The root names for the classification are shown in Figs. 2.13 and 2.14, while the field symbols are given in Fig. 2.15. The classification is easy to use as all that is required for most rocks are the values of Na_2O, K_2O and SiO_2. However, if the analysis falls in certain fields, additional calculations, such as the CIPW norm (see Appendix C), must be performed in order to arrive at the correct root name.

The TAS classification was originally constructed with the more common rock types in mind, using the following principles summarized by Le Bas & Streckeisen (1991):

(1) each field was chosen to accord as closely as possible with the current usage of the root name with the help of data from 24 000 analyses of named fresh volcanic rocks from the CLAIR and PETROS databases (Le Maitre, 1982)
(2) fresh rocks were taken to be those in which $H_2O+ < 2\%$ and $CO_2 < 0.5\%$
(3) each analysis was recalculated to 100% on an H_2O and CO_2 free basis
(4) wherever possible, the boundaries were located to minimize overlap between adjacent fields
(5) the vertical SiO_2 boundaries between the fields of basalt, basaltic andesite, andesite and dacite were chosen to be those in common use
(6) the boundary between the S (for silica Saturated) fields and the U (for silica Undersaturated) fields was chosen to be roughly parallel with the empirically

determined contours of 10% normative F in QAPF
(7) the boundary between the S fields and O (for silica Oversaturated) fields was chosen where there was a density minimum between volcanic rock series that were alkaline and those that were calc-alkaline
(8) the boundaries between fields S1–S2–S3–T were all made parallel to a pronounced edge found in the distribution of analyses of rocks that had been called trachyte
(9) similarly, the boundaries between fields U1–U2–U3–Ph were also drawn parallel to each other. They are not at right angles to the line separating fields S from U.

However, after the TAS classification was published, the Subcommission considered whether or not it was possible to include some of the olivine- and pyroxene-rich ("high-Mg") volcanic rocks, e.g. picrites, komatiites, meimechites and boninites, into the scheme. After lengthy discussions this has been done by using MgO and TiO_2 in conjunction with TAS (see Fig. 2.13, p.34). As a result these rocks must be considered first, as they are not the "normal" type of volcanic rocks for which the TAS classification was originally designed.

Similarly it has been found that nephelinites and melanephelinites both fall in fields F and U1 and must therefore be excluded before using the TAS classification.

It must also be stressed that the TAS classification is **purely descriptive**, and that **no genetic significance** is implied. Furthermore, analyses of rocks that are weathered, altered, metasomatized, metamorphosed or have undergone crystal accumulation should be used with caution, as spurious results may be obtained. As a general rule it is suggested that only analyses with $H_2O+ < 2\%$ and $CO_2 < 0.5\%$ should be used, unless the rock is a picrite, komatiite, meimechite or boninite, when this restriction is withdrawn. The application of TAS to altered rocks is discussed by Sabine *et* *al.* (1985), who found that many low-grade metavolcanic rocks could be satisfactorily classified.

Before using the classification the two following procedures must be adopted:

(1) analyses must be recalculated to 100% on an H_2O and CO_2 free basis

(2) if a CIPW norm has to be calculated to determine the correct root name, the amounts of FeO and Fe_2O_3 should be left as determined. If only total iron has been determined, it is up to the user to justify the method used for partitioning the iron between FeO and Fe_2O_3. One method that can be used to estimate what the FeO and Fe_2O_3 would have been is that of Le Maitre (1976). Remember, it is the feeling of the Subcommission that rocks should be named according to what they are, and not according to what they might have been.

As previously explained not all rock types fall neatly into the TAS fields so that one must check to see if the rock being classified is one of these types before using Fig. 2.14 directly. The rocks in question are the "high-Mg" volcanic rocks, i.e. picrite, komatiite, meimechite or boninite and the nephelinites and melanephelinites which fall in fields F and U1.

"High-Mg" volcanic rocks — these are may be distinguished by the following criteria as shown in Fig. 2.13:

(1) if $SiO_2 > 52\%$, $MgO > 8\%$ and $TiO_2 < 0.5\%$, the rock is a **boninite**

(2) if $52\% > SiO_2 > 30\%$, $MgO > 18\%$ and $(Na_2O + K_2O) < 2\%$, then the rock is a **komatiite** if $TiO_2 < 1\%$ or a **meimechite** if $TiO_2 > 1\%$

(3) if $52\% > SiO_2 > 30\%$, $MgO > 12\%$, and $(Na_2O + K_2O) < 3\%$, it is a **picrite**.

Note that this scheme is different from that

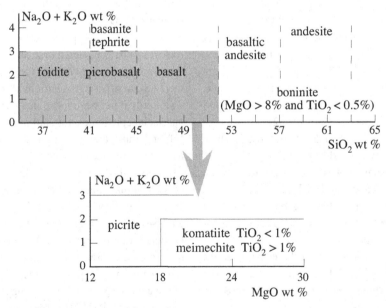

Fig. 2.13. Chemical classification and separation of the "high-Mg" volcanic rocks boninite, komatiite, meimechite and picrite prior to using the TAS classification. If a rock falls in the shaded rectangle of the TAS (upper) diagram, check in the lower diagram to see that it is not a komatiite, meimechite or picrite, before naming it as a foidite, picrobasalt or basalt. Similarly, a rock with $SiO_2 > 52\%$ should be checked to see that it is not a boninite (after Le Bas, 2000).

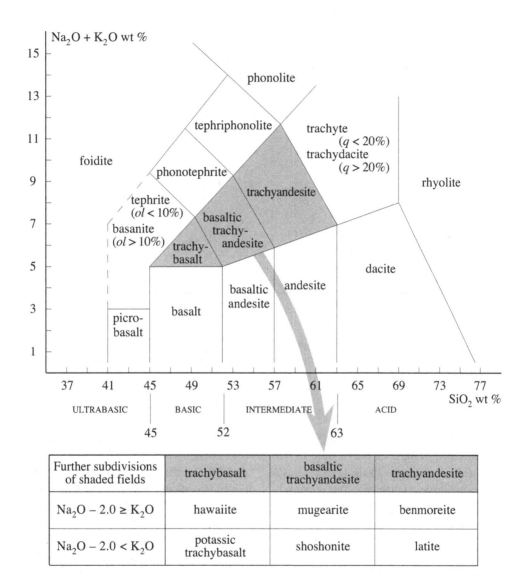

Fig. 2.14. Chemical classification of volcanic rocks using TAS (total alkali–silica diagram) (after Le Bas *et al.*, 1986, Fig. 2). Rocks falling in the shaded areas may be further subdivided as shown in the table pointed to by the arrow. The line between the foidite field and the basanite–tephrite field is dashed to indicate that further criteria must be used to separate these types. Abbreviations: *ol* = normative olivine; *q* = normative 100 * *Q* / (*Q* + *or* + *ab* + *an*).

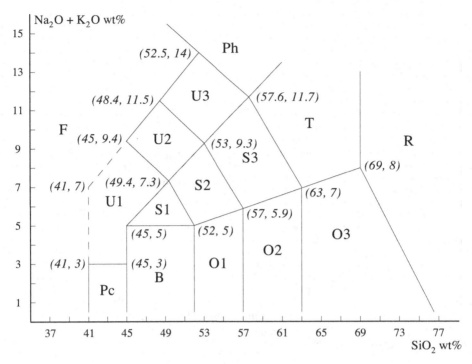

Fig. 2.15. Field symbols and coordinate points of TAS (after Le Bas *et al.*, 1986, Fig. 1). The numbers in brackets are the coordinates of the intersections of the lines.

published in Le Maitre *et al.* (1989, Fig. B.12). The lowering of MgO for picrite from 18% to 12% and increasing the alkalis from 2% to 3% makes many rocks into picrites that previously were classified as picrobasalt.

Nephelinites and melanephelinites — it has been found that nephelinites, melanephelinites and certain leucitites fall in both fields U1 and F, which is why the boundary between the two fields is dashed. They are distinguished by the following rules (after Le Bas, 1989):

(1) if normative *ne* > 20% the rock is a **nephelinite**.

(2) if normative *ne* < 20% and *ab* is present but is < 5% the rock is a **melanephelinite**.

If the rock is none of these six types you can use TAS diagram in Fig. 2.14 directly.

The field letters of the TAS diagram shown above are now described in further detail.

Field B (basalt) — the root name may be divided into **alkali basalt** and **subalkali basalt** according to the state of silica saturation – if the CIPW norm contains nepheline (*ne*) the rock is an alkali basalt, if not the rock is a subalkali basalt. This is based on the principle of the basalt tetrahedron (Yoder & Tilley, 1962)

The subalkali basalt group includes a large number of basalt varieties such as calc-alkali basalt (high-alumina basalt), mid-ocean ridge basalt, tholeiitic basalt, transitional basalt etc. Although none of these have been defined, but only categorized, the Subcommission does recommend that **tholeiitic basalt** should be used in preference to the term tholeiite (see Glossary).

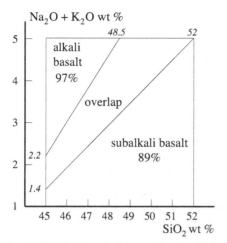

Fig. 2.16. Likelihood of correctly classifying alkali basalt and subalkali basalt using TAS (after Bellieni *et al.* 1983) assuming equal chances of a basalt being alkali or subalkali.

The TAS diagram has also been used many times to separate alkali basalt from subalkali basalt, resulting in numerable lines and curves being proposed. This is due to the fact that the undersaturation plane in the basalt tetrahedron consists of various planar surfaces (one for each normative type) none of which is perpendicular to the TAS surface, so that an exact correlation between TAS and silica saturation can never be achieved.

Bellieni *et al.* (1983) have investigated this problem using 7594 basalt analyses as defined by TAS. A synopsis of their results is shown in Fig. 2.16 which is divided into three fields. Assuming a basalt is equally likely to be alkali or subalkali, an analysis falling in the alkali basalt field has a 97% chance of being correctly classified, while one falling in the subalkali basalt field has an 89% chance of being correctly classified. An analysis falling in the overlap field is three times more likely to be an alkali basalt than a subalkali basalt.

Subdivision of fields B (basalt), O1 (basaltic andesite), O2 (andesite), O3 (dacite), R (rhyolite) — the root names may be qualified using the terms **low-K**, **medium-K**, and **high-K** as shown in Fig. 2.17. This is in accord with the concept developed by Peccerillo & Taylor (1976), but the lines have been slightly modified and simplified. It must be stressed that the

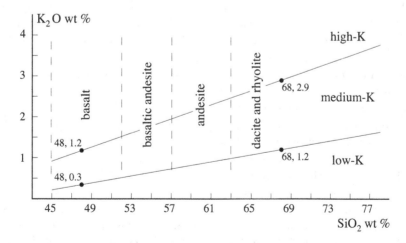

Fig. 2.17. Division of the basalt–rhyolite series into low-K, medium-K and high-K types. Note that high-K is **not** synonymous with potassic. The thick stippled lines indicate the equivalent position of some of the fields in the TAS diagram.

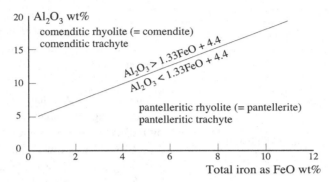

Fig. 2.18. Classification of trachytes and rhyolites into comenditic and pantelleritic types using the Al_2O_3 versus total iron as FeO diagram (after Macdonald, 1974). The coordinates of the bottom left of the line are (0.45, 5.0) and the top right are (10.98, 19.0).

term **high-K** is not synonymous with **potassic**, as high-K rocks can have more Na_2O than K_2O.

Field R (rhyolite) — the root name may be further subdivided into **peralkaline rhyolite**, if the **peralkaline index**, which is the molecular ratio $(Na_2O + K_2O) / Al_2O_3$, is greater than 1.

Field T (trachyte, trachydacite) — these two root names are separated by the function *100 * Q / (Q + an + ab + or)* which is the normative equivalent of Q in QAPF. If the value is less than 20% the rock is trachyte; if greater than 20% it is trachydacite. Trachytes may be further subdivided into **peralkaline trachytes**, if the peralkaline index > 1.

Peralkaline rhyolites and **trachytes** — the Subcommission has considered it useful to further subdivide these rocks into **comenditic rhyolite (= comendite)**, **comenditic trachyte**, **pantelleritic rhyolite (= pantellerite)**, and **pantelleritic trachyte** according to the method of Macdonald (1974), which is based on the relative amounts of Al_2O_3 versus total iron as FeO as shown in Fig. 2.18.

Field Ph (phonolite) — phonolites may be further subdivided into **peralkaline phonolites**, if the peralkaline index > 1.

Field S1 (trachybasalt) — the root name may be divided into **hawaiite** and **potassic trachyba-**

salt according to the relative amounts of Na_2O and K_2O. If $Na_2O – 2$ is greater than K_2O the rock is considered to be "sodic" and is called hawaiite; if $Na_2O – 2$ is less than K_2O the rock is considered to be "potassic" and is called potassic trachybasalt (see Fig. 2.14, p.35).

Field S2 (basaltic trachyandesite) — using the same criterion as for field S1, the root name may be divided into **mugearite** ("sodic") and **shoshonite** ("potassic").

Field S3 (trachyandesite) — using the same criterion as for field S1, the root name may be divided into **benmoreite** ("sodic") and **latite** ("potassic").

Field U1 (basanite, tephrite) — if normative *ol* > 10 % the rock is a basanite, if *ol* < 10% it is a tephrite.

Field F (foidite) — before deciding that the rock should be named a foidite check to see if it is a melilitite, using the following rules:

1) if the rock does not contain kalsilite but has normative *cs* (dicalcium silicate or larnite) > 10% and $K_2O < Na_2O$, then it is a **melilitite** (modal olivine < 10%) or an **olivine melilitite** (modal olivine > 10%)

2) if normative *cs* > 10%, $K_2O > Na_2O$ and $K_2O > 2\%$, then it is a **potassic melilitite** (modal olivine < 10%) or a **potassic olivine**

melilitite (modal olivine > 10%). The latter has been termed katungite, which mineralogically is a kalsilite-leucite-olivine melilitite

3) if normative *cs* is present but is < 10%, then the rock is a **melilite nephelinite** or a **melilite leucitite** according to the nature of the dominant feldspathoid mineral.

The rock should now be named a foidite but wherever possible this term should be replaced with a more specific term according to the dominant feldspathoid mineral.

2.12.3 PROVISIONAL "FIELD" CLASSIFICATION

The "field" classification of volcanic rocks should be used only as a provisional measure when neither an accurate mineral mode nor a chemical analysis is yet available. When either become available, the volcanic QAPF diagram or the TAS diagram should be used.

The classification is based on a simplified version of the volcanic QAPF diagram (Streckeisen, 1978, 1979) and is shown in Fig. 2.19. If the suffix "-oid" is felt to be linguistically awkward then the alternative adjectival form "-ic rock" may be used, i.e. use dacitic rock in place of dacitoid.

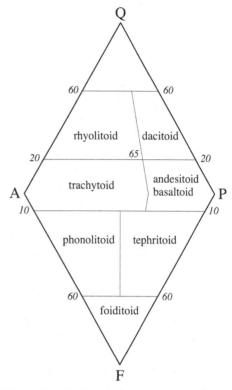

Fig. 2.19. Preliminary QAPF classification of volcanic rocks for field use (after Streckeisen, 1976, Fig. 6).

2.13 REFERENCES

BELLIENI, G., JUSTIN VISENTIN, E., LE MAITRE, R.W., PICCIRILLO, E.M. & ZANETTIN, B., 1983. Proposal for a division of the basaltic (B) field of the TAS diagram. *IUGS Subcommission on the Systematics of Igneous Rocks*. Circular No.38, Contribution No.102.

CLEMENT, C.R., SKINNER, E.M.W. & SCOTT SMITH, B.H., 1984. Kimberlite-defined. *Journal of Geology*. Vol.32, p.223–228.

DAWSON, J.B., 1980. Kimberlites and their Xenoliths. *Springer-Verlag, Berlin*. 252pp.

DUNWORTH, E.A. & BELL, K., 1998. Melilitolites: a new scheme of classification. *The Canadian Mineralogist*. Vol.36, p.895–903.

FISHER, R.V., 1966. Rocks composed of volcanic fragments. *Earth Science Reviews. International Magazine for Geo-Scientists. Amsterdam*. Vol.1, p.287–298.

FOLEY, S.F., VENTURELLI, G., GREEN, D.H. & TOSCANI, L., 1987. The ultrapotassic rocks: characteristics, classification and constraints for petrogenetic models. *Earth Science Reviews. International Magazine for Geo-Scientists. Amsterdam*. Vol.24, p.81–134.

GITTINS, J. & HARMER, R.E., 1987. What is ferrocarbonatite? A revised classification. *Journal of African Earth Sciences*. Vol.28(1), p.159–168.

LE BAS, M.J., 1989. Nephelinitic and basanitic rocks. *Journal of Petrology. Oxford*. Vol.30, p.1299–1312.

LE BAS, M.J., 1999. Sövite and alvikite: two chemically distinct calciocarbonatites C1 and C2. *South African Journal of Geology*. Vol.102(2), p.109–121.

LE BAS, M.J., 2000. IUGS reclassification of the high-Mg and picritic volcanic rocks. *Journal of Petrology. Oxford*. Vol.41(10), p.1467–1470.

LE BAS, M.J. & STRECKEISEN, A., 1991. The IUGS systematics of igneous rocks. *Journal*

of the Geological Society, London. Vol.148, p.825–833.

LE BAS, M.J., LE MAITRE, R.W., STRECKEISEN, A. & ZANETTIN, B., 1986. A chemical classification of volcanic rocks based on the total alkali – silica diagram. *Journal of Petrology. Oxford*. Vol.27, p.745–750.

LE BAS, M.J., LE MAITRE, R.W. & WOOLLEY, A.R., 1992. The construction of the total alkali-silica chemical classification of volcanic rocks. *Mineralogy and Petrology*. Vol.46, p.1–22.

LE MAITRE, R.W., 1976. Some problems of the projection of chemical data into mineralogical classifications. *Contributions to Mineralogy and Petrology*. Vol.56, p.181–189.

LE MAITRE, R.W., 1982. Numerical Petrology. *Elsevier, Amsterdam*. 281pp.

LE MAITRE, R.W. (Editor), BATEMAN, P., DUDEK, A., KELLER, J., LAMEYRE, M., LE BAS, M.J., SABINE, P.A., SCHMID, R., SØRENSEN, H., STRECKEISEN, A., WOOLLEY, A.R. & ZANETTIN, B., 1989. A Classification of Igneous Rocks and a Glossary of Terms. *Blackwell Scientific Publications, Oxford*. 193pp.

MACDONALD, R., 1974. Nomenclature and petrochemistry of the peralkaline oversaturated extrusive rocks. *Bulletin Volcanologique*. Vol.38, p.498–516.

MITCHELL, R.H., 1979. The alleged kimberlite-carbonatite relationship: additional contrary mineralogical evidence. *American Journal of Science. New Haven*. Vol.279, p.570–589.

MITCHELL, R.H., 1986. Kimberlites: Mineralogy, Geochemistry and Petrology. *Plenum Press, New York*. 442pp.

MITCHELL, R.H., 1994. Suggestions for revisions to the terminology of kimberlites and lamprophyres from a genetic viewpoint. In: Meyer, H.O.A. & Leonardos, O.H. (Editors) *Proceedings of the Fifth International Kimberlite Conference. 1. Kimberlites and related rocks and mantle xenoliths, Companhia* de Pesquisa de Recursos Minerais, Special Publication No.1/A, Brasilia. p.15–26.

MITCHELL, R.H., 1995. Kimberlites, Orangeites and Related Rocks. *Plenum Press, New York*. 410pp.

MITCHELL, R.H. & BERGMAN, S.C., 1991. Petrology of Lamproites. *Plenum Press, New York*. 447pp.

MITCHELL, R.H. & PUTNIS, A., 1988. Polygonal serpentine in segregation-textured kimberlite. *The Canadian Mineralogist*. Vol.26, p.991–997.

NIGGLI, P., 1931. Die quantitive mineralogische Klassifikation der Eruptivgesteine. *Schweizerische Mineralogische und Petrographische Mitteilungen, Zürich*. Vol.11, p.296–364.

PECCERILLO, A. & TAYLOR, S.R., 1976. Geochemistry of the Eocene calc-alkaline volcanic rocks from the Kastamonu area, northern Turkey. *Contributions to Mineralogy and Petrology*. Vol.58, p.63–81.

ROCK, N.M.S., 1991. Lamprophyres. *Blackie, London*. 285pp.

SABINE, P.A., HARRISON, R.K. & LAWSON, R.I., 1985. Classification of volcanic rocks of the British Isles on the total alkali oxide–silica diagram, and the significance of alteration. *British Geological Survey Report*. Vol.17, p.1–9.

SAHAMA, T.G., 1974. Potassium-rich alkaline rocks. In: Sørensen, H. (Editor) The Alkaline Rocks. *Wiley, New York*. p.96–109.

SCHMID, R., 1981. Descriptive nomenclature and classification of pyroclastic deposits and fragments: Recommendations of the IUGS Subcommission on the Systematics of Igneous Rocks. *Geology. The Geological Society of America. Boulder, Co*. Vol.9, p.41–43.

SCOTT SMITH, B.H., DANCHIN, R.V., HARRIS, J.W. & STRACKE, K.J., 1983. Kimberlites near Orroroo, South Australia. *In:* Kornprobst, J. (Editor) *Proceedings of the Third International Kimberlite Conference. Elsevier*.

Vol.1, p.121–142.

SKINNER, E.M.W., 1989. Contrasting Group-1 and Group-2 kimberlite petrology: towards a genetic model for kimberlites. *Proceedings of the Fourth International Kimberlite Conference, Perth, Australia. Geological Society of Australia*. Special Publication No.14, p.528–544.

SMITH, C.B., GURNEY, J.J., SKINNER, E.M.W., CLEMENT, C.R. & EBRAHIM, N., 1985. Geochemical character of southern African kimberlites: a new approach based upon isotopic constraints. *Transactions of the Geological Society of South Africa*. Vol.88, p.267–280.

STRECKEISEN, A., 1967. Classification and Nomenclature of Igneous Rocks (Final Report of an Inquiry). *Neues Jahrbuch für Mineralogie. Stuttgart, Abhandlungen*. Vol.107, p.144–240.

STRECKEISEN, A., 1973. Plutonic Rocks. Classification and nomenclature recommended by the IUGS Subcommission on the Systematics of Igneous Rocks. *Geotimes*. Vol.18(10), p.26–30.

STRECKEISEN, A., 1974. How should charnockitic rocks be named? *Centenaire de la Société Géologique de Belgique Géologie des Domaines Cristallins, Liège*. p.349–360.

STRECKEISEN, A., 1976. To each plutonic rock its proper name. *Earth Science Reviews. International Magazine for Geo-Scientists. Amsterdam*. Vol.12, p.1–33.

STRECKEISEN, A., 1978. IUGS Subcommission on the Systematics of Igneous Rocks. Classification and nomenclature of volcanic rocks, lamprophyres, carbonatites and melilitic rocks. Recommendations and suggestions. *Neues Jahrbuch für Mineralogie. Stuttgart, Abhandlungen*. Vol.134, p.1–14.

STRECKEISEN, A., 1979. Classification and nomenclature of volcanic rocks, lamprophyres, carbonatites, and melilitic rocks: Recommendations and suggestions of the IUGS Subcommission on the Systematics of

Igneous Rocks. *Geology. The Geological Society of America. Boulder, Co..* Vol.7, p.331–335.

TAINTON, K. M. & BROWNING, P., 1991. The group 2 kimberlite–lamproite connection: some constraints from the Barkly-West District, northern Cape Province, South Africa. *Fifth International Kimberlite Conference, Araxa. Extended Abstract. Companha de Pesquisa de Recursos Minerais, Special Publication No.2/91. Brasilia.* p.405–407.

WAGNER, P.A., 1914. The diamond fields of southern Africa. *Transvaal Leader, Johannesburg.*

WAGNER, P.A., 1928. The evidence of kimberlite pipes on the constitution of the outer part of the Earth. *South African Journal of Science.* Vol.25, p. 127–148.

WOOLLEY, A.R. & KEMPE, D.R.C., 1989. Carbonatites: nomenclature, average chemical compositions and element distribution. In: K. Bell (Editor), Carbonatite Genesis and Evolution. *Unwin Hyman, London.* p.1–14.

WOOLLEY, A.R., BERGMAN, S.C., EDGAR, A.D., LE BAS, M.J., MITCHELL, R.H., ROCK, N.M.S. & SCOTT SMITH, B.H. 1996. Classification of lamprophyres, lamproites, kimberlites, and the kalsilitic, melilitic, and leucitic rocks. *The Canadian Mineralogist.* Vol.34, p.175–186.

YODER, H.S. & TILLEY, C.E. 1962. Origin of basaltic magmas: an experimental study of natural and synthetic rock systems. *Journal of Petrology. Oxford.* Vol.3, p.342–532.

3 Glossary of terms

In order to make the glossary a standard reference for the future, every effort has been made to make the list of rock names as complete as possible. The initial list included all the names found in standard reference texts and glossaries such as Johannsen (1931, 1932, 1937, 1938, 1939), Tröger (1935, 1938), Sørensen (1974) and Tomkeieff *et al.* (1983), except for varietal names as outlined in section 3.1.1. This list was later extensively edited and modified by suggestions from contributors and particularly when the references were checked. The list published in the 1st edition contained 1586 terms which has now been extended to 1637.

3.1 DETAILS OF ENTRIES

Most of the glossary entries consist of five types of information:

(1) the term, with any alternative spellings in brackets

(2) a brief petrological description or comments on the term

(3) the author(s), year and page number of the source reference

(4) the origin of the term

(5) the location of the term in three standard texts.

Where given the last three are printed in italics and enclosed in brackets. Each of these five types of information is described in further detail below.

3.1.1 CHOICE OF TERMS

The principles with which names have been included in the glossary have been somewhat subjective. In general, varietal names have not included where they are self-explanatory, e.g. hornblende granite, biotite granite. However, varietal names have been included where they have assumed a special meaning (e.g. nepheline syenite, quartz monzonite) or are not self-evident (e.g. Oslo-essexite, Puys-andesite) or are in common usage (e.g. mid-ocean ridge basalt, I-type granite).

Additional terms in brackets are alternative spellings, usually due to different transliterations into English, often from German or Russian. Terms separated only by commas are discrete names but with the same derivations, e.g. pallioessexite, palliogranite.

The glossary of terms also contains all the root names recommended by the Subcommission together with some general, adjectival and chemical terms that are necessary for their definition and understanding. All these terms are listed in **BOLD CAPITALS** as they form part of the Subcommission classification. A total of 316 entries (or 19%) falls into this category. No attempt has been made to include all igneous textural terms as some petrological (as well as mineralogical) knowledge is expected of the reader.

Any of the others terms may also be used if they are felt to serve a useful purpose, although it is hoped that some of the terms described as "obsolete" would not be revived. Note, however, that the Subcommission has not defined any of these terms.

3.1.2 PETROLOGICAL DESCRIPTION

The petrological descriptions were designed to be concise and informative. Wherever possible the names have been described in terms of the root name that would have been used if the rock had been classified using the IUGS Subcommission recommendations as set out earlier in the book. In most cases this has been achieved by using the phrase "**a variety of**" followed by the root name. A total of 465 entries (or 28%) falls into this category.

The term "**obsolete**" is used when, to the best knowledge of members of the Subcommission, the term has not been used in publications for a long time. In many cases the terms were only used once in the original publication. It is hoped that these terms will not be revived. A total of 413 entries (or 25%) falls into this category.

The term "**local**" is used when, to the best knowledge of members of the Subcommission, the name has only been used for rocks from the particular local region or area where it was named. It is suggested that such names should not be used outside their local areas. A total of 312 entries (or 19%) falls into this category.

3.1.3 AMPHIBOLE AND PYROXENE NAMES

The International Mineralogical Association (IMA) has published several recommendations on the nomenclature of amphiboles (IMA, 1978; Leake *et al.*, 1997) and pyroxenes (Morimoto *et al.*, 1988). Where these changes are largely grammatical and involve no ambiguity they have been implemented, e.g. **titanaugit** has been changed to **titanian augite**. However, in many cases it was felt inappropriate to eliminate the "old" names not only for historical reasons but also because many identifications were based purely on well-known optical properties.

This is particularly true of the amphibole **barkevikite**, which petrologists have identified optically as commonly occurring in certain alkaline rocks. The IMA (1978, p.558) states that barkevikite is a "(sometimes sodian) ferroan

pargasitic hornblende" or a "(sometimes sodian) ferro-pargasitic hornblende". These names were later changed to "**(sometimes sodian) pargasite**" or "**(sometimes sodian) ferropargasite**" (Leake *et al.*, 1997, p.309). The IMA (p.558) also states that barkevikite has never been defined chemically and has also been used for other compositions. In this instance it seems rather pointless to try to guess what optically identified barkevikite might be.

For these reasons, all "old" names are followed by the IMA recommended names in brackets, e.g. **hypersthene (= enstatite)**, **torendrikite (= magnesio-riebeckite)**, with the exception of barkevikite which appears as **barkevikite (see p.44)**, referring to this page.

3.1.4 SOURCE REFERENCE

Given after the petrological description are the author(s) and date of the original source reference where it is thought the term was first used. This differs somewhat from many of the references given in some of the other standard texts where the source reference cited often refers to where the rock was first collected but not necessarily named. This explains some of the apparent discrepancies that will be found between these references and some of those in the standard texts, in particular Tröger and Tomkeieff.

The page number given with the reference is where the name is first used or where it is defined and described in detail.

3.1.5 ORIGIN OF NAME

Wherever possible the source reference is followed by information concerning the origin of the term, unless the derivation is obvious, e.g. pyroxenite. Most rocks and terms are named after a geographical locality or region, a Greek, Latin or other linguistic root or a person. In some cases where the derivation is obvious (i.e. mineralogical) the type locality is given instead. A total of 882 entries (or 57%) has this information.

All the geographical localities, except the very local, are given in their English spelling and conform to the style found in *The Times Atlas of the World*, Comprehensive Edition, 1998.

The frequency with which some countries and linguistic roots have been used is given in Table 3.1. As can be seen 49% of the terms come from only seven countries – USA, Italy, Russian Federation, Norway, Germany, UK and France. Similarly, 13% of the terms have a Greek or Latin derivation or have been named after people.

3.1.6 LOCATION IN STANDARD TEXTS

The last piece of information given is the location of the term in either Johannsen (1931, 1932, 1937, 1938 or 1939), Tröger (1935 or 1938) or Tomkeieff *et al.* (1983).

For example, *Johannsen v.4, p.165* indicates that the name can be found in Johannsen volume 4 on page 165. *Johannsen v.1 (2nd Edn.), p.238* indicates that the name can be found in Johannsen volume 1, 2nd edition on page 238. The dates of the various volumes are *v.1* = 1931; *v.1 (2nd Edn.)* = 1939; *v.2* = 1932; *v.3* = 1937; *v.4* = 1938.

Similarly, *Tröger 508* indicates that the name occurs as rock number 508 in Tröger (1935). *Tröger(38) 775⅔* indicates that the name occurs as rock number 775⅔ in Tröger (1938). Note that both of these references can be found in Tröger (1969), a reprinted facsimile edition.

Finally, *Tomkeieff p.261* indicates that the name can be found on page 261 of Tomkeieff *et al.* (1983).

The contribution of these texts has been considerable, although many discrepancies and errors have been found. Of the 1633 terms listed in this glossary, 833, or 51%, can be found in Tröger, 616 (38%) in Johannsen and 1179 (72%) in Tomkeieff. Only the relatively small number of 499, or 31%, of the terms can be found in all three texts. Comparing the texts with each other, Tomkeieff contains 339 terms that are not in Tröger or Johannsen; Tröger contains 39 terms that are not in Johannsen or Tomkeieff while Johannsen only contains 17 terms that are not in Tröger or Tomkeieff. Finally, a total of 371, or 23%, of the terms in this glossary is not contained in Tröger, Johannsen or Tomkeieff.

Table 3.1. *Countries and linguistic roots found 12 or more times in the origin of new rock terms*

	Number	Acc%[a]
USA	81	9.2
Italy	77	17.9
Russian Federation	76	26.5
from Greek	69	34.4
Norway	63	41.5
Germany	57	48.0
UK	42	52.7
France	39	57.1
Canada	31	60.7
named after people	28	63.8
Czech Republic	23	66.4
Sweden	23	69.0
from Latin	17	71.0
Madagascar	13	72.4
Australia	13	73.9
Portugal	12	75.3
South Africa	12	76.6

[a] Based on the 882 glossary entries for which information is available.

3.2 HISTORICAL PERSPECTIVE

For those interested in the historical perspective of igneous petrology, it is interesting to look at the distribution of new rock names and terms with time. For example, as can be seen in Table 3.2, of the 1560 terms for which the year in which the term was first used is known, only 36 are from the period prior to 1800 while 522 were defined in the period 1800–99 and a further 1002 in the period 1900–99.

Table 3.2. *Frequency of new rock terms and their references by century*

Period	Rocks Number	Rocks Acc%	References Number	References Acc%
pre 1800	36	2.3	21	2.7
1800–99	522	35.8	280	39.0
1900–99	1002	100.0	471	100.0
Totals	1560		772	

Table 3.3. *"Best" and "worst" periods since 1800 for the publication of new rock terms and their references*

	Rocks		References	
2-year periods				
Best	1937–38	71	1912–13	31
5-year periods				
Best	1909–13	121	1912–16	61
	1934–38	121	1913–17	61
Worst	1800–05	0	1800–05	0
	1801–06	0	1801–06	0
10-year periods				
Best	1911–20	205	1893–02	110
			1908–17	110
Worst	1985–94	7	1800–09	4
	1986–95	7		
	1987–96	7		
	1988–97	7		

Similarly, Table 3.3 shows the "best" and "worst" 2-, 5- and 10-year periods since 1800. Of the terms produced in the period 1934–1938, about half are from Tröger and Johannsen.

Table 3.4 gives a list of some of the more prolific individual years for the production of new terms and references containing new terms. The largest number of new rock terms produced in one year was 57 in 1973 which was mostly the work of the Subcommission – many of these, however, were newly defined terms which had previously been used as varietal names. The next three most prolific years, 1938, 1920 and 1911 are also mainly due to the efforts of Johannsen.

Table 3.4. *Years with 20 or more new rock terms and 10 or more references containing new rock terms*

Rocks Year	Rocks Num	Rocks Acc%[a]	References Year	References Num	References Acc%[a]
1973	57	3.7	1913	18	2.3
1938	50	6.9	1896	14	4.1
1920	37	9.2	1893	13	5.8
1911	36	11.5	1899	13	7.5
1913	32	13.6	1912	13	9.2
1935	30	15.5	1917	13	10.8
1898	28	17.3	1895	12	12.4
1983	27	19.0	1900	12	13.9
1978	27	20.8	1928	12	15.5
1917	27	22.5	1902	11	16.9
1811	24	24.0	1906	11	18.3
1931	23	25.5	1916	11	19.7
1926	22	26.9	1937	11	21.2
1921	22	28.3	1898	10	22.5
1901	22	29.7	1901	10	23.7
1937	21	31.1	1904	10	25.0
1912	21	32.4	1905	10	26.3
1900	21	33.8	1910	10	27.6
1933	20	35.1	1911	10	28.9
1896	20	36.3	1915	10	30.2
1895	20	37.6	1935	10	31.5

[a] Based on the 1560 terms for which the source references are known.

The largest number of references with new rocks names produced in one year was 18 in 1913, surprisingly due to 16 different authors.

Only 11 or about 50% of the years are common to both lists indicating that a large number of publications does not automatically correspond to a large number of rock names.

Since 1800 there have only been 23 years when no new rock terms were published. They were 1800–05, 1812, 1818, 1828–30, 1856, 1867, 1871, 1950–51, 1987–88, 1991, 1993-94, 1997 and 1999.

Figure 3.1 shows the frequency with which new rocks terms and references have appeared in the literature for 10-year periods since 1760, i.e. over the last 240 years. In order to see the influence of different "schools" of petrology, the contribution of new rock names from

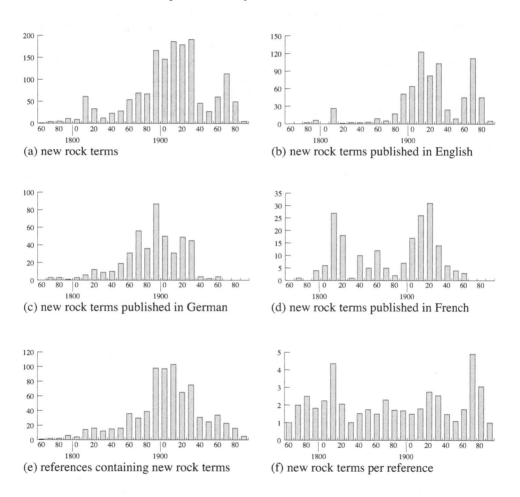

(a) new rock terms

(b) new rock terms published in English

(c) new rock terms published in German

(d) new rock terms published in French

(e) references containing new rock terms

(f) new rock terms per reference

Fig. 3.1. Frequency with which new rocks terms and their references have appeared in the literature over 10-year intervals from 1760 to 2000 for: (a) the number of new rock terms published; (b) the number of new rock terms published in English; (c) the number of new rock terms published in German; (d) the number of new rock terms published in French; (e) the number of references containing new rock terms; (f) the number of new rock terms per reference. The year values along the horizontal axis are the lower value for each class.

publications written in English (Fig. 3.1b), German (Fig. 3.1c) and French (Fig. 3.1d) are given as well.

The most prolific period for new rock terms (Fig. 3.1a) was from the 1890s to the 1930s, with minor peaks in the 1810s, mainly due to Brongniart, Cordier and Pinkerton, and in the 1970s, due to the Subcommission's work. The French school has two distinct peaks of activity, the earlier due to Brongniart and Cordier, while the later peak is due mainly to Lacroix. The major activity of the German school slightly preceded that of the English and the second peak of the French.

A slightly different picture is presented by the distribution of the number of references con-taining new rock names (Fig. 3.1e) where the early and late peaks shown in the distribution of new rock terms are missing.

Finally, the histograms of the number of new rock names per reference (Fig. 3.1f) is interest-ing. On average each publication produced about two new rock names. However, two distinct peaks are present, one reflecting the pioneering work of Brongniart, Cordier and Pinkerton in the 1810s, the other the work of the Subcommission in the 1970s. In this re-spect, the Subcommission's work can be said to be one of the most "productive" (to use a hackneyed political term) in the last 170 years. Whether we should be congratulated or not, only time will tell!

3.3 GLOSSARY

A-TYPE GRANITE. A general term for granitic rocks typically occurring in rift zones and in the interiors of stable continental plates. They are usually mildly alkaline granites with low CaO and Al_2O_3, high Fe / (Fe+Mg), high K_2O / Na_2O and K_2O values and consist of quartz, K-feldspar, minor plagioclase and Fe-rich biotite, and sometimes alkali amphibole. The prefix A stands for anorogenic. *(Loiselle & Wones, 1979, p.468)*

ABESSEDITE. An obsolete local term for a variety of peridotite composed of olivine, hornblende and phlogopite. *(Cotelo Neiva, 1947a, p.105; Abessédo Mine, Bragança district, Portugal; Tomkeieff p.3)*

ABSAROKITE. A collective term, related to banakite and shoshonite, for a trachyandesitic rock, containing phenocrysts of olivine and augite in a groundmass of augite, calcic plagioclase, alkali feldspar and sometimes leucite. Later defined chemically in several ways in terms of K_2O and SiO_2. *(Iddings, 1895a, p.938; Absaroka Range, Yellowstone National Park, Wyoming, USA; Tröger 271; Johannsen v.4, p.44; Tomkeieff p.4)*

ABYSSAL THOLEIITE. A term defined as a variety of tholeiitic basalt with K_2O < 0.4% occurring on the ocean floor; synonymous with mid-ocean ridge basalt. *(Miyashiro & Shido, 1975, p.268)*

ACHNAHAITE. An obsolete name used as a magma type for a biotite-bearing eucrite or variety of gabbro. *(Niggli, 1936, p.360; Achnaha, Ardnamurchan, Scotland, UK; Tröger(38) 362; Tomkeieff p.6)*

ACHNELITH. A term for pyroclastic fragments formed from solidified lava spray and whose external features are controlled by surface tension and not by fracturing. *(Walker & Croasdale, 1971, p.308; from the Greek achne = spray)*

ACID. A commonly used chemical term now defined in the TAS classification (Fig. 2.14, p.35) for rocks containing more than 63% SiO_2. See also intermediate, basic and ultrabasic. *(Abich, 1841, p.12; Tomkeieff p.6)*

ACIDITE. An obsolete general term applied to all acid rocks. *(Cotta, 1864a, p.824; Tröger 778; Tomkeieff p.6)*

ADAKITE. A term for a series of andesites, dacites and sodic rhyolites of unusual chemical composition erupted in some continent-based and island arc settings. They are characterized by high values of Sr, high Sr/Y and La/Yb ratios and negative Nb, Ti and Zr anomalies; the more mafic rocks of the series may contain high contents of transition metals such as Cr and Ni. They usually contain phenocrysts of plagioclase, amphibole, mica, very rare orthopyroxene, but no clinopyroxene; titanomagnetite, apatite, zircon and titanite are common. *(Defant & Drummond, 1990, p.662; Adak, Aleutian Islands, Alaska, USA)*

ADAM-GABBRO, ADAM-DIORITE, ADAM-TONALITE. A series of obsolete terms suggested for rocks intermediate between an adamellite and gabbro, diorite and tonalite respectively. *(Johannsen, 1920b, p.168; Tomkeieff p.7)*

ADAMELLITE. A term originally used for orthoclase-bearing tonalite of the Adamello Massif, but later used for granites with about equal amounts of alkali feldspar and plagioclase, which do not occur in the Adamello. The term should be avoided because of ambiguity and it is recommended that such rocks should be called monzogranite of QAPF field 3b (Fig. 2.4, p.22). *(Cathrein, 1890, p.74; Mt Adamello, Alto Adige, Italy; Tröger 779; Johannsen v.2, p.308; Tomkeieff p.6)*

AEGIAPITE. A mnemonic name suggested for a variety of pyroxenite consisting essentially of *aegi*rine and *apatite*. *(Belyankin & Vlodavets, 1932, p.63; Cape Turii, Kola Pe-*

ninsula, Russian Federation; Tröger(38) 775⅔; Tomkeieff p.8)

AEGINEITE. A mnemonic name suggested for a rock consisting essentially of *aegi*rine and *nephe*line. *(Belyankin, 1929, p.22)*

AEGIRINITE. An intrusive rock consisting essentially of aegirine-augite with minor nepheline and albite. *(Polkanov, 1940, p.303; Gremiakha-Vyrmes pluton, Kola Peninsula, Russian Federation; Tomkeieff p.8)*

AEGIRINOLITH (AIGIRINOLITH, EGIRINOLITH). An obsolete name for a variety of alkali clinopyroxenite containing aegirine-augite, titanite and magnetite. The original spelling was aigirinolith. *(Kretschmer, 1917, p.201; Tröger 780; Tomkeieff p.8)*

AEGISODITE. A mnemonic name suggested for a rock consisting of *aegi*rine and *sod*alite. *(Belyankin, 1929, p.22)*

AETNA-BASALT. A term from an obsolete chemical classification, based on feldspar composition rather than SiO_2 alone, for basaltic rocks in which CaO : Na_2O : $K_2O \approx$ 8 : 2.5 : 1 and $SiO_2 \approx$ 50%. *(Lang, 1891, p.236; Mt Etna, Sicily, Italy; Tomkeieff p.9)*

AFRIKANDITE (AFRICANDITE). A variety of melilitolite consisting of pyroxene, olivine, mica, perovskite, titanomagnetite and melilite. May be used as an optional term in the melilitic rocks classification if perovskite > 10% (section 2.4, p.11). *(Chirvinskii et al., 1940, p.31; Africanda railway station, Kola Peninsula, Russian Federation; Tomkeieff p.9)*

AGGLOMERATE. Now defined in the pyroclastic classification (section 2.2.2, p.7 and Fig. 2.1, p.8) as a pyroclastic rock in which bombs > 75%. Cf. Pyroclastic breccia. *(Leonhard, 1823b, p.685; from the Latin agglomerare = to join; Johannsen v.2, p.289; Tomkeieff p.9)*

AGGLUTINATE. A variety of agglomerate in which the ejecta were plastic when emplaced and are cemented together by thin skins of glass. *(Tyrrell, 1931, p.66; from the Latin gluten = glue; Tomkeieff p.9)*

AGPAITE (AGPAITIC). A general term for nepheline syenites characterized by the molecular ratio of (Na_2O + K_2O) / Al_2O_3 > 1 (originally > 1.2), high contents of Na, Fe, Cl and Zr with low Mg and Ca. Agpaite is now restricted (Sørensen, 1960) to peralkaline nepheline syenites characterized by complex Zr and Ti minerals, such as eudialyte, rather than simple minerals such as zircon and ilmenite. Agpaitic has commonly, but incorrectly, been used as a synonym for peralkaline. Cf. miaskitic. *(Ussing, 1912, p.341; Agpat (now Appat), Ilímaussaq, Greenland; Tröger 781; Johannsen v.1 (2nd Edn), p.238; Tomkeieff p.10)*

AIGIRINOLITH. See aegirinolith.

AILLIKITE. An ultramafic carbonate-rich lamprophyre consisting of various phenocrysts including olivine, diopsidic pyroxene, amphiboles and phlogopite in a matrix of similar minerals with at least partly primary carbonate and minor perovskite but no melilite. *(Kranck, 1939, p.75; Aillik, Labrador, Canada; Tomkeieff p.10)*

AILSYTE (AILSITE). A local name for a variety of microgranite or quartz microsyenite consisting of orthoclase, quartz and riebeckite. Cf. paisanite. *(Heddle, 1897, p.266; Ailsa Craig, Firth of Clyde, Scotland, UK; Tröger 32; Tomkeieff p.10)*

AIOUNITE. A local name for a variety of melteigite containing titanian augite and biotite in a cryptocrystalline groundmass of uncertain composition. *(Duparc, 1926, p.b119; El Aïoun, Morocco; Tröger(38) 404½; Tomkeieff p.10)*

AKENOBEITE. A local name for an aplitic dyke rock which is associated with granodiorite and composed of oligoclase, orthoclase, quartz, chloritized biotite and garnet. *(Kato, 1920, p.17; Higashiyama, Akénobé district, Tajima, Japan; Tröger 112; Johannsen v.2,*

p.360; Tomkeieff p.11)

AKERITE. A collective term for varieties of microsyenite and micromonzonite consisting of alkali feldspar, with more or less oligoclase, biotite, pyroxene and often quartz. They are characterized by rectangular oligoclase. *(Brögger, 1890, p.43; Vestre Aker, Oslo Igneous Province, Norway; Tröger 782; Johannsen v.3, p.62; Tomkeieff p.11)*

AKOAFIMITE. A leucocratic variety of hornblende-bearing quartz norite of QAPF field 10* (Fig. 2.4, p.22) belonging to the charnockitic rock series. It is suggested (Streckeisen, 1974, p.357) that this term should be abandoned. *(Schüller, 1949, p.574; Akoafim, Cameroon)*

ALABRADORITE. An obsolete name applied to rocks which contain alkali feldspar but no labradorite. *(Senft, 1857, Table II; Tomkeieff p.11)*

ALASKITE. A leucocratic variety of alkali feldspar granite consisting almost entirely of quartz and alkali feldspar. May be used as a synonym for leucocratic alkali feldspar granite of QAPF field 2 (p.23). *(Spurr, 1900a, p.189; named after Alaska, USA; Tröger 14; Johannsen v.2, p.106; Tomkeieff p.11)*

ALBANITE. A local name for a variety of leucitite composed essentially of leucite and clinopyroxene in about equal proportions with subordinate plagioclase. Minor amounts of alkali feldspar, nepheline and olivine may be present. The name was previously used for a bituminous material from Albania. *(Washington, 1920, p.47; Lake Albano, Alban Hills, near Rome, Italy; Tröger 784; Tomkeieff p.11)*

ALBASALT. An obsolete name proposed for a basalt with excess alumina and alkalis. *(Belyankin, 1931, p.30; from an abbreviation of aluminate basalt; Tomkeieff p.11)*

ALBITITE. A variety of alkali feldspar syenite consisting almost entirely of albite. *(Turner, 1896, p.728; Meadow Valley, Plumas County,*

California, USA; Tröger 169; Johannsen v.3, p.138; Tomkeieff p.12)

ALBITOPHYRE. An obsolete term for a porphyritic rock containing albite phenocrysts in a feldspathic groundmass. A variety of keratophyre. *(Coquand, 1857, p.78; Tröger 787; Johannsen v.3, p.139; Tomkeieff p.12)*

ALBORANITE. An obsolete local name for a variety of basalt containing phenocrysts of hypersthene (= enstatite), augite and calcic plagioclase in a groundmass of plagioclase, augite and glass. *(Becke, 1899, p.553; Alboran Island, near Cabo de Gata, Spain; Tröger 127; Johannsen v.3, p.284; Tomkeieff p.13)*

ALENTEGITE. An obsolete name suggested for a group of mica quartz "diorites" containing 16% to 33% of quartz. Erroneously spelt alentecite by Tomkeieff. *(Marcet Riba, 1925, p.293; Alentejo, Portugal; Tomkeieff p.13)*

ALEUTITE. A term proposed for a variety of porphyritic basaltic andesite with a fine-grained groundmass whose feldspars are intermediate in composition between those found in basalt and andesite. *(Spurr, 1900a, p.190; Katmai region, Aleutian Peninsula, Alaska, USA; Tröger 342; Johannsen v.3, p.179; Tomkeieff p.13)*

ALEXOITE. An obsolete local name for a variety of dunite composed of olivine, pyrrhotite and smaller amounts of magnetite and pentlandite. *(Walker, 1931, p.5; Alexo Mine, Dundonald Township, Ontario, Canada; Tröger 775; Johannsen v.4, p.407; Tomkeieff p.13)*

ALGARVITE. A name proposed for a variety of biotite melteigite. *(Lacroix, 1922, p.646; Navete, Caldas de Monchique, Algarve, Portugal; Tröger 610; Johannsen v.4, p.330; Tomkeieff p.14)*

ALKALI. A prefix given to a rock which contains either: (1) modal foids and/or alkali amphiboles or pyroxenes or (2) normative foids or acmite. *(Iddings, 1895b, p.183; Tomkeieff p.14)*

ALKALI ANDESITE. A term used as a synonym for

trachyandesite. *(Original reference uncertain; Tröger 1003; Tomkeieff p.14)*

ALKALI BASALT. A term originally used for basalts containing accessory foids. Such rocks generally contain a titanian augite and olivine as their main ferromagnesian phases. Now defined chemically as a variety of basalt in TAS field B (p.36) which contains normative nepheline. Cf. subalkali basalt. *(Hibsch, 1910, p.402; Böhmisches Mittelgebirge (now České Středohoří), N. Bohemia, Czech Republic; Tröger 381; Johannsen v.4, p.69; Tomkeieff p.14)*

ALKALI DIORITE. A variety of diorite with alkaline character due to the presence of such dark constituents as barkevikite (see p.44), kaersutite etc. *(Niggli, 1931, p.323)*

ALKALI FELDSPAR CHARNOCKITE. A member of the charnockitic rock series equivalent to orthopyroxene alkali feldspar granite QAPF field 2 (Table 2.10, p.20). *(Streckeisen, 1974, p.355)*

ALKALI FELDSPAR FOIDITE. A term suggested as an alternative to phonolitic foidite. *(Streckeisen, 1978, p.7)*

ALKALI FELDSPAR GRANITE. A special term for a variety of granite in which plagioclase is less than 10% of the total feldspar. Now defined modally in QAPF field 2 (Fig. 2.4, p.22). *(Streckeisen, 1973, p.26)*

ALKALI FELDSPAR RHYOLITE. A special term for a variety of rhyolite in which plagioclase is less than 10% of the total feldspar. Now defined modally in QAPF field 2 (Fig. 2.11, p.31). *(Streckeisen, 1978, p.4)*

ALKALI FELDSPAR SYENITE. A special term for a variety of syenite in which plagioclase is less than 10% of the total feldspar. Now defined modally in QAPF field 6 (Fig. 2.4, p.22). *(Streckeisen, 1973, p.26)*

ALKALI FELDSPAR TRACHYTE. A special term for a variety of trachyte in which plagioclase is less than 10% of the total feldspar. Now defined modally in QAPF field 6 (Fig. 2.11, p.31). *(Streckeisen, 1978, p.4)*

ALKALI GABBRO. A variety of gabbro in QAPF field 10', with alkaline character due to the presence of analcime or nepheline and such dark constituents as barkevikite (see p.44), kaersutite, titanian augite etc. *(Niggli, 1931, p.326; Tröger 789; Tomkeieff p.15)*

ALKALI GRANITE. A term used, but not recommended, as a synonym for peralkaline granite, i.e. a granite containing alkali amphibole or pyroxene. It should not be used as a synonym for alkali feldspar granite. *(Rosenbusch, 1896, p.56; Tröger 56; Tomkeieff p.15)*

ALKALI RHYOLITE. A term used, but not recommended, as a synonym for peralkaline rhyolite, i.e. a rhyolite containing alkali amphibole or pyroxene. It should not be used as a synonym for alkali feldspar rhyolite. *(Original reference uncertain)*

ALKALI SYENITE. A term used, but not recommended, as a synonym for peralkaline syenite, i.e. a syenite containing alkali amphibole or pyroxene. It should not be used as a synonym for alkali feldspar syenite. *(Rosenbusch, 1907, p.141; Tröger 176; Tomkeieff p.16)*

ALKALI TRACHYTE. A term used, but not recommended, as a synonym for peralkaline trachyte, i.e. a trachyte containing alkali amphibole or pyroxene. It should not be used as a synonym for alkali feldspar trachyte. *(Rosenbusch, 1908, p.916; Tröger 207; Tomkeieff p.16)*

ALKALIPLETE. An obsolete term for a melanocratic rock in which $(Na_2O + K_2O) > CaO$. *(Brögger, 1898, p.265; Tomkeieff p.16)*

ALKALIPTOCHE. An obsolete term for igneous rocks poor in alkalis. *(Loewinson-Lessing, 1900b, p.241; Tomkeieff p.16)*

ALKORTHOSITE. An obsolete name for a variety of syenite consisting predominantly of Na-orthoclase. *(Eckermann, 1942, p.403; Tomkeieff p.16)*

ALLALINITE. A local name for a saussuritized gabbro, consisting of uralite (= actinolite pseudomorph after pyroxene), talc, saussurite and altered olivine. Many original textures are preserved. *(Rosenbusch, 1896, p.328; Allalin, Zermatt, Switzerland; Johannsen v.3, p.229; Tomkeieff p.17)*

ALLGOVITE. An obsolete name temporarily proposed for a group of basaltic rocks which could not be classified as melaphyres or trapps as their age was unknown. *(Winkler, 1859, p.669; Allgovia, Allgäuer Alps, Bavaria, Germany; Tröger 790; Johannsen v.3, p.299; Tomkeieff p.17)*

ALLIVALITE. A variety of troctolite composed of olivine and highly calcic plagioclase. *(Harker, 1908, p.71; Allival, now Hallival, Island of Rhum, Scotland, UK; Tröger 364; Johannsen v.3, p.348; Tomkeieff p.17)*

ALLOCHETITE. A porphyritic fine-grained variety of nepheline monzosyenite containing phenocrysts of labradorite, orthoclase, titanian augite and nepheline in a felted groundmass of augite, biotite, hornblende, nepheline and orthoclase. *(Ippen, 1903, p.133; Allochet Valley, Mt Monzoni, Alto Adige, Italy; Tröger 518; Johannsen v.4, p.176; Tomkeieff p.17)*

ALLOITE. An obsolete term for a tuff consisting of fragments of feldspathic glass and some crystals. *(Cordier, 1816, p.366; Tomkeieff p.18)*

ALNÖITE. An ultramafic rock with phenocrysts of phlogopite-biotite, olivine and augite in a groundmass of melilite (often altered to calcite), augite and/or biotite with minor perovskite, garnet and calcite. *(Rosenbusch, 1887, p.805; Alnö Island, Västernorrland, Sweden; Tröger 746; Johannsen v.4, p.385; Tomkeieff p.19)*

ALPHA GRANITE. A possible term suggested for granites falling into QAPF field 3a. *(Streckeisen, 1973, p.28)*

ALSBACHITE. A local name for a porphyritic dyke rock containing phenocrysts of quartz, feldspar and garnet in a fine-grained groundmass of quartz, feldspar and colourless mica. *(Chelius, 1892, p.2; Melibocus, Alsbach, Odenwald, Germany; Tröger 111; Johannsen v.2, p.359; Tomkeieff p.20)*

ALVIKITE. A special term in the carbonatite classification for the medium- to fine-grained variety of calcite carbonatite consisting principally of calcite (section 2.3, p.10). *(Eckermann, 1942, p.403; Alvik, Alnö Island, Västernorrland, Sweden; Tomkeieff p.21)*

AMBONITE. A local collective name for a series of andesites and dacites containing cordierite possibly formed by assimilation of sillimanite-cordierite gneiss. *(Verbeek, 1905, p.100; Ambon Island, Moluccas, Indonesia; Tröger 791; Johannsen v.3, p.176; Tomkeieff p.21)*

AMHERSTITE. A leucocratic variety of quartz monzodiorite of QAPF field 9* (Fig. 2.4, p.22) belonging to the charnockitic rock series. The rock consists essentially of andesine-microcline antiperthite with minor quartz and hypersthene (= enstatite). It is suggested (Streckeisen, 1974, p.357) that this term should be abandoned. *(Although attributed to Watson & Taber, 1913, the reference does not contain the name although it does describe the type rock; Amherst County, Virginia, USA; Tröger 294; Tomkeieff p.22)*

AMIATITE. A term from an obsolete chemical classification, based on feldspar composition rather than SiO_2 alone, for a class of rocks in which $CaO : Na_2O : K_2O \approx 1.1 : 1 : 1.8$ and $SiO_2 \approx 63\%$. *(Lang, 1891, p.226; Mt Amiata, Tuscany, Italy; Tröger 792; Tomkeieff p.22)*

AMNEITE. A mnemonic name suggested for a rock consisting of *am*phibole and *ne*pheline. *(Belyankin, 1929, p.22)*

AMPASIMENITE. A local name for a porphyritic variety of ijolite or nephelinite with a glassy matrix. *(Lacroix, 1922, p.647; Ampasimena, Madagascar; Tröger 793; Tomkeieff p.22)*

AMPHIBOLDITE. An erroneous spelling of amphibololite. *(Tomkeieff et al., 1983, p.23)*

AMPHIBOLEID. An obsolete field term for a coarse-grained igneous rock consisting almost entirely of amphibole. *(Johannsen, 1911, p.320; Tomkeieff p.22)*

AMPHIBOLIDE. A revised spelling recommended to replace the field term amphiboleid. Now obsolete. *(Johannsen, 1926, p.182; Johannsen v.1, p.57; Tomkeieff p.22)*

AMPHIBOLITE. A term originally used for any rock with a groundmass of hornblende in which other minerals are disseminated. Now solely used for metamorphic rocks consisting of hornblende and plagioclase. *(Brongniart, 1813, p.40; Tröger 700; Johannsen v.4, p.442; Tomkeieff p.22)*

AMPHIBOLOLITE. A general term used in France for medium- to coarse-grained igneous rocks composed almost entirely of amphibole. *(Lacroix, 1894, p.270; Tröger 700; Tomkeieff p.23)*

AMPHIGENITE. An obsolete term originally used for leucite-tephrites but later used for intrusive rocks composed of more than 90% modal leucite. *(Cordier, 1842, vol.1, p.388; from the French amphigène = leucite; Tröger 794; Johannsen v.4, p.337; Tomkeieff p.23)*

AMYGDALITE. An old term for a coarse-grained amygdaloidal basalt with "nodules or kernals of chalcedony, agate, calcareous spar". Cf. mandelstein. *(Pinkerton, 1811a, p.89; Tomkeieff p.24)*

AMYGDALOID. A term proposed for a rock containing amygdales. *(Cronstedt, 1758, p.228; Tomkeieff p.24)*

AMYGDALOPHYRE. An obsolete term for a porphyritic rock containing amygdales. *(Jenzsch, 1853, p.395; Tröger 795; Tomkeieff p.24)*

ANABOHITSITE. A local name for a variety of websterite consisting of hypersthene (= enstatite), augite, ilmenite and magnetite with minor olivine and hornblende. *(Lacroix, 1914,* *p.419; Anabohitsy, Madagascar; Tröger 680; Johannsen v.4, p.464; Tomkeieff p.24)*

ANALCIME BASALT. A term used for an alkaline mafic volcanic rock consisting of olivine, titanian augite and analcime with minor feldspars. The name should not be used as the term basalt is now restricted to a rock containing essential plagioclase. As the rock is a variety of foidite it should be given the appropriate name, e.g. olivine analcimite. *(Lindgren, 1886, p.727; Highwood Mts, Montana, USA; Tröger 655; Johannsen v.4, p.345; Tomkeieff p.25)*

ANALCIME BASANITE. Now defined in QAPF field 14 (Fig. 2.11, p.31) as a variety of basanite in which analcime is the most abundant foid. *(Although this term is attributed to Hibsch, 1920, p.69, he does not use the name, but only describes it in the group of nepheline basanites; Tröger 599)*

ANALCIME DIORITE. Now defined in QAPF field 14 (Fig. 2.4, p.22) as a variety of foid diorite in which analcime is the most abundant foid.

ANALCIME GABBRO. Now defined in QAPF field 14 (Fig. 2.4, p.22) as a variety of foid gabbro in which analcime is the most abundant foid. The special term teschenite may be used as an alternative.

ANALCIME MONZODIORITE. Now defined in QAPF field 13 (Fig. 2.4, p.22) as a variety of foid monzodiorite in which analcime is the most abundant foid.

ANALCIME MONZOGABBRO. Now defined in QAPF field 13 (Fig. 2.4, p.22) as a variety of foid monzogabbro in which analcime is the most abundant foid.

ANALCIME MONZOSYENITE. Now defined in QAPF field 12 (Fig. 2.4, p.22) as a variety of foid monzosyenite in which analcime is the most abundant foid. The term is synonymous with analcime plagisyenite.

ANALCIME PHONOLITE. Now defined in QAPF field 11 (Fig. 2.11, p.31) as a variety of phonolite in which analcime is the most abundant foid. *(Pelikan, 1906, p.118; Radzein (now Radejčn), between Lovosice and Teplice, N. Bohemia, Czech Republic; Johannsen v.4, p.132)*

ANALCIME PLAGISYENITE. Now defined in QAPF field 12 (Fig. 2.4, p.22) as a variety of foid plagisyenite in which analcime is the most abundant foid. The term is synonymous with analcime monzosyenite.

ANALCIME SYENITE. Now defined in QAPF field 11 (Fig. 2.4, p.22) as a variety of foid syenite in which analcime is the most abundant foid. *(Hibsch, 1899, p.72; Schlossberg, near Velké Březno, Böhmisches Mittelgebirge (now České Středohoří), N. Bohemia, Czech Republic; Johannsen v.4, p.114; Tomkeieff p.24)*

ANALCIMITE. An alkaline volcanic rock composed essentially of analcime, titanian augite, opaques and only minor olivine. Now defined as a variety of foidite in QAPF field 15c (Fig. 2.11, p.31) in which analcime is the most abundant foid. *(Gemmellaro, 1845, p.318, ; the reference cited by Johannsen and Tomkeieff is erroneous; Cyclope Island, Catania, Sicily, Italy; Tröger 654; Johannsen v.4, p.336; Tomkeieff p.24)*

ANALCIMOLITH. An obsolete term proposed for a monomineralic volcanic rock consisting of analcime, i.e. a leucocratic analcimite. *(Johannsen, 1938, p.336; Tomkeieff p.25)*

ANALCITITE. As analcime is the mineral spelling recommended by the International Mineralogical Association analcimite should be used in preference to analcitite. *(Pirsson, 1896, p.690; Johannsen v.4, p.337)*

ANAM-AEGISODITE. A mnemonic name suggested for a rock consisting of *an*alcime, *am*phibole, *aegi*rine and *soda*lite. *(Belyankin, 1929, p.22)*

ANAMESEID. An obsolete field term for dark coloured, non-porphyritic aphanitic igneous rock. Also called melano-aphaneid.

(Johannsen, 1911, p.321; Tomkeieff p.25)

ANAMESEID PORPHYRY. An obsolete field term for a dark-coloured porphyritic igneous rock with an aphanitic groundmass. Also called melanophyreid. *(Johannsen, 1911, p.321)*

ANAMESITE. An obsolete term for basaltic rocks which are between basalt and dolerite in texture. *(Leonhard, 1832, p.150; from the Greek anamesos = in the middle; Tröger 796; Johannsen v.3, p.291; Tomkeieff p.25)*

ANATECTITE (ANATEXITE). An igneous rock produced by the remelting of crustal rocks. Cf. prototectite and syntectite. *(Loewinson-Lessing, 1934, p.7; Tomkeieff p.26)*

ANCHORITE. An obsolete local name for a diorite containing occasional syenite veins and scattered mafic segregation patches. *(Lapworth, 1898, p.419; Anchor Inn, Nuneaton, Warwickshire, England, UK; Tröger 797; Tomkeieff p.26)*

ANDELATITE. An obsolete term suggested for an extrusive rock intermediate between andesite and latite. *(Johannsen, 1920b, p.174; Tröger 798; Tomkeieff p.27)*

ANDENDIORITE. A local name suggested for diorites that are younger than the usual ones. *(Stelzner, 1885, p.201; Tröger 799)*

ANDENGRANITE. A local name suggested for granitic rocks that are younger than the usual ones. *(Stelzner, 1885, p.201; Tröger 799)*

ANDENNORITE. A local name suggested for norites that are younger than the usual ones. *(Wolff, 1899, p.482; Tröger 799; Tomkeieff p.27)*

ANDERSONITE. An obsolete name suggested for a group of amphibole tonalites. *(Marcet Riba, 1925, p.293; named after W. Anderson)*

ANDESIBASALT. A term used in the USSR as a synonym for basaltic andesite. *(Original reference uncertain)*

ANDESILABRADORITE. A name used in France for rocks transitional between andesite and labradorite (an old French group name which includes basalt). *(Tröger, 1935, p.322; Tröger*

890; Tomkeieff p.27)

ANDESINE BASALT. A term for a dark-coloured olivine-bearing andesite. *(Iddings, 1913, p.191; Lava of 1910, Mt Etna, Sicily, Italy; Tröger 327; Tomkeieff p.27)*

ANDESINITE. A term proposed for a coarse-grained rock consisting essentially of andesine. *(Turner, 1900, p.110; Tröger 293; Johannsen v.3, p.146; Tomkeieff p.27)*

ANDESITE. An intermediate volcanic rock, usually porphyritic, consisting of plagioclase (frequently zoned from labradorite to oligoclase), pyroxene, hornblende and/or biotite. Now defined modally in QAPF fields 9 and 10 (Fig. 2.11, p.31) and, if modes are not available, chemically in TAS field O2 (Fig. 2.14, p.35). *(Buch, 1836, p.190; Andes Mts, South America; Tröger 324; Johannsen v.3, p.160; Tomkeieff p.27)*

ANDESITE-BASALT. An obsolete term originally used as a synonym for basanite and later for rocks intermediate in composition between andesite and basalt. *(Bořický, 1874, p.43; Tröger 801; Tomkeieff p.27)*

ANDESITE-TEPHRITE. An obsolete term for a foid-bearing variety of trachyandesite composed of plagioclase, sanidine, augite and haüyne. Olivine and nepheline may also be present. *(Colony & Sinclair, 1928, p.307; Tröger 267; Tomkeieff p.27)*

ANDESITOID. Originally used as a variety of andesite containing considerable amounts of sanidine in the groundmass. Now proposed for preliminary use in the QAPF "field" classification (Fig. 2.19, p.39) for volcanic rocks tentatively identified as andesite. *(Sigmund, 1902, p.282; Tröger 802; Tomkeieff p.28)*

ANGARITE. A local name for Siberian basalts and dolerites. *(Loewinson-Lessing et al., 1932, p.71; River Angara, Siberia, Russian Federation; Tomkeieff p.28)*

ANGOLAITE. A peralkaline syenite consisting of alkali feldspar (87% in type material), titanaugite with aegirine-augite rims (5%),

aegirine (0.7%), fayalite (3.6%) and accessory magnetite, apatite, aenigmatite, hastingsite and cancrinite. *(Andrade, 1954; Chamaco complex, Angola)*

ANKARAMITE. A porphyritic melanocratic basanite with abundant phenocrysts of pyroxene and olivine. *(Lacroix, 1916a, p.182; Ankaramy, Ampasindava, Madagascar; Tröger 408; Johannsen v.3, p.338; Tomkeieff p.29)*

ANKARANANDITE. A group name for varieties of orthopyroxene alkali feldspar syenite to orthopyroxene syenite of QAPF fields 6–6* to 7–7* (Fig. 2.4, p.22) belonging to the charnockitic rock series. It is suggested (Streckeisen, 1974, p.357) that this term should be abandoned. *(Giraud, 1964, p.49; Betafo-Ankaranando, Madagascar)*

ANKARATRITE. A melanocratic variety of olivine nephelinite containing biotite. *(Lacroix, 1916b, p.256; Ankaratra, Madagascar; Tröger 623; Johannsen v.4, p.366; Tomkeieff p.29)*

ANORTHITISSITE. An obsolete name given to a melanocratic variety of gabbro consisting of 70% hornblende and 22% anorthite. *(Tröger, 1935, p.175; Koswinski Mts, Urals, Russian Federation; Tröger 403; Johannsen v.3, p.350; Tomkeieff p.30)*

ANORTHITITE. A term proposed for a variety of anorthosite consisting essentially of anorthite. In the original description the composition of the plagioclase was probably incorrectly determined. *(Turner, 1900, p.110; Tröger 300; Johannsen v.3, p.338; Tomkeieff p.30)*

ANORTHOBASE. According to Tomkeieff et al. (1983) this rock is a diabase containing calcic plagioclase. However, the term does not appear in the reference cited. The same incorrect reference is also cited by Loewinson-Lessing (1932). *(Belyankin, 1911, p.363; Tomkeieff p.30)*

ANORTHOCLASE BASALT. A variety of trachybasalt containing phenocrysts of anorthoclase, augite and enstatite in a groundmass of these minerals, labradorite and opaques. *(Skeats & Summers, 1912, p.33; Sugarloaf Hill, Mt Macedon, Victoria, Australia; Tröger 273; Tomkeieff p.30)*

ANORTHOCLASITE. A variety of alkali feldspar syenite composed almost entirely of anorthoclase. *(Loewinson-Lessing, 1901, p.114; Tröger 805; Johannsen v.3, p.5; Tomkeieff p.30)*

ANORTHOLITE. An erroneous term stated to be synonymous with anorthosite. However, the reference cited (Hunt, 1864) does not contain the term, only the term anorthosite. *(Tomkeieff et al., 1983, p.30)*

ANORTHOPHYRE. An obsolete name for a porphyritic anorthoclase syenite. Cf. pilandite. *(Loewinson-Lessing, 1900a, p.174; Tröger 804; Tomkeieff p.30)*

ANORTHOSITE. A leucocratic plutonic rock consisting essentially of plagioclase often with small amounts of pyroxene. Now defined modally in QAPF field 10 (Fig. 2.4, p.22). The term is synonymous with plagioclasite. *(Hunt, 1862, p.62; Laurentian Mts, Quebec, Canada; Tröger 291; Johannsen v.3, p.196; Tomkeieff p.31)*

ANORTHOSYENITE. An obsolete term for a variety of syenite with phenocrysts of anorthoclase. *(Loewinson-Lessing, 1900a, p.174; Tröger 805; Tomkeieff p.31)*

ANOTERITE. An obsolete name for a variety of rapakivi granite with euhedral quartz thought to have crystallized at a high level. *(Sederholm, 1891, p.21; from the Greek anoteros = higher up; Tomkeieff p.31)*

ANTHRAPHYRE. An obsolete term proposed for igneous rocks intruding anthracite-rich formations. *(Ebray, 1875, p.291)*

ANTIFENITEPEGMATITE. A sodic syenite pegmatite composed mainly of antiperthite with minor biotite and opaques. *(Barth, 1927, p.97; Island of Seiland, Finnmark, Norway; Tröger 173)*

ANTIRAPAKIVI. A term originally used for a texture in which ovoids of plagioclase are mantled by orthoclase. Tomkeieff *et al.* use it as a variety of rapakivi granite with this texture. *(Vakar, 1931, p.1026; Umylymnan Mt, Kolyma region, Siberia, Russian Federation; Tomkeieff p.32)*

ANTSOHITE. A local name for a lamprophyric dyke rock consisting of phenocrysts of biotite in a groundmass of biotite, hornblende and interstitial quartz. *(Lacroix, 1922, p.431; Antsohy, Tsaratanana, Madagascar; Tröger 147; Tomkeieff p.32)*

APACHITE. A variety of peralkaline phonolite rich in sodic amphiboles and containing sodic pyroxene and aenigmatite. *(Osann, 1896, p.402; Apache (now Davis) Mts, Texas, USA; Tröger 466; Johannsen v.4, p.129; Tomkeieff p.33)*

APANEITE. A mnemonic name from *apa*tite and *ne*pheline, used for alkaline intrusive rocks consisting mainly of apatite with variable amounts of nepheline and minor aegirine and biotite, i.e. a nepheline-bearing apatitolite. *(Vlodavets, 1930, p.34; Khibina complex, Kola Peninsula, Russian Federation; Tröger(38) 775⅓; Tomkeieff p.33)*

APATITOLITE. A rock of magmatic origin composed essentially of apatite. *(Loewinson-Lessing, 1936, p.989; Khibina complex, Kola Peninsula, Russian Federation; Tomkeieff p.33)*

APHANEID. An obsolete field term for non-porphyritic rocks which contain a megascopically indeterminable component. *(Johannsen, 1911, p.318; from the Greek aphanes = invisible; Tomkeieff p.33)*

APHANIDE. A revised spelling recommended to replace the field term aphaneid. Now obsolete. *(Johannsen, 1926, p.182; from the Greek aphanes = invisible; Johannsen v.1, p.56; Tomkeieff p.33)*

APHANITE. A term originally suggested by Haüy for fine-grained rocks, including trapp, ophite and variolite, of dioritic composition.

Now generally applied to all fine-grained igneous rocks. *(D'Aubuisson de Voisins, 1819, p.147; Tröger 806; Tomkeieff p.33)*

APLITE.   A term used both for fine-grained granitic rocks, consisting only of feldspar and quartz, and as a group name for any leucocratic fine-grained to aphanitic dyke rock. *(Leonhard, 1823a, p.51;from the Greek haploos = simple;Tröger 807;Johannsen v.2, p.91; Tomkeieff p.34)*

APLO-.   A prefix used for light-coloured rocks with simple mineralogy and few ferromagnesian minerals, e.g. aplodiorite (Tröger 105), aplogranite (Tröger 808). Cf. haplo-. Bailey & Maufe (1960, p. 211) later abandoned the term aplogranite in favour of binary granite. *(Bailey & Maufe, 1916, p.160; Tomkeieff p.34)*

APLOID.   An obsolete term for a nepheline-bearing aplite. *(Shand, 1910, p.377; Tröger 809; Tomkeieff p.34)*

APLOSYENITE.   An obsolete term used as a family name for syenites with less than 5% mafics. *(Tröger, 1935, p.79; Tröger 163–165)*

APO-.   A prefix used to denote that one rock has been derived from another by a specific alteration process. This can include devitrification, e.g. aporhyolite (Tröger 810), apoperlite and apoobsidian would be acid volcanic rocks whose structure was once glassy. *(Bascom, 1893, p.828;from the Greek apo = from, off; Tomkeieff p.34)*

APOGRANITE.   An albitized or greisenized granite located in the apices of intrusions and frequently mineralized in Sn, W, Be, Nb–Ta, Li and B. *(Beus et al., 1962, p.5)*

APOTROCTOLITE.   A term used for a coarse-grained rock consisting essentially of alkali feldspar and olivine, with minor pyroxene, biotite and chlorite. *(Barth, 1944, p.57; Tomkeieff p.35)*

APPINITE.   A local, general term for medium- to coarse-grained, meso- to melanocratic rocks with conspicuous hornblende in a base of oligoclase-andesine and/or orthoclase with or without quartz. The plutonic equivalent of vogesite and spessartite. *(Bailey & Maufe, 1916, p.167;Appin district, near Ballachulish, Scotland, UK; Tröger 811; Tomkeieff p.35)*

ARAPAHITE.   A local name for a variety of basalt containing over 50% magnetite. *(Washington & Larsen, 1913, p.452; named after Arapaho Indians, Colorado, USA; Tröger 777; Johannsen v.3, p.303; Tomkeieff p.36)*

ARENDALITE.   A comprehensive term for rocks of the charnockitic series found in the Arendal area. It is suggested (Streckeisen, 1974, p.357) that this term should be abandoned. *(Bugge, 1940, p.81; Arendal, Norway)*

ARGEINITE.   A local name for a variety of olivine hornblendite composed of hornblende and olivine. *(Lacroix, 1933, p.194; Argeins, Pyrénées, France; Tröger 707; Tomkeieff p.37)*

ARIEGITE.   A local group name for a variety of websterite composed of clinopyroxene, orthopyroxene, abundant spinel and often pyrope-rich garnet and hornblende. *(Lacroix, 1901a, p.360; Lac de Lherz, now Lhers, Ariège, Pyrénées, France; Tröger 684; Johannsen v.4, p.461; Tomkeieff p.38)*

ARIZONITE.   A local name for a dyke rock consisting of 80% quartz with 18% orthoclase and 2% muscovite. *(Spurr & Washington, 1917, p.34; Helvitia, Arizona, USA; Tröger 7; Johannsen v.2, p.32; Tomkeieff p.38)*

ARKESINE.   An obsolete name for hornblende biotite granodiorites and granites in the Dent Blanche nappe of the Swiss and Italian Alps. *(Jurine, 1806, p.373; Mont Blanc, France; Tomkeieff p.39)*

ARKITE.   A local name for a leucocratic variety of nepheline fergusite consisting of leucite or pseudoleucite in a matrix of nepheline, melanite garnet, biotite, sodic pyroxene, amphibole and minor alkali feldspar. *(Washington, 1901, p.617; named after the abbre-*

viation of Arkansas, USA; Tröger 629; Johannsen v.4, p.274; Tomkeieff p.39)

ARSOITE. A local name for a variety of trachyte consisting of phenocrysts of sanidine, diopside, andesine and a little olivine in a groundmass of sanidine, oligoclase and diopside with minor amounts of leucite. The same rock had previously been called ciminite by Washington. *(Reinisch, 1912, p.121; Arso flow of 1302, Epomeo, Ischia, Italy; Tröger 252; Johannsen v.4, p.36; Tomkeieff p.40)*

ASCHAFFITE. A local name for a variety of kersantite (a lamprophyre) with abundant biotite in a matrix of plagioclase and quartz. *(Gümbel, 1865, p.206; Aschaffenburg, Bavaria, Germany; Tröger 812; Johannsen v.3, p.190; Tomkeieff p.40)*

ASCLERINE. An obsolete term for a tuff consisting of fragments of altered feldspathic glass and some crystals. *(Cordier, 1816, p.372; Tomkeieff p.40)*

ASH, ASH GRAIN. Now defined in the pyroclastic classification (section 2.2.1, p.7) as a pyroclast with a mean diameter < 2 mm. *(Schmid, 1981, p.42; Tomkeieff p.41)*

ASH TUFF. Now defined in the pyroclastic classification (section 2.2.2, p.8 and Fig. 2.1, p.8) as a pyroclastic rock in which ash > 75%. The term is synonymous with tuff. See also coarse (ash) tuff, fine (ash) tuff and dust tuff. *(Hibsch, 1896, p.234; Tomkeieff p.41)*

ASH-STONE. A term used for a lithified volcanic ash. *(Williams, 1941, p.279; Tomkeieff p.40)*

ASO LAVA. Aso lava is the same as ash-stone in Japan. *(Williams, 1941, p.279; Aso Volcano, Japan; Tomkeieff p.41)*

ASPERITE. An obsolete field term for a variety of andesite with trachytic character. *(Becker, 1888, p.151; from the Latin asper = rough; Tröger 813)*

ASPRONE. An alternative name for sperone. *(Gmelin, 1814; Tomkeieff p.41)*

ASSYNTITE. A local name for a variety of nepheline syenite composed of abundant

orthoclase, smaller amounts of sodalite, and nepheline with aegirine-augite, biotite and large crystals of titanite. *(Shand, 1910, p.403; Borralan complex, Assynt, Scotland, UK; Tröger 439; Johannsen v.4, p.101; Tomkeieff p.42)*

ASTRIDITE. A green ultrabasic rock composed mainly of chromo-jadeite and picotite. *(*Willems, 1934, p.120; named after Astrid, Queen of Belgium; Tomkeieff p.42)*

ATATSCHITE. An obsolete term originally used for a glassy orthophyre containing sillimanite and cordierite, but later shown to be contact metamorphosed tuffs and agglomerates with porphyry boulders. The correct transliteration should be atachite. *(Morozewicz, 1901, p.16; Atatch Ridge, Magnitnaya Mts, S. Urals, Russian Federation; Tröger 258; Tomkeieff p.42)*

ATLANTITE. An obsolete term for a melanocratic variety of basanite. *(Lehmann, 1924, p.118; named after islands of the Atlantic Ocean; Tröger 577; Johannsen v.4, p.240; Tomkeieff p.42)*

AUGANITE. A collective name for augite andesites composed essentially of augite and plagioclase. *(Winchell, 1912, p.657; Tröger 814; Johannsen v.3, p.280; Tomkeieff p.44)*

AUGITITE. A volcanic rock composed essentially of augite and opaque phenocrysts in an indeterminate dark coloured matrix which may be analcime. *(Rosenbusch, 1883, p.404; Tröger 580; Johannsen v.4, p.21; Tomkeieff p.45)*

AUGITOPHYRE. An obsolete name for an augite porphyry. *(Palmieri & Scacchi, 1852, p.67; Johannsen v.3, p.322; Tomkeieff p.45)*

AVEZACITE. A local name for a variety of pyroxene hornblendite which occurs as dykes and consists of phenocrysts of hornblende in a groundmass of hornblende, augite and abundant ilmenite. *(Lacroix, 1901b, p.826; Avezac-Prat, S.W. Lannemezan, Pyrénées, France; Tröger 714; Johannsen v.4, p.449;*

Tomkeieff p.48)

BAHIAITE. An obsolete local name for a variety of hornblende orthopyroxenite consisting of hypersthene (= enstatite), hornblende and small amounts of olivine and spinel. *(Washington, 1914a, p.86; near Maracas, Bahia, Brazil; Tröger 679; Johannsen v.4, p.432; Tomkeieff p.50)*

BAJAITE. Originally used as "bajaite series rocks" the term is now used as a rock name for a variety of boninite with MgO ≈ 8%, SiO_2 ≈ 56% and characterized by high contents of Sr (> 1000 ppm) and high K/Rb ratios (> 1000). It was originally described as a magnesian andesite. *(Rogers et al., 1985, p.392; Baja California, Mexico)*

BALDITE. A variety of analcimite composed essentially of analcime and titanian augite with minor olivine and opaques. The rock was originally called analcime basalt. *(Johannsen, 1938, p.393; Big Baldy Mt, Little Belt Mts, Montana, USA; Tröger(38) 640½; Tomkeieff p.51)*

BALGARITE. A local name for a variety of trachyte with marked spheroidal structure composed essentially of perthite with minor amounts of biotite and aegirine-augite. *(Borisov, 1963, p.225; Balgarovo, Bulgaria)*

BALTORITE. A local comprehensive term for potassic vogesites and potassic minettes. *(Comucci, 1937, p.737; Alto Baltoro, Karakorum, Kashmir, India)*

BANAKITE. A collective term, related to absarokite and shoshonite, for a trachyandesitic rock containing phenocrysts of augite and sometimes olivine in a groundmass of sanidine mantling labradorite–andesine, augite, biotite, analcime and opaques. It is similar to absarokite but contains less olivine and augite. *(Iddings, 1895a, p.947; named after Bannock Indians, Yellowstone National Park, Wyoming, USA; Tröger 527; Johannsen v.4, p.44; Tomkeieff p.51)*

BANATITE. A term used for a series of rocks ranging from granite to diorite (but mainly granodiorite) that were intruded in Upper Cretaceous time in the Banat and adjacent areas of Hungary and Yugoslavia. *(Cotta, 1864b, p.13; Banat district, Transylvania, Romania; Tröger 816; Johannsen v.2, p.348; Tomkeieff p.51)*

BANDAITE. A variety of glassy andesite composed of labradorite, sanidine, hypersthene (= enstatite) and augite in a base of glass containing potential quartz and feldspar. *(Iddings, 1913, p.111; Bandai San, Iwasciro Province, Japan; Tröger 158; Johannsen v.2, p.416; Tomkeieff p.52)*

BARAMITE. Altered ultramafic plutonic rock composed of serpentine, magnesite and opal. *(Hume et al., 1935, p.26; Baramia Mine, Upper Egypt, Egypt; Tröger(38) 677½; Tomkeieff p.52)*

BARNEITE. A mnemonic name suggested for a rock consisting of *bar*kevikite and *ne*pheline. *(Belyankin, 1929, p.22)*

BARSHAWITE. A melanocratic variety of analcime nepheline monzosyenite consisting of kaersutite, titanian augite, alkali feldspar, andesine, nepheline and analcime. *(Johannsen, 1938, p.283; Barshaw, Paisley, Scotland, UK; Tröger(38) 564½; Tomkeieff p.53)*

BASALATITE. A term suggested for an extrusive rock intermediate between basalt and latite. *(Johannsen, 1920c, p.212; Tröger 1004)*

BASALT. A volcanic rock consisting essentially of calcic plagioclase and pyroxene. Olivine and minor foids or minor interstitial quartz may also be present. Now defined modally in QAPF fields 9 and 10 (Fig. 2.11, p.31) and, if modes are not available, chemically in TAS field B (Fig. 2.14, p.35). For some other varieties of basalt see also alkali basalt, high-alumina basalt, island arc basalt, mid-ocean ridge basalt (MORB), olivine basalt, olivine tholeiite, subalkali basalt, tholeiite, tholeiitic basalt, transitional basalt. *(A term of great antiquity, probably Egyptian in origin, usually attributed to Pliny,[†] AD 77 – see Johannsen for further discussion; Tröger 378; Johannsen v.3, p.246; Tomkeieff p.54)*

BASALT-TRACHYTE. An obsolete name for a trachyte containing augite and hornblende. *(Vogelsang, 1872, p.541; Tomkeieff p.55)*

BASALT-WACKE. An obsolete term for an altered basalt. *(Cotta, 1855, p.44; Tomkeieff p.55)*

BASALTIC ANDESITE. A term for a volcanic rock which has plagioclase of variable composition with the ferromagnesian minerals more commonly found in basalts, e.g. olivine. Now defined chemically in TAS field O1 (Fig. 2.14, p.35). *(Anderson, 1941, p.389; one of the Modoc basalts, Medicine Lake, California, USA)*

BASALTIC KOMATIITE (KOMATIITIC BASALT). A member of the komatiite series with MgO in the range 5% to 15% typically displaying spinifex texture and evidence of rapid quenching of high-temperature magma. Chemically they are intermediate between tholeiitic basalt and boninite and are thought to have been komatiite magmas contaminated by crustal material. *(Viljoen & Viljoen, 1969, p.61; Komati River, Barberton, Transvaal, South Africa)*

BASALTIC TRACHYANDESITE. A group term introduced for rocks intermediate between trachyandesite and trachybasalt, i.e. an analogous name to basaltic andesite in the oversaturated rocks. Now defined chemically in TAS field S2 (Fig. 2.14, p.35). *(Le Bas et al., 1986, p.747)*

BASALTIN. An obsolete term for a fine-grained variety of basalt. *(Pinkerton, 1811a, p.32; Tomkeieff p.54)*

BASALTINE. An obsolete term for a porphyritic or fine-grained basalt. *(Born, 1790, p.395; Tomkeieff p.54)*

BASALTITE. An obsolete term originally used for a variety of melaphyre, but later used in several other senses ranging from specific types of basalt to a group name for rocks consisting of labradorite, augite, nepheline and leucite. *(Raumer, 1819, p.101; Tröger 817; Tomkeieff p.54)*

BASALTOID. The term was originally used as the name for a black Egyptian basalt, but later as a collective term for black basic volcanic rocks. Now proposed for preliminary use in the QAPF "field" classification (Fig. 2.19, p.39) for volcanic rocks tentatively identified as basalt. *(Haüy, 1822, p.542; Tomkeieff p.55)*

BASALTON. An obsolete term for a coarse-grained basalt, e.g. dolerite. *(Pinkerton, 1811a, p.72; Tomkeieff p.55)*

BASANITE. A term originally used for a porphyritic basalt containing pyroxene phenocrysts, but later used as a group name for rocks composed of clinopyroxene, plagioclase, essential foids and olivine. Now defined modally in QAPF field 14 (Fig. 2.11, p.31) and, if modes are not available, chemically in TAS field U1 (Fig. 2.14, p.35). *(A term of great antiquity usually attributed to Theophrastus, 320 BC – see Johannsen for further discussion; from the Greek basanos = touchstone; Tröger 818; Johannsen v.4, p.230; Tomkeieff p.55)*

BASANITIC FOIDITE. A collective term for alkaline volcanic rocks consisting of foids with some plagioclase as defined modally in QAPF field 15b (Fig. 2.11, p.31). It is distinguished from tephritic foidite by having more than 10% modal olivine. If possible the most abundant foid should be used in the name, e.g. basanitic nephelinite, basanitic leucitite etc. *(Streckeisen, 1978, p.7)*

[†] There is considerable confusion with the derivation of the word basalt as some versions of Pliny contain the phrase "quem vocant basalten" while others contain "quem vocant basanites".

BASANITOID. A term originally used for volcanic rocks intermediate between olivine basalt and basanite. Has also been used for rocks with the composition of basanite but without modal nepheline. *(Bücking, 1881, p.154; Johannsen v.4, p.69; Tomkeieff p.55)*

BASIC. A commonly used chemical term now defined in the TAS classification (Fig. 2.14, p.35) as a rock with SiO_2 between 45% and 52%. See also ultrabasic, intermediate and acid. *(Abich, 1841, p.12; Tomkeieff p.56)*

BASITE. An obsolete group name for all basic igneous rocks. *(Cotta, 1864a, p.824; Tröger 819; Tomkeieff p.56)*

BASITE-PORPHYRY. An obsolete term for a porphyritic nepheline and leucite rock. *(Vogelsang, 1872, p.542; Tomkeieff p.56)*

BATUKITE. A melanocratic variety of leucitite essentially composed of clinopyroxene with small amounts of olivine and leucite. *(Iddings & Morley, 1917, p.595; Batuku, Sulawesi, Indonesia; Tröger 647; Johannsen v.4, p.369; Tomkeieff p.57)*

BAUCHITE. A local name for varieties of fayalite-bearing orthopyroxene granite and quartz monzonite of QAPF fields 3, 7* and 8* (Fig. 2.4, p.22). It is suggested (Streckeisen, 1974, p.357) that this term should be abandoned. *(Oyawoye, 1965, p.689; Bauchi, Jos Plateau, Nigeria; Tomkeieff p.57)*

BAULITE. An obsolete term for a variety of rhyolite. *(Bunsen, 1851, p.199; Mt Baula, Faeroe Islands; Tomkeieff p.57)*

BEBEDOURITE. A local name for a variety of biotite pyroxenite composed essentially of aegirine-augite and biotite with perovskite and opaques. *(Tröger, 1928, p.202; Bebedouro, Minas Gerais, Brazil; Tröger 692; Johannsen v.4, p.452; Tomkeieff p.58)*

BEFORSITE. A medium- to fine-grained variety of dolomite carbonatite which often occurs as dykes and consists principally of dolomite. *(Eckermann, 1928a, p.386; Bergeforsen, near Alnö Island, Väster-*

norrland, Sweden; Tröger 759; Tomkeieff p.59)

BEKINKINITE. A melanocratic variety of theralite consisting of abundant brown amphibole and lesser titanian augite with minor olivine, nepheline, labradorite and alkali feldspar. *(Rosenbusch, 1907, p.441; Mt Bekinkina, Madagascar; Tröger 515; Johannsen v.4, p.332; Tomkeieff p.59)*

BELOEILITE. A local name for a variety of sodalitolite consisting of about 70% sodalite with nepheline, alkali feldspar, oligoclase and a little aegirine, originally described as a feldspathic variety of tawite. *(Johannsen, 1920b, p.163; Beloeil, now Mt St Hilaire, Quebec, Canada; Tröger 506; Johannsen v.4, p.282; Tomkeieff p.60)*

BELUGITE. An obsolete term for a plutonic rock whose feldspars are intermediate in composition between those typical of diorite and gabbro. *(Spurr, 1900a, p.189; Beluga River, Alaska, USA; Tröger 330; Johannsen v.3, p.159; Tomkeieff p.60)*

BENMOREITE. Originally described as a member of the sodic volcanic rock series falling in Daly's compositional gap between mugearite and trachyte, with a differentiation index between 65 and 75. It consists essentially of anorthoclase or sodic sanidine, Fe-olivine and ferroaugite (= augite). Now defined chemically as the sodic variety of trachyandesite in TAS field S3 (Fig. 2.14, p.35). *(Tilley & Muir, 1963, p.439; Ben More, Island of Mull, Scotland, UK)*

BENTONITE. A clayey material, composed largely of smectite and montmorillonite, formed by the alteration and devitrification of volcanic rocks, often with shard and phenocryst outlines preserved. *(Knight, 1898, p.491; Fort Benton Formation, Wyoming, USA)*

BERESHITE. See bjerezite, which is an earlier transliteration.

BERESITE. A miners' term for a kaolinized dyke

rock which was originally a quartz porphyry but now looks like a greisen. *(Rose, 1837, p.186; Beresovsk, Urals, Russian Federation; Tröger 9; Johannsen v.2, p.23; Tomkeieff p.61)*

BERGALITE. A local name for an ultramafic lamprophyre similar to alnöite except that the matrix contains essential feldspathoids as well as melilite. The rock could be described as a haüyne melilitite. The term is essentially synonymous with polzenite. *(Söllner, 1913, p.415; Oberbergen, Kaiserstuhl, Baden, Germany; Tröger 662; Johannsen v.4, p.379; Tomkeieff p.61)*

BERGMANITE. An obsolete term for a variety of serpentinite. *(Pinkerton, 1811b, p.53; named after T. Bergman, 1780; Tomkeieff p.61)*

BERINGITE. An local name for a melanocratic variety of trachyte containing nearly 50% of barkevikite (see p.44). *(Starzynski, 1912, p.671; Bering Island, Bering Sea, Russian Federation; Tröger 231; Johannsen v.3, p.106; Tomkeieff p.61)*

BERMUDITE. A local name for a lamprophyric rock consisting of biotite and minor titanian augite in a groundmass of analcime (probably after nepheline). Possibly equivalent in composition to ouachitite. *(Pirsson, 1914, p.340; named after Bermuda; Tröger 621; Johannsen v.4, p.345; Tomkeieff p.61)*

BERONDRITE. A melanocratic variety of theralite with kaersutite, titanian augite and labradorite rimmed with alkali feldspar and small amounts of nepheline. *(Lacroix, 1920, p.21; Beronda Valley, Bezavona Massif, Madagascar; Tröger 545; Johannsen v.4, p.198; Tomkeieff p.62)*

BESCHTAUITE. A local name for a variety of porphyritic rhyolite that contains phenocrysts of sanidine and oligoclase. *(*Bayan, 1866; Mt Beschtau, Pyatigorsk, N. Caucasus, Russian Federation; Tröger 95; Tomkeieff p.62)*

BETA GRANITE. A possible term suggested for granites falling into QAPF field 3b. *(Streckeisen, 1973, p.28)*

BIELENITE. An obsolete local name for a variety of olivine websterite consisting of diallage (= altered diopside), enstatite and olivine. *(Kretschmer, 1917, p.153; Biele River (now Bělá), Moravia, Czech Republic; Tröger 820; Johannsen v.4, p.437; Tomkeieff p.62)*

BIGWOODITE. A variety of alkali feldspar syenite consisting essentially of orthoclase, microcline and albite with traces of foids and sodic pyroxene and/or amphibole. *(Quirke, 1936, p.179; Bigwood Township, Ontario, Canada; Tröger(38) 163½; Johannsen v.4, p.31; Tomkeieff p.62)*

BIMSSTEIN (BYNFTEIN, BYMSFTEIN). An old German term for pumice. *(Boetius de Boot, 1636, p.576; Tröger 821; Johannsen v.2, p.282)*

BINARY GRANITE. An obsolete term applied to a variety of granite containing both muscovite and biotite. *(Keyes, 1895, p.714; Johannsen v.2, p.225; Tomkeieff p.63)*

BINEITE. A mnemonic name suggested for a rock consisting of *bi*otite and *ne*pheline. *(Belyankin, 1929, p.22)*

BINEMELITE. A mnemonic name suggested for a rock consisting of *bi*otite, *ne*pheline and *mel*ilite. *(Belyankin, 1929, p.22)*

BIOTITITE. A rock consisting almost entirely of biotite. *(Washington, 1927, p.187; Villa Senni, near Rome, Italy; Tröger 720; Johannsen v.4, p.441; Tomkeieff p.66)*

BIQUAHORORTHANDITE. An unwieldy name constructed by Johannsen to illustrate some of the possibilities of the mnemonic classification of Belyankin (1929) for a granitic rock consisting of *bi*otite, *qua*rtz, *hor*nblende, *ortho*clase and *and*esine. Cf. hobiquandorthite and topatourbiolilepiquorthite. *(Johannsen, 1931, p.125)*

BIRKREMITE (BJERKREIMITE). A member of the charnockitic rock series, this term was originally used for an orthopyroxene alkali feldspar granite in QAPF field 2 (Fig. 2.4, p.22), but as it was later shown to contain mesoperthite, it became a mesoperthite charnockite of QAPF field 3. It is suggested (Streckeisen, 1974, p.358) that this term should be abandoned. *(Kolderup, 1903, p.117; Birkrem, Egersund district, Norway; Tröger 15; Johannsen v.2, p.46; Tomkeieff p.66)*

BISTAGITE. An obsolete local name for a variety of clinopyroxenite composed essentially of diopside. *(Yachevskii, 1909, p.46; Bis-Tag Range, Yenisey Province, Siberia, Russian Federation; Tröger(38) 821½; Tomkeieff p.67)*

BIZARDITE. A local name applied to an ultramafic lamprophyre or aillikite. *(Stansfield, 1923b, p.551; Bizard Island, Montreal, Quebec, Canada; Tröger 665; Tomkeieff p.68)*

BIZEUL. According to Tomkeieff a local French name for diabase or diorite. *(Tomkeieff et al., 1983, p.68)*

BJEREZITE. A term for a porphyritic micromonzonite or monzogabbro rich in nepheline phenocrysts and containing analcime, titanian augite rimmed by aegirine. Later transliterated as bereshite. *(Erdmanns-dörffer, 1928, p.88; Bjerez River, Yenisey River, Siberia, Russian Federation; Tröger 562; Johannsen v.4, p.292; Tomkeieff p.68)*

BJERKREIMITE. An alternative spelling of birkremite. *(Original reference uncertain)*

BJÖRNSJÖITE. A local name for a variety of trachyte consisting essentially of albite and small amounts of aegirine-augite. *(Brögger, 1932, p.34; Lake Björnsjö, Oslo district, Norway; Tröger 205; Tomkeieff p.68)*

BLACOLITE. An obsolete term for a variety of serpentinite. *(Pinkerton, 1811b, p.53; named after J. Black, 1760; Tomkeieff p.69)*

BLAIRMORITE. A leucocratic variety of analcimite with phenocrysts of analcime in a groundmass of analcime and aegirine-augite with minor nepheline, sanidine and melanite garnet. *(Knight, 1905, p.275; Blairmore, Alberta, Canada; Tröger 653; Johannsen v.4, p.256; Tomkeieff p.70)*

BLATTERSTEIN. A local German name for a variolite. *(Leonhard, 1823a, p.108; Tröger 822; Tomkeieff p.71)*

BLOCK. Now defined in the pyroclastic classification (section 2.2.1, p.7) as a pyroclast with a mean diameter > 64 mm and whose angular to subangular shape indicates that it was solid during its formation Cf. bomb. *(Wentworth, 1935, p.227; Tomkeieff p.71)*

BODERITE. A term proposed for a melanocratic variety of nosean phonolite, originally called a nosean sanidine biotite augitite, with more alkali feldspar than nosean, biotite and augite. Cf. riedenite and rodderite. *(Taylor et al., 1967, p.409, and Frechen, 1971, p.41 for the description; Boder Wald, near Rieden, Eifel district, near Koblenz, Germany)*

BOGUEIRITE. An obsolete local name for a variety of hornblende orthopyroxenite composed of enstatite and hornblende. *(Cotelo Neiva, 1947b, p.118; Bogueiro Hill, Bragança district, Portugal; Tomkeieff p.73)*

BOGUSITE. A term synonymous with teschenite or analcime gabbro. *(Johannsen, 1938, p.220; Boguszow, Silesia, Poland; Tröger(38) 565⅓; Tomkeieff p.73)*

BOJITE. A plutonic rock originally described as consisting of labradorite, brown hornblende, minor augite and a little biotite, i.e. a variety of gabbro. Tröger (1935), however, showed the plagioclase to be andesine and the rock to be a diorite. The term was later adopted by Johannsen (1938) in its original sense as an alternative to hornblende gabbro. *(Weinschenk, 1899, p.541; named after Boii, a Celtic tribe, Pfaffenreuth district, Bavaria, Germany; Tröger 309; Johannsen v.3, p.226; Tomkeieff p.73)*

BOLSENITE. A term from an obsolete chemical classification, based on feldspar composition rather than SiO_2 alone, for a class of rocks in which $CaO : Na_2O : K_2O \approx 1.9 : 1 : 4.8$ and $SiO_2 \approx 60\%$. *(Lang, 1891, p.221; Lake Bolsena, Umbria, Italy; Tomkeieff p.74)*

BOMB. Now defined in the pyroclastic classification (section 2.2.1, p.7) as a pyroclast with a mean diameter > 64 mm and whose shape or surface indicates that it was in a wholly or partly molten condition during its formation and subsequent transport. Cf. block. *(Wentworth & Williams, 1932, p.46; Tomkeieff p.74)*

BONINITE. A high magnesia, low alkali, andesitic rock consisting of phenocrysts of protoenstatite (which inverts to clinoenstatite), orthopyroxene, clinopyroxene and olivine in a glassy base full of crystallites. The rock exhibits textures characteristic of rapid growth and was originally described as an hyaloandesite. Now defined chemically in the TAS classification (Fig. 2.13, p.34). *(Petersen, 1890a, p.25, and Petersen, 1890b for the description; Bonin Islands, now Ogaswara-gunto Islands, Japan; Tröger 160; Johannsen v.3, p.174; Tomkeieff p.74)*

BORENGITE. A dyke rock consisting essentially of K-feldspar with minor fluorite. *(Eckermann, 1960, p.519; Bäräng, Alnö Island, Västernorrland, Sweden)*

BORNITE. An obsolete term for an altered porphyry, predating its use as a mineral name. *(Pinkerton, 1811b, p.239; named after Baron von Born; Tomkeieff p.75)*

BOROLANITE. A local name for a coarse-grained variety of nepheline syenite consisting of alkali feldspar, nepheline (or its alteration products), melanite and biotite, and sometimes characterized by the presence of pseudoleucite. *(Horne & Teall, 1892, p.163; Loch Borolan, now Borralan, Assynt, Scotland, UK; Tröger 420; Johannsen v.4, p.139; Tomkeieff p.75)*

BORZOLITE. An obsolete term for a hornblende-bearing rock with carbonate amygdales associated with serpentinite. *(Issel, 1880, p.187; Borzoli, Genoa, Italy; Tomkeieff p.75)*

BOSTONITE. A variety of fine-grained leucocratic alkali feldspar syenite consisting almost entirely of alkali feldspar and characterized by a bostonitic texture, i.e. irregular, subparallel laths of feldspar arranged in a divergent manner. *(Hunter & Rosenbusch, 1890, p.447; Boston, Massachusetts, USA; Tröger 171; Johannsen v.4, p.22; Tomkeieff p.75)*

BOTTLEITE. A local Irish name for a tachylyte. *(Kinahan, 1875, p.426; Tomkeieff p.75)*

BOWRALITE. A variety of peralkaline syenite pegmatite composed of idiomorphic feldspars with interstitial arfvedsonite and aegirine-augite. *(Mawson, 1906, p.606; Bowral, New South Wales, Australia; Tröger 195; Johannsen v.3, p.32; Tomkeieff p.76)*

BOXITE. A name given to a hypothetical set of data used to illustrate the problems of the statistical interpretation of rock compositions. *(Aitchison, 1984, p.534)*

BRACCIANITE. An aphyric variety of tephritic leucitite essentially composed of leucite and clinopyroxene in about equal proportions with small amounts of plagioclase. Minor olivine and nepheline may be present. *(Lacroix, 1917e, p.1030; Lake Bracciano, Sabatini Volcanoes, near Rome, Italy; Tröger 583; Johannsen v.4, p.307; Tomkeieff p.77)*

BRAGANÇAITE. A local name for a variety of hornblende harzburgite composed of olivine, enstatite and hornblende. *(Cotelo Neiva, 1947a, p.112; Bragança district, Portugal; Tomkeieff p.77)*

BRANDBERGITE. A local name for a peralkaline granite containing biotite and small amounts of arfvedsonite. *(Chudoba, 1930, p.389; Brandberg, Erongo Mts, Namibia; Tröger 27; Johannsen v.2, p.99; Tomkeieff p.77)*

BRECCIOLE. A term originally used for piperno-like basaltic tuffs. *(Brongniart, 1824, p.3; Val Nera, Chiampo Valley, near Arzignano, Italy; Tomkeieff p.77)*

BRONZITITE. A variety of orthopyroxenite composed almost entirely of bronzite (= enstatite). *(Williams, 1890, p.47; Tröger 675; Johannsen v.4, p.459; Tomkeieff p.78)*

BUCHNERITE. An obsolete name for lherzolite. *(Wadsworth, 1884, p.85; named after Dr Otto Buchner for writings on meteorites; Tröger 823)*

BUCHONITE. A local name for a variety of melanocratic tephrite containing labradorite, hornblende, titanian augite, biotite, nepheline and minor orthoclase. *(Sandberger, 1872, p.208; Buchonia, now Rhön Mts, Germany; Tröger 579; Johannsen v.4, p.202; Tomkeieff p.79)*

BUGITE. A term applied to a series of dark fine-grained rocks, belonging to the charnockitic rock series, with 50% to 75% SiO_2 and consisting mainly of plagioclase, hypersthene (= enstatite) and quartz. The series is characterized by the absence of orthoclase and a complete transition in mineral composition from the basic to acid end members. It is suggested (Streckeisen, 1974, p.358) that this term should be abandoned. *(Bezborod'ko, 1931, p.129; Bug River, Podolia, Ukraine; Tröger(38) 131½; Tomkeieff p.80)*

BURGASITE. A local name for a potassic volcanic rock with marked spheroidal structure composed of almost equal amounts of alkali feldspar and various zeolites with minor augite, biotite and aegirine-augite. It may be a metasomatized andesitic rock. *(Borisov, 1963, p.225; Burgas, Bulgaria)*

BUSONITE. An erroneous spelling of busorite. *(Tomkeieff et al., 1983, p.80)*

BUSORITE. A local name for a coarse-grained cancrinite syenite with primary calcite, alkali feldspar, lepidomelane and aegirine. *(Béthune, 1956, p.394; Busori, Lueshe, Kivu, Democratic Republic of Congo; Tomkeieff p.80)*

BYTOWNITITE. A variety of anorthosite composed mainly of highly calcic plagioclase. Cf. allivalite. *(Johannsen, 1920a, p.60; Tröger 299; Johannsen v.3, p.196; Tomkeieff p.80)*

C-TYPE GRANITE. A general term for granitic rocks with charnockitic features, i.e. gener-ally lacking hornblende and containing inverted pigeonite, exceptionally calcic alkali feldspar and potassic plagioclase. They are characterized by distinctively high abundances of K_2O, TiO_2, P_2O_5 and LILE and low CaO at a given SiO_2 level compared to metamorphic charnockites, I-, S- and A-type granites. The prefix C stands for charnockitic. *(Kilpatrick & Ellis, 1992, p.155)*

CABYTAUITE. A mnenonic name from *ca*lcite, *by*townite and *au*gite, given to an altered dolerite composed mainly of highly calcic plagioclase, minor augite, calcite and biotite with secondary iron oxides. *(Vlodavets, 1952, p.659; Tomkeieff p.81)*

CALC-ALKALI BASALT. A basalt not defined by its mineralogy but by its association with rocks of the basalt–andesite–dacite suite of the orogenic belts and island arcs. *(Original reference uncertain)*

CALC-TONALITE. An obsolete term for a tonalite in which the alkali feldspar to plagioclase ratio is less than 1 to 6.5. *(Bailey & Maufe, 1916, p.160; Tomkeieff p.85)*

CALCICLASITE. A term for varieties of anorthosite containing calcic plagioclase. *(Johannsen, 1937, p.338; Tröger 300; Tomkeieff p.83)*

CALCIOCARBONATITE. A chemically defined variety of carbonatite in which wt % $CaO/(CaO+MgO+FeO+Fe_2O_3+MnO) > 0.8$ (Fig. 2.2, p.10). *(Woolley & Kempe, 1989)*

CALCIOPLETE. An obsolete term for a melanocratic rock in which CaO > (Na_2O + K_2O). *(Brögger, 1898, p.265; Tomkeieff p.83)*

CALCIPTOCHE. An obsolete adjectival term for igneous rocks poor in calcium. *(Loewinson-Lessing, 1901, p.118; Tomkeieff p.84)*

CALCITE-CARBONATITE. A variety of carbonatite in which the main carbonate is calcite (section 2.3, p.10). *(Streckeisen, 1978, p.12)*

CALCITFELS. An obsolete term for a calcite-rich member of the ijolite–melteigite series. *(Brögger, 1921, p.136; Tomkeieff p.84)*

CALCITITE. A term proposed for a calcareous carbonatite or calcite carbonatite. *(Belyankin & Vlodavets, 1932, p.53; Tomkeieff p.84)*

CALCOGRANITONE. An obsolete term for a gabbroic rock impregnated with calcite. *(Mazzuoli & Issel, 1881, p.326; Tomkeieff p.85)*

CALTONITE. An obsolete local name for a variety of analcime basanite consisting of olivine, titanian augite, labradorite and analcime. *(Johannsen, 1938, p.242; Calton Hill, Derbyshire, England, UK; Tröger(38) 599½; Tomkeieff p.86)*

CAMPANITE. A variety of leucite tephrite with large leucite crystals associated with essential plagioclase and clinopyroxene with minor sanidine, nepheline and haüyne. *(Lacroix, 1917b, p.207; Mt Somma, Naples, Campania, Italy; Tröger 573; Johannsen v.4, p.190; Tomkeieff p.86)*

CAMPTONITE. A variety of lamprophyre composed of phenocrysts of combinations of olivine, kaersutite, titanian augite, and Ti-rich biotite in a matrix of the same minerals (minus olivine) with plagioclase and sometimes subordinate alkali feldspar and feldspathoids. A volatile-rich basanite or alkali basalt. Now defined in the lamprophyre classification (Table 2.9, p.19). *(Rosenbusch, 1887, p.333; Campton Falls, New Hampshire, USA; Tröger 322; Johannsen v.4, p.63; Tomkeieff p.86)*

CAMPTOSPESSARTITE. A local term used for a basic spessartite containing titanian augite. *(Tröger, 1931, p.139; Golenz, S.W. Bautzen, Lausitz, Saxony, Germany; Tröger 322; Johannsen v.3, p.193; Tomkeieff p.86)*

CAMPTOVOGESITE. A name suggested for a basic variety of vogesite containing basaltic hornblende. *(Loewinson-Lessing, 1905a, p.15; Tagil River, Ural Mts, Russian Federation; Tröger 250)*

CANADITE. A coarse-grained variety of nepheline syenite in which the feldspar is essentially albite or sodic plagioclase, with abundant mafic minerals, notably biotite and amphibole. *(Quensel, 1913, p.177; Monmouth Township, Ontario, Canada; Tröger 484; Johannsen v.4, p.185; Tomkeieff p.86)*

CANCALITE. A variety of potassic lamproite containing normative olivine but no normative leucite. Mineralogically is an enstatite-sanidine-phlogopite lamproite (Table 2.7, p.17). *(Fuster et al., 1967, p.35; Cancarix and Calasparra, Spain)*

CANCARIXITE. A local name for a variety of peralkaline quartz syenite consisting essentially of sanidine and lesser amounts of aegirine-augite and quartz. *(Parga-Pondal, 1935, p.66; Cancarix, Sierra de las Cabras, Spain; Tröger(38) 233½; Tomkeieff p.87)*

CANCRINITE DIORITE. Now defined in QAPF field 14 (Fig. 2.4, p.22) as a variety of foid diorite in which cancrinite is the most abundant foid. *(Streckeisen, 1976, p.21)*

CANCRINITE GABBRO. Now defined in QAPF field 14 (Fig. 2.4, p.22) as a variety of foid gabbro in which cancrinite is the most abundant foid. *(Streckeisen, 1976, p.21)*

CANCRINITE MONZODIORITE. Now defined in QAPF field 13 (Fig. 2.4, p.22) as a variety of foid monzodiorite in which cancrinite is the most abundant foid. *(Streckeisen, 1976, p.21)*

CANCRINITE MONZOGABBRO. Now defined in QAPF field 13 (Fig. 2.4, p.22) as a variety of foid monzogabbro in which cancrinite is the most abundant foid. *(Streckeisen, 1976, p.21)*

CANCRINITE MONZOSYENITE. Now defined in QAPF field 12 (Fig. 2.4, p.22) as a variety of foid monzosyenite in which cancrinite is the most abundant foid. The term is synonymous with cancrinite plagisyenite. *(Streckeisen, 1976, p.21)*

CANCRINITE PLAGISYENITE. Now defined in QAPF field 12 (Fig. 2.4, p.22) as a variety of foid plagisyenite in which cancrinite is the most abundant foid. The term is synonymous with cancrinite monzosyenite. *(Streckeisen, 1976, p.21)*

CANCRINITE SYENITE. Now defined in QAPF field 11 (Fig. 2.4, p.22) as a variety of foid syenite in which cancrinite is the most abundant foid. *(Törnebohm, 1883, p.399; Tröger 431; Johannsen v.4, p.108)*

CANTALITE. An obsolete term used in several ways. It was originally defined as an extremely silica-rich rock but later redefined for two compositions of pitchstone and a tephritic trachyte. *(Leonhard, 1821, p.122; Verrières, Plomb du Cantal, Auvergne, France; Tröger 49; Tomkeieff p.87)*

CARBONATITE. A collective term for an igneous rocks in which the modal amount of primary carbonate minerals > 50% (section 2.3, p.10). *(Brögger, 1921, p.350; Fen Complex, Telemark, Norway; Tröger 752; Tomkeieff p.88)*

CARBOPHYRE. An obsolete term proposed for igneous rocks intruding Carboniferous formations. *(Ebray, 1875, p.291; Tomkeieff p.89)*

CARMELOITE. A local name for an iddingsite-bearing basalt. *(Lawson, 1893, p.8; Carmelo Bay, California, USA; Tröger 326; Johannsen v.3, p.282; Tomkeieff p.89)*

CARROCKITE. An obsolete term for devitrified tachylyte veins in granophyre. *(Groom, 1889, p.43; Carrock Fell, Cumbria, England, UK; Tomkeieff p.89)*

CARVOEIRA. A local Brazilian name for a quartz-tourmaline rock. *(Eschwege, 1832, p.178; Villa Rica, Minas Gerais, Brazil; Johannsen v.2, p.22; Tomkeieff p.90)*

CASCADITE. A local name for a sodic melanocratic lamprophyre having phenocrysts of biotite, olivine and augite in a matrix essentially of alkali feldspar with patches of what could have been leucite. *(Pirsson, 1905, p.55; Cascade Creek, Highwood Mts, Montana, USA; Tröger 229;*

Johannsen v.4, p.30; Tomkeieff p.90)

CAUCASITE (KAUKASITE). A local name for a variety of granite containing anorthoclase or sanidine. *(Belyankin, 1924, p.422; Caucasus Mts, Central Ridge, Russian Federation; Tröger 880; Tomkeieff p.91)*

CAVALORITE. An obsolete local name proposed in place of the term oligoclasite for a leucocratic plutonic rock consisting of orthoclase and lesser oligoclase with hornblende. *(Capellini, 1878, p.124; Mt Cavaloro, Bologna, Italy; Tröger(38) 824½; Johannsen v.3, p.145; Tomkeieff p.92)*

CECILITE. An aphyric variety of melilite leucitite consisting of essential leucite and clinopyroxene with small amounts of plagioclase, melilite and nepheline. *(Cordier, 1868, p.117; Tomb of Cecilia Metella, Capo di Bove, Rome, Italy; Tröger 671; Johannsen v.4, p.306; Tomkeieff p.92)*

CEDRICITE. A melanocratic diopside-leucite lamproite (Table 2.7, p.17) essentially composed of leucite and diopside with minor olivine pseudomorphs, amphibole, phlogopite and rutile. *(Wade & Prider, 1940, p.64; Mt Cedric, West Kimberley district, West Australia, Australia; Tomkeieff p.92)*

CERAMICITE. A spelling of keramikite attributed to Kotô (1916a) but not found in the reference cited. *(Tomkeieff et al., 1983, p.94)*

CHARNO-ENBERBITE. A member of the charnockitic rock series equivalent to orthopyroxene granodiorite of QAPF field 4 (Table 2.10, p.20). The term is synonymous with opdalite. *(Tobi, 1972)*

CHARNOCKITE. A term now applied to any orthopyroxene granite of QAPF field 3 (Table 2.10, p.20). The term was originally defined as a specific type of orthopyroxene granite found associated with other members of what was called the charnockitic rock series, ranging from granitic and intermediate members through to noritic and orthopyroxene peridotites and pyroxenites. *(Holland, 1900,* *p.134; tombstone of Job Charnock, St John's Churchyard, Calcutta, India; Tröger 16; Johannsen v.2, p.234; Tomkeieff p.96)*

CHIBINITE. See khibinite.

CHLOROPHYRE. An obsolete term originally applied to green porphyries from Egypt. *(Rozière, 1826, p.303; Tröger 825; Tomkeieff p.100)*

CHRISTIANITE. A term from an obsolete chemical classification, based on feldspar composition rather than SiO_2 alone, for granitic rocks in which $CaO : Na_2O : K_2O \approx 1 : 2.3 : 2.4$ and $SiO_2 \approx 70\%$. Later (Brögger, 1921) the spelling was changed to kristianite and the term was used as a local name for a biotite granite with approximately equal amount of Na_2O and K_2O. *(Lang, 1891, p.226; Christiania (now Oslo), Norway; Tröger 826; Tomkeieff p.100)*

CHROMITITE. An ultramafic rock consisting essentially of chromite. *(Johannsen, 1920c, p.225; Tröger 771; Johannsen v.4, p.466; Tomkeieff p.100)*

CIMINITE. A variety of trachyte characterized by phenocrysts of sanidine, augite, labradorite and olivine in a trachytic groundmass of orthoclase, labradorite, augite, olivine and magnetite. The same rock was later called arsoite by Reinisch. *(Washington, 1896b, p.838; Fontana di Fiescoli, Cimino Volcano, near Viterbo, Italy; Tröger 254; Johannsen v.3, p.90; Tomkeieff p.101)*

CINERITE. A basic volcanic ash consisting of glass and pyroxene fragments. Later extended to include all types of formations derived from volcanic ash. *(Cordier, 1816, p.385; from the Latin cineres = ashes; Tomkeieff p.101)*

CIZLAKITE. A melanocratic variety of quartz monzodiorite or monzogabbro consisting of abundant augite with smaller amounts of hornblende and plagioclase with minor alkali feldspar, quartz and biotite. *(Nikitin & Klemen, 1937, p.172; Cizlak, Slovenia)*

CLINKSTONE. See klingstein.

CLINOPYROXENE NORITE. A basic plutonic rock consisting mainly of calcic plagioclase, orthopyroxene and minor clinopyroxene. Now defined modally in the gabbroic rock classification (Fig. 2.6, p.25). *(Johannsen, 1937, p.238)*

CLINOPYROXENITE. An ultramafic plutonic rock consisting almost entirely of clinopyroxene. Now defined modally in the ultramafic rock classification (Fig. 2.9, p.28). *(Wyllie, 1967, p.2)*

COARSE (ASH) GRAIN. Now defined in the pyroclastic classification (Table 2.3, p.9) as a variety of ash grain in which the mean diameter is between 2 mm and 1/16 mm. *(Wentworth & Williams, 1932, p.46)*

COARSE (ASH) TUFF. Now defined in the pyroclastic classification (Table 2.3, p.9) as a variety of tuff in which the mean diameter is between 2 mm and 1/16 mm. *(Wentworth & Williams, 1932, p.46)*

COCCITE. See kokkite.

COCCOLITE. An obsolete name applied to granular pyroxene rocks from Sweden. *(Cordier, 1842, vol.4, p.45; from the Greek kokkos = grain or kernel; Tomkeieff p.108)*

COCITE. A melanocratic leucite-rich lamprophyre largely composed of clinopyroxene and subordinate olivine in a matrix of alkali feldspar, leucite and clinopyroxene with minor biotite. Similar to tjosite but with leucite instead of nepheline. *(Lacroix, 1933, p.195; Coc Pia, Upper Tonkin, Vietnam; Tröger 500; Johannsen v.4, p.279; Tomkeieff p.108)*

COLORADOITE. A variety of quartz latite that corresponds to the opdalitic magma-type of Niggli. *(Niggli, 1923, p.118; San Cristobal Quadrangle, Colorado, USA; Tröger 99; Tomkeieff p.111)*

COLOUR INDEX (COLOUR RATIO). The volume percentage or ratio of dark-coloured or mafic minerals to light-coloured or felsic minerals in a rock. In terms of the QAPF terminology the colour index is denoted by the parameter M' (Table 2.2, p.5). *(Shand, 1916, p.403; Tomkeieff p.111)*

COLUMBRETITE. A variety of tephritic leucite phonolite essentially composed of alkali feldspar and plagioclase with subordinate leucite/analcime, clinopyroxene and amphibole. *(Johannsen, 1938, p.168; Bauzá, Columbrete Islands, Spain; Tröger(38) 533; Tomkeieff p.111)*

COMENDITE. Originally described as a variety of porphyritic leucocratic alkali rhyolite containing phenocrysts of quartz, alkali feldspar, aegirine, arfvedsonite or riebeckite and minor biotite. It is now defined and distinguished from pantellerite as a variety of peralkaline rhyolite of TAS field R in which $Al_2O_3 > 1.33 \times$ total iron as FeO + 4.4 (Fig. 2.18, p.38). Synonymous with comenditic rhyolite. *(Bertolio, 1895, p.50; Comende, San Pietro Island, Sardinia, Italy; Tröger 48; Johannsen v.2, p.65; Tomkeieff p.111)*

COMENDITIC RHYOLITE. Now defined as a variety of peralkaline rhyolite of TAS field R in which $Al_2O_3 > 1.33 \times$ total iron as FeO + 4.4 (Fig. 2.18, p.38). Synonymous with comendite.

COMENDITIC TRACHYTE. Now defined as a variety of peralkaline trachyte of TAS field T in which $Al_2O_3 > 1.33 \times$ total iron as FeO + 4.4 (Fig. 2.18, p.38).

COMMENDITE. An erroneous spelling of comendite. *(Irvine & Baragar, 1971, p.530)*

CONCHAITE. An obsolete name for a variety of hornblende orthopyroxenite composed of bronzite (= enstatite) and hornblende. *(Cotelo Neiva, 1947b, p.122; Concha Farm, São Antonio, Portugal; Tomkeieff p.114)*

CONGRESSITE. A local name for a coarse-grained variety of urtite consisting of about 70% nepheline with minor alkali feldspar, biotite, occasional sodalite and accessory minerals. *(Adams & Barlow, 1913, p.90; Congress*

Bluff, Craigmont Hill, Ontario, Canada; Tröger 605; Johannsen v.4, p.280; Tomkeieff p.116)*

COPPAELITE. A variety of melilitite consisting of almost equal amounts of pyroxene and melilite with variable amounts of phlogopite and kalsilite but no olivine. A kamafugitic rock and now regarded as a kalsilite-phlogopite melilitite (Tables 2.5 and 2.6, p.12). *(Sabatini, 1903, p.378; Coppaeli di Sotto, near Rieti, Umbria, Italy; Tröger 751; Johannsen v.4, p.371; Tomkeieff p.119)*

CORCOVADITE. An obsolete local name for the hypabyssal equivalent of andengranite (granitic rocks that are younger than the usual ones) consisting of phenocrysts of plagioclase, hornblende, minor quartz and biotite in a fine-grained groundmass of plagioclase, quartz and orthoclase. *(Scheibe, 1933, p.160; Marmato, Cauca River, Colombia; Tröger 114; Tomkeieff p.120)*

CORONITE. A term for any igneous rock containing coronas. *(Shand, 1945, p.251; Tomkeieff p.121)*

CORSILITE. An old term for corsite. *(Pinkerton, 1811b, p.78; Tomkeieff p.122)*

CORSITE. An orbicular variety of gabbro composed essentially of calcic plagioclase with hornblende, minor hypersthene (= enstatite) and quartz. Spheroids are of variable composition but the matrix is richer in hornblende. This rock has also been called kugeldiorite, miagite and napoleonite. *(Zirkel, 1866b, p.133; named after Corsica; Tröger 361; Johannsen v.3, p.230; Tomkeieff p.122)*

CORTLANDTITE. A variety of pyroxene olivine hornblendite composed of large crystals of hornblende poikilitically enclosing olivine, hypersthene (= enstatite) and augite. The rock had previously been called hudsonite, which was objected to as hudsonite (= hastingsite) was already in use as a mineral name. *(Williams, 1886, p.30; named after Cortlandt, New York, USA; Tröger 708;*

Johannsen v.4, p.425; Tomkeieff p.123)

COSCHINOLITE. An obsolete term for a variety of brownish borzolite (hornblende-bearing rock) with cavities indicating former amygdales. *(Issel, 1880, p.187; from the Latin coccineus = scarlet-red; Tomkeieff p.123)*

COVITE. A local name for a melanocratic nepheline syenite with sodic pyroxene and amphibole. *(Washington, 1901, p.615; Magnet Cove, Arkansas, USA; Tröger 485; Johannsen v.4, p.103; Tomkeieff p.124)*

COXITE. A name given to a hypothetical set of data used to illustrate the problems of the statistical interpretation of rock compositions. *(Aitchison, 1984, p.534)*

CRAIGMONTITE. A coarse-grained leucocratic variety of nepheline diorite consisting of approximately 2/3 nepheline and 1/3 oligoclase with small amounts of corundum. *(Adams & Barlow, 1910, p.313; Craigmont Hill, Ontario, Canada; Tröger 548; Johannsen v.4, p.299; Tomkeieff p.125)*

CRAIGNURITE. A local Hebridian name for a series of volcanic rocks occurring as cone sheets and ranging in composition from intermediate to acid. They are characterized by acicular and skeletal crystals of plagioclase, augite and hornblende in a felsic groundmass. *(Thomas & Bailey, 1924, p.224; Craignure, Island of Mull, Scotland, UK; Tröger 98; Tomkeieff p.125)*

CRINANITE. A variety of olivine analcime dolerite or gabbro composed of olivine, titanian augite, and labradorite with minor analcime. Although it has less analcime and more olivine than teschenite the two names have been used interchangeably. *(Flett, 1911, p.117; Loch Crinan, Sound of Jura, Scotland, UK; Tröger 566; Johannsen v.4, p.70; Tomkeieff p.127)*

CRISTULITE. An obsolete name for a porphyritic rock containing phenocrysts of quartz and feldspar. *(Cordier, 1868, p.85; Tomkeieff p.127)*

CROMALTITE. A local name for an alkali pyroxenite consisting of aegirine-augite, melanite, biotite, iron ore and apatite. *(Shand, 1910, p.394; Ledmore Lodge, near Cromalt Hills, Assynt, Scotland, UK; Tröger 694; Johannsen v.4, p.454; Tomkeieff p.128)*

CRYPTODACITE. A term for a dacite in which the silica phase is present in a glassy groundmass. *(Belyankin, 1923, p.100; Tomkeieff p.130)*

CRYSTAL TUFF. Now defined in the pyroclastic classification (section 2.2.2, p.8) as a variety of tuff in which crystal fragments predominate. *(Pirsson, 1915, p.193)*

CUMBERLANDITE. A variety of dunite composed of olivine, ilmenite, magnetite and small amounts of labradorite and spinel. *(Wadsworth, 1884, p.75; Cumberland, Rhode Island, USA; Tröger 766; Johannsen v.3, p.335; Tomkeieff p.134)*

CUMBRAITE. An obsolete local name for a dyke rock composed of phenocrysts of calcic plagioclase, enstatite and augite in a groundmass of glass. *(Tyrrell, 1917, p.306; Great Cumbrae, Firth of Clyde, Scotland, UK; Tröger 125; Tomkeieff p.134)*

CUMULATE. A general term proposed for igneous rocks in which most of the crystals are thought to have accumulated by gravitational settling from a magma. *(Wager et al., 1960, p.73; from the Latin cumulus = heap; Tomkeieff p.134)*

CUMULOPHYRE. An obsolete term for a porphyritic rock in which the clusters of phenocrysts resemble cumulus clouds. *(Iddings, 1909, p.224; Tomkeieff p.134)*

CUSELITE (KUSELITE). A local name for an altered lamprophyric rock containing phenocrysts of plagioclase and uralitized diopside in a granular groundmass of plagioclase, orthoclase, chloritized biotite, hornblende, diopside and interstitial quartz. *(Rosenbusch, 1887, p.503; Cusel (now Kusel), Rheinland-Pfalz, Germany; Tröger 218; Johannsen v.3, p.190; Tomkeieff p.135)*

CUYAMITE. A local name for an olivine-free or

olivine-poor variety of teschenite. *(Johannsen, 1938, p.243; Cuyamas Valley, California, USA; Tröger(38) 565⅔; Tomkeieff p.136)*

DACITE. A volcanic rock composed of quartz and sodic plagioclase with minor amounts of biotite and/or hornblende and/or pyroxene. The volcanic equivalent of granodiorite and tonalite. Now defined modally in QAPF fields 4 and 5 (Fig. 2.11, p.31) and, if modes are not available, chemically in TAS field O3 (Fig. 2.14, p.35). *(Hauer & Stache, 1863, p.72; Dacia, a Roman province, Transylvania, Romania; Tröger 148; Johannsen v.2, p.390; Tomkeieff p.137)*

DACITE-ANDESITE. An obsolete term for an andesitic rock containing olivine and quartz. *(Dannenberg, 1900, p.239; Caucasus, Russian Federation; Tomkeieff p.137)*

DACITOID. Originally used as a collective name for acidic volcanic rocks having the chemical composition of dacite but containing no quartz. Now proposed for preliminary use in the QAPF "field" classification (Fig. 2.19, p.39) for volcanic rocks tentatively identified as dacite. *(Lacroix, 1919, p.297; Tomkeieff p.137)*

DAHAMITE. An obsolete local term for a fine-grained variety of peralkaline granite containing phenocrysts of albite in a groundmass of equal amounts of alkali feldspar, albite and quartz. Minor riebeckite is present. *(Pelikan, 1902, p.397; Dahamis, Socotra Island, Indian Ocean; Tröger 30; Tomkeieff p.137)*

DAMKJERNITE (DAMTJERNITE). A melanocratic variety of nepheline lamprophyre containing phenocrysts of biotite and titanian augite in a groundmass of the same minerals with some nepheline, calcite and orthoclase. As pointed out by Sæther (1957) the name was actually misspelt by Brögger in the belief that the locality was named Damkjern instead of Damtjern. The strictly correct spelling should have been damtjernite but damkjernite is in common usage. *(Brögger, 1921, p.276;*

Damkjern, Fen Complex, Telemark, Norway; Tröger 499; Johannsen v.4, p.277; Tomkeieff p.138)

DANCALITE. A variety of analcime tephrite with phenocrysts of oligoclase, augite with aegirine rims and rare amphibole in a groundmass of plagioclase laths with interstitial analcime. *(De-Angelis, 1925, p.82; Dankalia, Ethiopia; Tröger 570; Johannsen v.4, p.189; Tomkeieff p.138)*

DANUBITE. An obsolete term for an amphibole hypersthene (= enstatite) andesite in which the labradorite phenocrysts were thought to be nepheline. *(Krenner, 1910, p.130; Tröger 828)*

DAVAINITE. An obsolete term for a granular variety of hornblendite, probably metamorphic, consisting essentially of brown hornblende with cores of pyroxene. Hypersthene (= enstatite) and feldspar are also present but in small amounts. *(Wyllie & Scott, 1913, p.501; Garabal Hill, Loch Beinn Damhain or Ben Davain, Scotland, UK; Tröger 702; Johannsen v.4, p.448)*

DELDORADITE (DELDORADOITE). A leucocratic variety of cancrinite syenite containing biotite and aegirine. *(Johannsen, 1938, p.77; Deldorado Creek, Uncompahgre Quadrangle, Colorado, USA; Tröger(38) 412½)*

DELLENITE. A name proposed for a rock intermediate between dacite and rhyolite. Cf. toscanite. *(Brögger, 1895, p.59; Lake Dellen, Gävleborg, Sweden; Tröger 96; Johannsen v.2, p.309; Tomkeieff p.140)*

DELLENITOID. A rock which is chemically a dellenite but contains no visible orthoclase. *(Lacroix, 1923, p.9)*

DEVONITE. A temporary name suggested by A. Johannsen for a variety of porphyritic diabase. *(Clarke, 1910, p.40; Mt Devon, Missouri, USA; Tröger 315; Johannsen v.3, p.320)*

DIABASE. A term for medium-grained rocks of basaltic composition that has been used in two distinct ways. British usage implied heavy alteration, but French, German and American usage implied an ophitic texture. The original definition included a transitional texture between that of basalts and that of coarse-grained rocks. Now regarded as being synonymous with dolerite and an approved synonym for microgabbro of QAPF field 10 (Fig. 2.4, p.22). *(Brongniart, 1807, p.456; from the Greek diabasis = crossing over; Tröger 390; Johannsen v.3, p.311; Tomkeieff p.144)*

DIABASITE. A term proposed, but never used, for a diabase aplite. *(Polenov, 1899, p.464; Johannsen v.3, p.61)*

DIALLAGITE. Although originally suggested for a rock composed of labradorite and diallage (= altered diopside), it was more commonly used for rocks consisting essentially of diallage. Now obsolete. *(Cloizeaux, 1863, p.108; Tröger 681; Johannsen v.4, p.457)*

DIASCHISTITE. A term for a rock from a minor intrusion which has a different composition from the main intrusion. *(Johannsen, 1931, p.5; from the Greek diaschistos = cleaved; Tomkeieff p.147)*

DIATECTITE (DIATEXITE). A term used for a rock formed by diatexis, i.e. complete or almost complete melting of pre-existing rocks. *(Fiedler, 1936, p.493; Tomkeieff p.148)*

DIMELITE. A mnemonic name suggested for a rock consisting essentially of *di*opside and *me*lilite. *(Belyankin, 1929, p.22)*

DIOPSIDITE. A variety of pyroxenite consisting almost entirely of diopside (or chrome-diopside). *(Lacroix, 1895, p.752; Lhers, Ariège, Pyrénées, France; Tröger 682; Johannsen v.4, p.455)*

DIOREID. An obsolete field term for a coarse-grained igneous rock containing more than 50% amphibole; feldspar is subordinate. *(Johannsen, 1911, p.320; Tomkeieff p.150)*

DIORIDE. A revised spelling recommended to replace the field term dioreid. Now obsolete. *(Johannsen, 1926, p.181; Johannsen v.1, p.57; Tomkeieff p.150)*

DIORITE. A plutonic rock consisting of intermediate plagioclase, commonly with hornblende and often with biotite or augite. Now defined modally in QAPF field 10 (Fig. 2.4, p.22). *(D'Aubuisson de Voisins, 1819, p.146; from the Greek diorizein = to distinguish; Tröger 308; Johannsen v.3, p.146)*

DIORITITE. An obsolete name for a variety of diorite or diorite-aplite which occurs as dykes. *(Polenov, 1899, p.464; Johannsen v.3, p.61; Tomkeieff p.150)*

DIORITOID. Originally used as a group name for igneous rocks consisting mainly of plagioclase, amphibole and biotite. Now proposed for preliminary use in the QAPF "field" classification (Fig. 2.10, p.29) for plutonic rocks tentatively identified as diorite or monzodiorite. *(Gümbel, 1888, p.87; Tomkeieff p.150)*

DIORITOPHYRITE. An obsolete term for a porphyritic diorite aplite. *(Polenov, 1899, p.464; Tomkeieff p.150)*

DIOROID. An obsolete mnemonic name for a *dior*ite containing feldspath*oid*s. *(Shand, 1910, p.378; Tomkeieff p.150)*

DISSOGENITE. A term applied to igneous rocks containing two different sets of minerals, such as might be created by assimilation. *(Lacroix, 1922, p.374; from the Greek dissos = twofold, genos = family; Tröger 832; Tomkeieff p.153)*

DITRO-ESSEXITE. A comprehensive term for various plutonic rocks of essexitic and theralitic chemistry from the alkaline complex of Ditro, Romania. The term includes such types as diorite, monzonite, essexite with large amounts of barkevikite (see p.44) and kaersutite. *(Streckeisen, 1952, p.274, and 1954, p.394 for the description; Ditro (now Ditrau), Transylvania, Romania)*

DITROITE. A biotite-bearing variety of nepheline syenite with cancrinite and primary calcite; sodalite penetrates along fractures and intergranular boundaries. Used by Brögger

(1890) as a general term for nepheline syenites of granular texture. *(Zirkel, 1866a, p.595; Ditro (now Ditrau), Transylvania, Romania; Tröger 427; Johannsen v.4, p.98; Tomkeieff p.153)*

DOLEREID. An obsolete field term for a rock containing more than 50% of ferromagnesian minerals when the species cannot be determined megascopically. Cf. gabbreid and dioreid. *(Johannsen, 1911, p.320; Tomkeieff p.154)*

DOLERINE. An obsolete name for a rock with a feldspathic groundmass containing chlorite. *(Jurine, 1806, p.374; Mont Blanc, France; Tomkeieff p.154)*

DOLERITE. A rock intermediate in grain size between basalt and gabbro and composed essentially of plagioclase, pyroxene and opaque minerals; often with ophitic texture. If olivine is present the rock may be called olivine dolerite – if quartz, quartz dolerite. Now regarded as being synonymous with diabase and an approved synonym for microgabbro of QAPF field 10 (Fig. 2.4, p.22). *(D'Aubuisson de Voisins, 1819, p.556; from the Greek doleros = deceptive; Tröger 833; Johannsen v.3, p.291; Tomkeieff p.154)*

DOLEROID. An obsolete mnemonic name for a *dolerite* containing *foids*. *(Shand, 1910, p.378; Tomkeieff p.154)*

DOLOMITE-CARBONATITE. A variety of carbonatite in which the main carbonate is dolomite (section 2.3, p.10). Cf. beforsite and rauhaugite. *(Tröger, 1935, p.303; Tröger 758)*

DOMITE. A local name for a variety of trachyte that contains a few phenocrysts of oligoclase and biotite in a potentially quartz-bearing glassy groundmass. *(Buch, 1809, p.244; Puy de Dôme, Auvergne, France; Tröger 103; Johannsen v.3, p.66; Tomkeieff p.156)*

DOREITE. A local name for a variety of trachyandesite characterized by microphenocrysts of andesine and augite. *(Lacroix, 1923, p.328; Mont Dore, Auvergne, France;*

Tröger 268; Tomkeieff p.157)

DORGALITE. An obsolete local name for a variety of olivine basalt containing phenocrysts of olivine only. *(Amstutz, 1925, p.300; Mt Pirische, near Dorgali, Sardinia, Italy; Tröger 328; Johannsen v.3, p.282; Tomkeieff p.157)*

DRAKONTITE (DRAKONITE). A local name for a variety of trachyte containing biotite and alkali amphibole. *(Reinisch, 1912, p.121; Drachenfels, Siebengebirge, near Bonn, Germany; Tröger 834; Johannsen v.4, p.35; Tomkeieff p.157)*

DRUSITE. An obsolete name for a gabbroic rock containing concentric coronas of minerals (drusy structure). *(*Fedorov, 1896, p.168; Tomkeieff p.158)*

DUCKSTEIN. A local German name for a nonstratified tuff (trass). *(Although attributed to Wolff, 1914, p.404, the term had been used before; Tomkeieff p.159)*

DUMALITE. A local name for a variety of trachyandesite characterized by intersertal texture and a glassy groundmass of nepheline composition. *(Loewinson-Lessing, 1905b, p.276; Dumala Gorge, N. Caucasus, Russian Federation; Tröger 556; Johannsen v.4, p.190; Tomkeieff p.159)*

DUNCANITE. An oceanic volcanic rock of the icelandite group, in which groundmass plagioclase is andesine, Cf. jervisite. *(Stewart & Thornton, 1975, p.568; named after Duncan Island, Galapagos Islands, Pacific Ocean)*

DUNGANNONITE. A local name for a variety of nepheline-bearing diorite containing corundum and scapolite. *(Adams & Barlow, 1908, p.67; Dungannon, Renfrew County, Ontario, Canada; Tröger 550; Johannsen v.4, p.59)*

DUNITE. An ultramafic plutonic rock consisting essentially of olivine. Now defined modally in the ultramafic rock classification (Fig. 2.9, p.28). *(Hochstetter, 1859, p.275; Dun Mountain, Nelson, New Zealand; Tröger 724; Johannsen v.4, p.405; Tomkeieff p.159)*

DURBACHITE. A fine- to medium-grained melanocratic variety of syenite consisting of large flakes of biotite with hornblende and megacrysts of orthoclase in a groundmass of oligoclase and a little quartz. *(Sauer, 1892, p.247; Durbach, near Offenburg, Black Forest, Germany; Tröger 243; Johannsen v.3, p.86; Tomkeieff p.160)*

DUST GRAIN. Now defined in the pyroclastic classification (section 2.2.2, p.7) as a variety of ash grain in which the mean diameter is < 1/16 mm. Synonymous with fine (ash) grain. *(Schmid, 1981, p.42)*

DUST TUFF. Now defined in the pyroclastic classification (section 2.2.2, p.8) as a variety of tuff in which the mean diameter is < 1/16 mm. Synonymous with fine (ash) tuff. *(Schmid, 1981, p.43)*

EGERINOLITH. An erroneous spelling of aigirinolith or aegirinolith. *(Tomkeieff et al., 1983, p.165)*

EHRWALDITE. A lamprophyric rock of basanitic composition composed of titanian augite, biotite, serpentinized olivine and glass. *(Pichler, 1875, p.927; Wetterschroffen, Ehrwald, Tyrol, Austria; Tröger 347; Tomkeieff p.165)*

EICHSTÄDTITE. An obsolete name suggested for a group of quartz norites. *(Marcet Riba, 1925, p.293; named after F. Eichstädt)*

EKERITE. A local name for a peralkaline granite containing anorthoclase microperthite and small amounts of arfvedsonite and aegirine. *(Brögger, 1906, p.136; Eker, Oslo district, Norway; Tröger 18; Tomkeieff p.166)*

ELKHORNITE. A variety of alkali feldspar syenite containing some labradorite. *(Johannsen, 1937, p.92; Elkhorn, Montana, USA; Tröger(38) 244½; Tomkeieff p.167)*

ELVAN. A local term for dyke rocks of granitic composition containing phenocrysts of quartz and orthoclase with tourmaline occurring as isolated crystals and in radiating groups. *(Conybeare, 1817, p.401; from the Celtic el = rock, and van = white; Tröger 113; Johannsen v.2, p.300; Tomkeieff p.168)*

ELVANITE. An alternative term for elvan. *(Cotta, 1866, p.214; Johannsen v.2, p.300; Tomkeieff p.168)*

ENDERBITE. A member of the charnockitic rock series consisting essentially of quartz, antiperthite, orthopyroxene and magnetite. It is equivalent to orthopyroxene tonalite of QAPF field 5 (Table 2.10, p.20). *(Tilley, 1936, p.312; named after Enderby Land, Antarctica; Johannsen v.1 (2nd Edn), p.250; Tomkeieff p.169)*

ENGADINITE. A term for granites with low quartz content, that correspond to the engadinitic magma-type of Niggli (1923, p.110). *(Tröger, 1935, p.20; Platta mala granite, near Ramosch, Engadine, Switzerland; Tröger 13; Tomkeieff p.171)*

ENGELBURGITE. A variety of hybrid biotite granodiorite containing leucocratic spots consisting of alkali feldspar and abundant titanite. *(Frentzel, 1911, p.145; Engelburg, Bavaria, Germany; Tröger(38) 107½; Tomkeieff p.171)*

ENSTATITITE. A variety of orthopyroxenite composed almost entirely of enstatite. *(Streng, 1864, p.260; Tröger 674; Tomkeieff p.171)*

ENSTATOLITE. An obsolete term for a variety of orthopyroxenite composed almost entirely of enstatite. *(Pratt & Lewis, 1905, p.30; Tröger 836; Johannsen v.4, p.459; Tomkeieff p.171)*

EORHYOLITE. An old name proposed for a devitrified acid igneous rock of Archean age. *(Nordenskjöld, 1893, p.153; Tomkeieff p.172)*

EPIBASITE. An obsolete term suggested for non-porphyritic lamprophyres – somewhat a contradiction of terms. *(Sederholm, 1926, p.14; Tomkeieff p.173)*

EPIBUGITE. A leucocratic variety of enderbite of QAPF field 5 (Fig. 2.4, p.22). It is a member of the bugite series with 68% to 72% SiO_2, 0% to 8% hypersthene (= enstatite), oligoclase-andesine antiperthite and quartz. It is suggested (Streckeisen, 1974, p.358) that this term should be abandoned. *(Bezborod'ko, 1931, p.137; Bug River, Podolia, Ukraine; Tröger(38) 129½; Tomkeieff p.173)*

EPIDIABASE. An obsolete name for epidiorite. *(Issel, 1892, p.324; Tröger 395; Tomkeieff p.173)*

EPIDIORITE. A term originally applied to a diorite in which the pyroxene had been altered to amphibole. Current usage as a field term implies rocks of basaltic composition recrystallized by regional metamorphism. *(Gümbel, 1874, p.10; Tröger 395; Tomkeieff p.174)*

EPISYENITE. A term coined by uranium exploration geologists for an igneous-looking rock of syenite composition, displaying rounded cavities produced by hydrothermal dissolution of quartz crystals, that can ultimately host uranium ore deposits. Contrary to the common usage (epidiorite), the prefix epi- is used here as a qualifier to the name of the rock resulting from the alteration of granite. *(Lameyre, 1966, p.170)*

ERCALITE. An obsolete local name for a variety of granophyre composed of clusters of quartz and sodic plagioclase surrounded by a graphic intergrowth of quartz and feldspar. *(Watts, 1925, p.327; Ercall, near Wrekin, Shropshire, England, UK; Tomkeieff p.177)*

ESBOITE. A local name for a variety of diorite, often orbicular, consisting of abundant sodic plagioclase with small amounts of orthoclase and biotite. *(Sederholm, 1928, p.72; Esbo, now Espoo, near Helsinki, Finland; Tröger 306; Tomkeieff p.179)*

ESMERALDITE. A local name for a granular variety of quartzolite consisting predominantly of quartz with some muscovite. It is thought to be an extreme differentiate but, unlike greisen, not produced by pneumatolytic action. Cf. northfieldite. *(Spurr, 1906, p.382; Esmeralda County, Nevada, USA; Tröger 5;*

Johannsen v.2, p.17; Tomkeieff p.179)

ESPICHELLITE. A local name for a variety of camptonite (a lamprophyre) containing analcime. *(Souza-Brandão, 1907, p.2; Cape Espichel, Lisbon, Portugal; Tröger 559; Johannsen v.4, p.207; Tomkeieff p.179)*

ESSEXITE. A variety of nepheline monzogabbro or nepheline monzodiorite containing titanian augite, kaersutite and/or biotite with labradorite, lesser alkali feldspar and nepheline. May be used as a synonym for nepheline monzogabbro or nepheline monzodiorite of QAPF field 13 (p.24). *(Sears, 1891, p.146; Salem Neck, Essex County, Massachusetts, USA; Tröger 542; Johannsen v.4, p.191; Tomkeieff p.179)*

ESSEXITE-AKERITE. A term suggested for a dark plagioclase-rich akerite. *(Brögger, 1933, p.35; Lortlauptal, W. Vikersund, Oslo district, Norway; Tröger 276)*

ESSEXITE-BASALT. An obsolete name for a volcanic rock consisting of serpentinized olivine phenocrysts in a groundmass of olivine, augite, labradorite, sanidine, biotite, ore and interstitial nepheline. Johannsen (1938) renamed it westerwaldite. *(Lehmann, 1924, p.175; Utanjilua, Patagwa, Malawi; Tröger 382; Johannsen v.4, p.203; Tomkeieff p.180)*

ESSEXITE-DIABASE. An obsolete name for a variety of nepheline tephrite or essexite containing small amounts of orthoclase. *(Erdmannsdörffer, 1907, p.22; Tröger 601; Tomkeieff p.180)*

ESSEXITE-DIORITE. An obsolete term for a variety of diorite containing titanian augite similar to rongstockite and kauaiite. *(Tröger, 1935, p.124; Tröger 282, 284)*

ESSEXITE-FOYAITE. An obsolete term for a variety of nepheline monzosyenite, synonymous with husebyite. *(Tröger, 1935, p.214; Tröger 508)*

ESSEXITE-GABBRO. An obsolete term for a variety of essexite rich in plagioclase and poor in feldspathoids. *(Lacroix, 1909, p.543;*

Jordanne Valley, Cantal, Auvergne, France; Tröger 283; Tomkeieff p.180)

ESSEXITE-MELAPHYRE. An obsolete term for a geologically old altered essexite-basalt. *(Brögger, 1906, p.120; Holmestrand, Oslo Igneous Province, Norway; Tröger 389; Tomkeieff p.180)*

ESSEXITE-PORPHYRITE. A variety of essexite containing plagioclase phenocrysts. *(Brögger, 1906, p.121; Tröger 557; Tomkeieff p.180)*

ESTERELLITE. A local name for a porphyritic variety of quartz diorite consisting of zoned plagioclase and hornblende in a groundmass of quartz, orthoclase and plagioclase. *(Michel-Lévy, 1897, p.39; Esterel, Provence, France; Tröger 143; Johannsen v.2, p.400; Tomkeieff p.180)*

ETINDITE. A variety of leucite nephelinite essentially composed of nepheline, clinopyroxene and leucite. *(Lacroix, 1923, p.65; Etinde Volcano, Cameroon; Tröger 619; Johannsen v.4, p.367; Tomkeieff p.180)*

ETNAITE. A local term used for trachyandesites from Mt Etna. *(Rittmann, 1960, p.117; Lava of 1669, Catania, Mt Etna, Sicily, Italy)*

EUCRITE (EUCRYTE, EUKRITE). A term originally and properly applied to meteorites composed of anorthite and augite (Rose, 1863), it was extended to include varieties of gabbro in which the plagioclase was bytownite. The term should no longer be used as a rock name. *(Cotta, 1866, p.148; from the Greek eukritos = clear, distinct; Tröger 358; Johannsen v.3, p.347; Tomkeieff p.181)*

EUDIALYTITE. A melanocratic plutonic rock consisting essentially of eudialyte with minor variable amounts of microcline, arfvedsonite and aegirine. *(Eliseev, 1937, p.1100; Lovozero complex, Kola Peninsula, Russian Federation; Tomkeieff p.181)*

EUKTOLITE. A name given by Rosenbusch to a melanocratic volcanic rock largely composed of melilite, leucite and olivine, when he was unaware that the rock had already been called

venanzite. *(Rosenbusch, 1899, p.110; from the Greek euktos = desired; San Venanzo, Umbria, Italy; Tröger 838; Johannsen v.4, p.361)*

EULYSITE. An obsolete term for a variety of peridotite composed essentially of olivine, diopside, and opaques with anthophyllite. *(Erdmann, 1849, p.837; from the Greek eulytos = easy to dissolve or break; Tröger 726; Johannsen v.4, p.412; Tomkeieff p.181)*

EUPHOTIDE. An obsolete term originally suggested by Haüy for a saussuritized gabbro. *(Brongniart, 1813, p.42; Tröger 839; Johannsen v.3, p.229; Tomkeieff p.181)*

EURITE. A name originally suggested by d'Aubuisson for a compact felsitic rock and later extended to cover all aphanitic rocks of granitic composition. Cf. felsite. Although this name appears in Johannsen's index it was not found in the text. *(Brongniart, 1813, p.43; from the Greek eurys = broad; Rocher de Sanadoire, Auvergne, France; Tröger 840; Tomkeieff p.182)*

EURYNITE. An obsolete term for a porphyritic eurite (felsitic rock). *(Cordier, 1868, p.81; Tomkeieff p.182)*

EUSTRATITE. A local name for a lamprophyric dyke rock with rare phenocrysts of olivine, corroded hornblende and occasional augite in a groundmass of augite, titanomagnetite, feldspar and glass. *(Kténas, 1928, p.1632; Haghios Eustratios, now Ayios Evstrátios Island, Greece; Tröger 521; Johannsen v.4, p.178; Tomkeieff p.182)*

EUTECTITE. According to Tomkeieff an igneous rock formed by the crystallization of residual liquids. However, the reference cited (Bowen, 1914) does not contain the name. *(Tomkeieff et al., 1983, p.183)*

EUTECTOFELSITE (EUTEKTOFELSITE, EUTECTO-PHYRE). An obsolete term for a variety of quartz porphyry with a whitish earthy tufflike appearance which cleaves into imperfect tablets. The description given by Tomkeieff appears to be erroneous. *(Kotô, 1909, p.189; Tröger 841; Tomkeieff p.183)*

EVERGREENITE. A local name for a variety of alkali feldspar granite containing wollastonite, which was originally identified as enstatite. *(Ritter, 1908, p.751; Evergreen Mine, Apex, Oregon, USA; Tröger 71; Tomkeieff p.184)*

EVISITE. A local comprehensive term for peralkaline granites and syenites with aegirine and/or riebeckite corresponding to the evisitic magma-type of Niggli (1923, p.148). *(Tröger, 1935, p.317; Evisa, Corsica, France; Tröger 842; Tomkeieff p.184)*

FARRISITE. A local name for a melanocratic variety of melilite lamprophyre containing augite, barkevikite (see p.44), biotite and olivine but no feldspar. *(Brögger, 1898, p.64; Lake Farris, Larvik, Oslo Igneous Province, Norway; Tröger 522; Johannsen v.4, p.389; Tomkeieff p.189)*

FARSUNDITE. A member of the charnockitic rock series equivalent to orthopyroxene monzogranite of QAPF field 3b (Table 2.10, p.20), consisting of oligoclase, microcline, quartz, hornblende, hypersthene (= enstatite) and a little clinopyroxene. The term is also frequently used as a comprehensive name for granitic rocks of the Farsund area. It is suggested (Streckeisen, 1974, p.354) that this term should be abandoned, to avoid ambiguity. *(Kolderup, 1903, p.110; Farsund, Egersund, Norway; Tröger 106; Tomkeieff p.189)*

FASIBITIKITE. A fine-grained variety of peralkaline granite, similar to rockallite, containing large amounts of aegirine and riebeckite and characterized by the presence of eucolite. *(Lacroix, 1915, p.257; Ampasibitika, Madagascar; Tröger 64; Tomkeieff p.190)*

FASINITE. A variety of melteigite consisting of 60%–70% titanian augite and some nepheline, with minor olivine, biotite and orthoclase. *(Lacroix, 1916b, p.257; from the Madagascan*

fasina = sand, referring to long sand beaches; Tröger 611; Johannsen v.4, p.252; Tomkeieff p.190)

FELDSPAR-BASALT. An obsolete name proposed for a fine-grained rock consisting essentially of plagioclase and augite to distinguish it from leucite- and nepheline-basalt. *(Zirkel, 1870, p.108; Tröger 888; Tomkeieff p.190)*

FELDSPATHIDOLITE. An obsolete term for an igneous rock composed mainly of feldspathoids. *(Loewinson-Lessing, 1901, p.114; Tomkeieff p.191)*

FELDSPATHINE. An obsolete name for a feldspar rock. *(Cordier, 1868, p.74; Tomkeieff p.191)*

FELDSPATHITE. An obsolete collective name for monzonite and quartz monzonite aplites. *(Kolenec, 1904, p.162; Tröger 844)*

FELDSPATHOIDITE. A general term for extrusive rocks composed essentially of feldspathoids, now superseded by the term foidite. *(Johannsen, 1938, p.336; Tomkeieff p.192)*

FELDSPATHOLITE. An obsolete term for an igneous rock composed mainly of feldspar. *(Loewinson-Lessing, 1901, p.114; Tomkeieff p.192)*

FELDSPATTAVITE. An obsolete term originally used for a feldspathic variety of tawite (a sodalite aegirine rock). *(Tröger, 1935, p.318; Tröger 846; Tomkeieff p.192)*

FELSEID. An obsolete field term for an aphanitic, non-porphyritic light-coloured igneous rock. Also called leuco-aphaneid. *(Johannsen, 1911, p.321; Tomkeieff p.192)*

FELSEID PORPHYRY. An obsolete field term for a porphyritic light-coloured igneous rock with an aphanitic groundmass. Also called leucophyreid. *(Johannsen, 1911, p.321)*

FELSIC. A collective term for modal quartz, feldspars and feldspathoids which was introduced to stop the normative term salic incorrectly being used for that purpose. See also mafic and femic. *(Cross et al., 1912, p.561)*

FELSIDE. A revised spelling recommended to replace the field term felseid. Now obsolete.

(Johannsen, 1926, p.182; Johannsen v.1, p.57; Tomkeieff p.192)

FELSITE (FELSITIC). As a rock term initially used for the microcrystalline groundmass of porphyries but now more commonly used for microcrystalline rocks of granitic composition. The name had previously been used by Kirwan (1794) for a microcrystalline variety of feldspar. *(Gerhard, 1815, p.18; from feldspar; Tröger 847; Tomkeieff p.192)*

FELSOANDESITE. An obsolete term used for andesites in which the groundmass is cryptocrystalline. *(Johannsen, 1937, p.170; Tomkeieff p.193)*

FELSOGRANOPHYRE. An obsolete name for a porphyry in which the texture of the groundmass is between granitic and felsitic. *(Zirkel, 1894a, p.168; Tomkeieff p.193)*

FELSOPHYRE. An obsolete group name for a porphyritic rock with a microcrystalline groundmass containing some glass. *(Vogelsang, 1872, p.534; from the German Fels = rock, and phyre = porphyritic; Tröger 848; Johannsen v.2, p.275; Tomkeieff p.193)*

FELSOPHYRITE. A term originally proposed for non-porphyritic rocks with microcrystalline texture and some glass. Later used for effusive rocks of diorite composition. Now obsolete. *(Vogelsang, 1872, p.534; Tomkeieff p.194)*

FELSOVITROPHYRE. A term proposed for a porphyry in which the texture of the groundmass is between felsitic and glassy. *(Vogelsang, 1875, p.160; Tomkeieff p.194)*

FELSPARITE. An obsolete term proposed for a granite rich in feldspar. *(Boase, 1834, p.17; Tomkeieff p.194)*

FELSTONE. An obsolete term for a felsite. *(Leonhard, 1823a, p.210; Tomkeieff p.194)*

FEMIC. A name used in the CIPW normative classification for one of two major groups of normative minerals which includes the Fe and Mg silicates such as olivine and pyroxene as well as the Fe and Ti oxides, apatite and fluorite. The other group is named salic. Cf. mafic. *(Cross et al., 1902, p.573; Johannsen v.1, p.173; Tomkeieff p.194)*

FENITE. A metasomatic rock, normally associated with carbonatites or ijolites and occasionally with nepheline syenites and peralkaline granites, composed of alkali feldspar, sodic pyroxene and/or alkali amphibole. Some varieties are monomineralic alkali feldspar rocks. *(Brögger, 1921, p.156; Fen Complex, Telemark, Norway; Tröger 187; Johannsen v.4, p.32; Tomkeieff p.194)*

FERGANITE. An obsolete name for a variety of clinopyroxenite consisting essentially of clinopyroxene with minor olivine, plagioclase and magnetite. *(Lebedev & Vakhrushev, 1953, p.119; Fergana, E.S.E. of Tashkent, Uzbekistan; Tomkeieff p.195)*

FERGUSITE. A plutonic rock consisting of roughly 70% pseudoleucite (alkali feldspar, nepheline, kalsilite and minor analcime) and 30% pyroxene. Now defined modally as a variety of foidite in QAPF field 15 (Fig. 2.8, p.27). *(Pirsson, 1905, p.74; Fergus County, Highwood Mts, Montana, USA; Tröger 628; Johannsen v.4, p.325; Tomkeieff p.195)*

FERRILITE. An obsolete local name for a dolerite. *(Kirwan, 1794, p.229; Rowley Regis, Staffordshire, England, UK; Tomkeieff p.195)*

FERROCARBONATITE. A term now used in two senses: (1) modally as a variety of carbonatite in which the main carbonate is iron-rich (p.10) and (2) chemically as a variety of carbonatite in which weight percent $CaO / (CaO+MgO+FeO+Fe_2O_3+MnO) < 0.8$ and $MgO < (FeO+Fe_2O_3+MnO)$ (Fig. 2.2, p.10). *(Le Bas, 1977, p.37)*

FERRODIORITE. A term for rocks of gabbroic appearance containing Fe-rich pyroxenes and olivines but with plagioclase more sodic than An_{50}. Many such rocks were previously called ferrogabbro (e.g. from Skaergaard). *(Wager & Vincent, 1962, p.26)*

FERROGABBRO. A variety of gabbro in which the pyroxenes and olivines are Fe-rich varieties. Since the plagioclase of gabbros should be more calcic than An_{50}, most of the rocks originally given this name (e.g. from Skaergaard) have been renamed as ferrodiorites. *(Wager & Deer, 1939, p.98; Tomkeieff p.196)*

FERROLITE. An obsolete term originally proposed for iron ore rocks, but later restricted to iron ore of magmatic origin. *(Wadsworth, 1893, p.92; Tomkeieff p.196)*

FERROPICRITE. A chemically defined variety of picrite in which the amount of total iron as FeO exceeds 14 wt %. This value was later changed to 13% (Hanski, 1992). *(Hanski & Smolkin, 1989, p.71; Pechenga, Kola Peninsula, Russian Federation)*

FERROPLETE. An obsolete term for an igneous rock rich in iron oxides. *(Brögger, 1898, p.266; Tomkeieff p.196)*

FIASCONITE. A local name for a variety of leucite basanite in which the plagioclase is rich in anorthite. *(Johannsen, 1938, p.307; Mt Fiascone, Vulsini district, near Viterbo, Italy; Tröger(38) 595¾; Tomkeieff p.196)*

FINANDRANITE. A K-rich variety of alkali feldspar syenite consisting essentially of microcline and torendrikite (= magnesio-riebeckite), with some biotite and ilmenite. *(Lacroix, 1922, p.378; Ambatofinandrahana, Madagascar; Tröger 178; Johannsen v.3, p.13; Tomkeieff p.197)*

FINE (ASH) GRAIN. Now defined in the pyroclastic classification (section 2.2.2, p.7) as a variety of ash grain in which the mean diameter is < 1/16 mm. Synonymous with dust grain.

FINE (ASH) TUFF. Now defined in the pyroclastic classification (section 2.2.2, p.8) as a variety of tuff in which the mean diameter is < 1/16 mm. Synonymous with dust tuff.

FINLANDITE. A name suggested for a group of gabbros *sensu stricto. (Marcet Riba, 1925,* *p.293; named after Finland)*

FIRMICITE. An obsolete term for rocks consisting of hornblende and feldspar. *(Pinkerton, 1811b, p.42; named after Firmicus, who first mentioned alchemy; Tomkeieff p.197)*

FITZROYITE. A leucite-phlogopite lamproite (Table 2.7, p.17) consisting essentially of leucite and phlogopite in a chlorite-rich groundmass. Erroneously spelt fitzroydite by Sørensen (1974). *(Wade & Prider, 1940, p.59; Fitzroy River Basin, Kimberley district, West Australia, Australia; Tomkeieff p.198)*

FLOOD BASALT. A general term for basaltic lavas occurring in continental regions as vast accumulations of subhorizontal flows that have been erupted, probably from fissures, in rapid succession on a regional scale. Many of the lavas are transitional basalts, but tholeiitic basalts are also common. The term is synonymous with plateau basalt. *(Tyrrell, 1937, p.94; Tomkeieff p.201)*

FLORIANITE. An obsolete term for a variety of granite containing red plagioclase, white orthoclase, quartz and pinite. However, the source reference cited by Tomkeieff does not contain the name although it does mention granites from the Florian Hills. *(Tomkeieff et al., 1983, p.201; Florian Hills (Swabian Alb), Germany)*

FLORINITE. A local name for a melanocratic variety of monchiquite (a lamprophyre) with phenocrysts of olivine and augite and smaller augites and biotite in an altered matrix. *(Lacroix, 1933, p.197; Sainte-Florine, Brassac, Auvergne, France; Tröger 406; Tomkeieff p.201)*

FLUOLITE. An obsolete term, originally used by Lampadius, for a variety of pitchstone from Iceland. *(Glocker, 1831, p.720; Tomkeieff p.204)*

FOID DIORITE. A collective term for alkaline plutonic rocks consisting of feldspathoids (10%–60% of the felsic minerals), intermediate plagioclase and large amounts of mafic minerals. Now defined modally in QAPF field 14 (Fig. 2.4, p.22). *(Streckeisen, 1973, p.26)*

FOID DIORITOID. Proposed for preliminary use in the QAPF "field" classification (Fig. 2.10, p.29) for plutonic rocks thought to contain essential foids and in which plagioclase is thought to be more abundant than alkali feldspar. *(Streckeisen, 1973, p.28)*

FOID GABBRO. A collective term for basic alkaline plutonic rocks consisting of feldspathoids (10%–60% of the felsic minerals), calcic plagioclase and large amounts of mafic minerals. Now defined modally in QAPF field 14 (Fig. 2.4, p.22). *(Streckeisen, 1973, p.26)*

FOID GABBROID. Proposed for preliminary use in the QAPF "field" classification (Fig. 2.10, p.29) for plutonic rocks thought to contain essential foids and in which plagioclase is thought to be more abundant than alkali feldspar. *(Streckeisen, 1973, p.28)*

FOID MONZODIORITE. A collective term for alkaline plutonic rocks consisting of feldspathoids (10%–60% of the felsic minerals), intermediate plagioclase with subordinate alkali feldspar and large amounts of mafic minerals. Now defined modally in QAPF field 13 (Fig. 2.4, p.22). *(Streckeisen, 1973, p.26)*

FOID MONZOGABBRO. A collective term for basic alkaline plutonic rocks consisting of feldspathoids (10%–60% of the felsic minerals), calcic plagioclase with subordinate alkali feldspar and large amounts of mafic minerals. Now defined modally in QAPF field 13 (Fig. 2.4, p.22). *(Streckeisen, 1973, p.26)*

FOID MONZOSYENITE. A collective term for rare alkaline plutonic rocks consisting of feldspathoids, alkali feldspar, plagioclase and mafic minerals. Now defined modally in QAPF field 12 (Fig. 2.4, p.22). The term is synonymous with foid plagisyenite.

(Streckeisen, 1973, p.26)

FOID PLAGISYENITE. May be used as a synonym for foid monzosyenite of QAPF field 12 (p.24). *(Streckeisen, 1973, p.26)*

FOID SYENITE. A collective term for leucocratic alkaline plutonic rocks consisting of feldspathoids (10%–60% of the felsic minerals), alkali feldspar and mafic minerals. Now defined modally in QAPF field 11 (Fig. 2.4, p.22). *(Streckeisen, 1973, p.26)*

FOID SYENITOID. Proposed for preliminary use in the QAPF "field" classification (Fig. 2.10, p.29) for plutonic rocks thought to contain essential foids and in which alkali feldspar is thought to be more abundant than plagioclase. *(Streckeisen, 1973, p.28)*

FOID-BEARING ALKALI FELDSPAR SYENITE. A collective term for alkali feldspar syenites containing small amounts of feldspathoids (less than 10% of the felsic minerals). Now defined modally in QAPF field 6' (Fig. 2.4, p.22). *(Streckeisen, 1973, p.26)*

FOID-BEARING ALKALI FELDSPAR TRACHYTE. A collective term for alkali feldspar trachytes containing small amounts of feldspathoids (less than 10% of the felsic minerals). Now defined modally in QAPF field 6' (Fig. 2.11, p.31). *(Streckeisen, 1978, p.4)*

FOID-BEARING ANORTHOSITE. A collective term for anorthosites containing small amounts of feldspathoids (less than 10% of the felsic minerals). Now defined modally in QAPF field 10' (Fig. 2.4, p.22). *(Streckeisen, 1973, p.26)*

FOID-BEARING DIORITE. A collective term for diorites containing small amounts of feldspathoids (less than 10% of the felsic minerals). Now defined modally in QAPF field 10' (Fig. 2.4, p.22). *(Streckeisen, 1973, p.26)*

FOID-BEARING GABBRO. A collective term for gabbros containing small amounts of

82 3 Glossary of terms

feldspathoids (less than 10% of the felsic minerals). Now defined modally in QAPF field 10' (Fig. 2.4, p.22). *(Streckeisen, 1973, p.26)*

FOID-BEARING LATITE. A collective term for latites containing small amounts of feldspathoids (less than 10% of the felsic minerals). Now defined modally in QAPF field 8' (Fig. 2.11, p.31). *(Streckeisen, 1978, p.4)*

FOID-BEARING MONZODIORITE. A collective term for monzodiorites containing small amounts of feldspathoids (less than 10% of the felsic minerals). Now defined modally in QAPF field 9' (Fig. 2.4, p.22). *(Streckeisen, 1973, p.26)*

FOID-BEARING MONZOGABBRO. A collective term for monzogabbros containing small amounts of feldspathoids (less than 10% of the felsic minerals). Now defined modally in QAPF field 9' (Fig. 2.4, p.22). *(Streckeisen, 1973, p.26)*

FOID-BEARING MONZONITE. A collective term for monzonites containing small amounts of feldspathoids (less than 10% of the felsic minerals). Now defined modally in QAPF field 8' (Fig. 2.4, p.22). *(Streckeisen, 1973, p.26)*

FOID-BEARING SYENITE. A collective term for syenites containing small amounts of feldspathoids (less than 10% of the felsic minerals). Now defined modally in QAPF field 7' (Fig. 2.4, p.22). *(Streckeisen, 1973, p.26)*

FOID-BEARING TRACHYTE. A collective term for trachytes containing small amounts of feldspathoids (less than 10% of the felsic minerals). Now defined modally in QAPF field 7' (Fig. 2.11, p.31). *(Streckeisen, 1978, p.4)*

FOIDITE. A general term for volcanic rocks defined in QAPF field 15 (Fig. 2.11, p.31), i.e. rocks containing more than 60% foids in total light-coloured constituents. If modes

are not available, chemically defined in TAS field F (Fig. 2.14, p.35). If possible the most abundant foid should be used in the name, e.g. nephelinite, leucitite etc. *(Streckeisen, 1965; Tomkeieff p.205)*

FOIDITOID. Proposed for preliminary use in the QAPF "field" classification (Fig. 2.19, p.39) for volcanic rocks tentatively identified as foidite. *(Streckeisen, 1978, p.3)*

FOIDOLITE. A general term for plutonic rocks defined in QAPF field 15 (Fig. 2.4, p.22), i.e. rocks containing more than 60% foids in total light-coloured constituents. If possible the most abundant foid should be used in the name, e.g. nephelinolite, leucitolite etc. *(Streckeisen, 1973, p.26)*

FORELLENSTEIN. A local German name applied to a spotted troctolite. *(Rath, 1855, p.551; from the German Forelle =trout; Tröger 849; Johannsen v.3, p.225; Tomkeieff p.206)*

FORTUNITE. Originally described as a variety of trachyte characterized by the presence of phlogopite and bronzite (= enstatite). Now regarded as an hyalo-enstatite-phlogopite lamproite (Table 2.7, p.17). *(Adan de Yarza, 1893, p.350; Fortuna, Murcia, Spain; Tröger 233; Johannsen v.3, p.20; Tomkeieff p.207)*

FOURCHITE. A melanocratic analcime lamprophyre containing abundant augite but devoid of olivine and feldspar. *(Williams, 1891, p.107; Fourche Mts, Arkansas, USA; Tröger 376; Johannsen v.4, p.391; Tomkeieff p.207)*

FOYAITE. A hypersolvus nepheline syenite sometimes used as a group name for nepheline syenites. Now used as a term for nepheline syenites having a foyaitic (= trachytic) texture caused by the platy alkali feldspar crystals. *(Blum, 1861, p.426; Mt Foia, Monchique, Portugal; Tröger 414; Johannsen v.4, p.100; Tomkeieff p.208)*

FRAIDRONITE (FRAIDONITE). A local French name for a variety of biotite lamprophyre which was later described as a variety of kersantite consisting of biotite, chlorite, plagioclase and minor quartz. *(Dumas, 1846, p.572; Tomkeieff p.208)*

GABBREID. An obsolete field term for a coarse-grained igneous rock which contains more than 50% pyroxene. Feldspar is subordinate in amount. *(Johannsen, 1911, p.320; Tomkeieff p.214)*

GABBRIDE. A revised spelling recommended to replace the field term gabbreid. Now obsolete. *(Johannsen, 1926, p.182; Johannsen v.1, p.57; Tomkeieff p.214)*

GABBRITE. An obsolete name for an aplitic gabbro with only minor ferromagnesian minerals. *(Polenov, 1899, p.464; Tröger(38) 849½; Johannsen v.3, p.308; Tomkeieff p.214)*

GABBRO. A coarse-grained plutonic rock composed essentially of calcic plagioclase, pyroxene and iron oxides. If olivine is an essential constituent it is olivine gabbro – if quartz, quartz gabbro. Now defined modally in QAPF field 10 (Fig. 2.4, p.22). *(Targioni-Tozzetti, 1768, p.432; from an old Florentine name; Tröger 348; Johannsen v.3, p.205; Tomkeieff p.214)*

GABBRO-NELSONITE. A local name for a melanocratic variety of dolerite which occurs as dykes and consists of labradorite, pyroxene, ilmenite and apatite. *(Watson & Taber, 1910, p.206; Nelson County, Virginia, USA; Tröger 770; Tomkeieff p.215)*

GABBRODIORITE. A term originally proposed for a gabbro in which the pyroxene is altered to uralitic hornblende. Now used for rocks which are mineralogically on the boundary of gabbro and diorite, i.e. their plagioclase is An_{50}. *(Törnebohm, 1877b, p.391; Tröger 331; Johannsen v.3, p.238; Tomkeieff p.214)*

GABBROGRANITE. An obsolete term for rocks intermediate between gabbro and granite and synonymous with farsundite when used in the sense of a comprehensive name for granitic rocks of the Farsund area. *(Törnebohm, 1881, p.21; Tröger 850)*

GABBROID. A term originally used as a group name for rocks of the gabbro–norite groups, although used by Shand (1927) for rocks of theralitic composition. Now proposed for preliminary use in the QAPF "field" classification (Fig. 2.10, p.29) for plutonic rocks tentatively identified as gabbro or monzogabbro. *(Gümbel, 1888, p.87; Tomkeieff p.214)*

GABBRONORITE. A collective name for a plutonic rock consisting of calcic plagioclase and roughly equal amounts of clinopyroxene and orthopyroxene. Now defined modally in the gabbroic rock classification (Fig. 2.6, p.25). *(Streckeisen, 1973, p.27)*

GABBROPHYRE. A dyke rock of basaltic composition consisting of labradorite and rare augite phenocrysts in a felted groundmass of hornblende needles with occasional quartz. Cf. malchite. Odinite was suggested as an alternative term. *(Chelius, 1892, p.3; Tröger 319; Johannsen v.3, p.326; Tomkeieff p.215)*

GABBROPHYRITE. An obsolete name for a porphyritic gabbro-aplite. *(Polenov, 1899, p.464; Tomkeieff p.215)*

GABBROPORPHYRITE. An obsolete name for a porphyritic rock containing phenocrysts of plagioclase and some pyroxene in a fine-grained crystalline groundmass of calcic plagioclase. *(Rosenbusch, 1898, p.205; Tröger 371; Johannsen v.3, p.309; Tomkeieff p.215)*

GABBROSYENITE. An obsolete name proposed for a rock mineralogically between monzonite and an orthoclase-bearing gabbro. *(Tarasenko, 1895, p.15; Tröger 852; Johannsen v.3, p.212)*

GALLINACE (GALLINAZO). A local Peruvian name for a variety of obsidian, but later restricted by Cordier (1816) to basic glasses. *(Faujas de Saint-Fond, 1778, p.172; Tomkeieff p.215)*

GAMAICU. A local South American name for a variolite. *(Tomkeieff et al., 1983, p.215)*

GAREWAITE. An obsolete local name for an

ultramafic lamprophyre composed of phenocrysts of corroded diopside in a groundmass of olivine, pyroxene, chromite, magnetite and kaolinized labradorite. *(Duparc & Pearce, 1904, p.154; Springs of Garewaïa River, Tilai Range, Urals, Russian Federation; Tröger 741; Johannsen v.3, p.336; Tomkeieff p.216)*

GARGANITE. A local name for a variety of lamprophyre containing olivine, hornblende, augite, orthoclase and plagioclase found in the centre of a kersantite dyke. *(Viola & Stefano, 1893, p.135; Punta delle Pietre Nere, Foggia, Italy; Tröger 853; Johannsen v.3, p.40; Tomkeieff p.216)*

GARRONITE. An obsolete name for a variety of melanocratic diorite consisting mainly of biotite, augite and andesine with a little hornblende showing a reaction relationship with augite. Minor orthoclase and quartz occur. *(Reynolds, 1937, p.476; Slievegarron, County Down, Northern Ireland, UK; Tomkeieff p.216)*

GAUSSBERGITE. A variety of lamproite containing phenocrysts of leucite, augite and olivine in a glass matrix. *(Lacroix, 1926, p.599; Gaussberg, Kaiser Wilhelm II Land, Antarctica; Tröger 504; Johannsen v.4, p.262; Tomkeieff p.217)*

GAUTEITE. A porphyritic variety of syenite consisting of phenocrysts of hornblende, pyroxene, plagioclase and occasionally biotite, set in a groundmass composed essentially of alkali feldspar. Analcime may occur. *(Hibsch, 1898, p.84; Gaute (now Kout), S. of Děčín, N. Bohemia, Czech Republic; Tröger 245; Johannsen v.4, p.39; Tomkeieff p.218)*

GEBURITE-DACITE. An obsolete name for a variety of dacite containing hypersthene (= enstatite). *(Gregory, 1902, p.203; from the Aboriginal gebur = Mt Macedon, Victoria, Australia; Tröger 854; Tomkeieff p.218)*

GESTELLSTEIN. An old German term for greisen. *(Brückmanns, 1778, p.214; Johannsen v.2,*

p.18; Tomkeieff p.221)*

GHIANDONE. A local Italian name for a porphyritic granite or augen gneiss. *(Tomkeieff et al., 1983, p.221)*

GHIZITE. An obsolete name for a volcanic rock consisting of phenocrysts of titanian augite, biotite and olivine in a groundmass of colourless glass with analcime and microlites of augite and olivine. Cf. scanoite. *(Washington, 1914b, p.753; Ghizu, Mt Ferru, Sardinia, Italy; Tröger 589; Johannsen v.4, p.216; Tomkeieff p.221)*

GIBELITE. A local name for a variety of trachyte consisting essentially of Na-microcline with small amounts of augite, aenigmatite and quartz. *(Washington, 1913, p.691; Mt Gibelé, Pantelleria Island, Italy; Tröger 211; Johannsen v.3, p.18; Tomkeieff p.222)*

GIUMARRITE. A local name for a variety of monchiquite (a lamprophyre) containing hornblende. *(Viola, 1901, p.309; Giumarra, Sicily, Italy; Tröger 377; Tomkeieff p.222)*

GLADKAITE. An obsolete name for a diorite aplite composed of plagioclase, quartz, hornblende, and biotite with minor epidote. With less quartz it passes into plagiaplite. *(Duparc & Pearce, 1905, p.1614; Gladkaïa-Sopka Ridge, N. Urals, Russian Federation; Tröger 140; Johannsen v.2, p.405; Tomkeieff p.223)*

GLEESITES. A collective name for subvolcanic haüyne rocks. *(Kalb & Bendig, 1938; Glees, Laacher See, near Koblenz, Germany; Tomkeieff p.225)*

GLENMUIRITE. A local name for a variety of analcime gabbro or monzogabbro consisting of labradorite and augite with analcime, olivine and alkali feldspar. The rock was originally described as a teschenite. *(Johannsen, 1938, p.194; Glenmuir Water, near Lugar, Scotland, UK; Tröger(38) 566½; Tomkeieff p.225)*

GLIMMERITE. An ultramafic rock consisting almost entirely of biotite. The term was suggested as an alternative to biotitite to avoid the rather discordant sound of consecutive "tites". *(Larsen & Pardee, 1929, p.104; from the German Glimmer = mica; Tröger 719; Johannsen v.4, p.441; Tomkeieff p.226)*

GOODERITE. A coarse-grained variety of alkali feldspar syenite consisting of about 80% albite with a little nepheline and biotite. *(Johannsen, 1938, p.57; Gooderham, Glamorgan Township, Ontario, Canada; Tröger(38) 179½; Tomkeieff p.228)*

GORDUNITE. An obsolete name for a variety of wehrlite composed of olivine, diopside and pyrope garnet. *(Grubenmann, 1908, p.129; Gordunotal, Bellinzona, Switzerland; Tröger 737; Johannsen v.4, p.421; Tomkeieff p.228)*

GORYACHITE. A local name for a leucocratic plutonic rock composed of up to 70% nepheline, with calcic to intermediate plagioclase, augite zoned to titanian augite, with small amounts of alkali feldspar and olivine. Sporadic aegirine, alkali amphibole and fluorite may also be present. *(Luchitskii, 1960, p.114; Mt Goryachaya, Minusinsk Depression, Siberia, Russian Federation)*

GOUGHITE. An oceanic volcanic rock equivalent to high-K trachybasalt, chemically defined by a K_2O/Na_2O ratio higher than 1:2 and a differentiation index less than 62.5. Cf. tristanite. *(Stewart & Thornton, 1975, p.566; named after Gough Island, South Atlantic Ocean)*

GRAMMITE. An obsolete term for graphic granite. *(Langius, 1708, p.42; Tomkeieff p.230)*

GRANATINE. An obsolete general term for rocks consisting of three minerals, usually quartz and feldspar plus a ferromagnesian. *(Kirwan, 1794, p.342; Tomkeieff p.230)*

GRANATITE. A group name for ultramafic rocks containing abundant garnet. *(Tröger, 1935, p.289; Tröger 716–718)*

GRANEID. An obsolete field term for a coarse-grained igneous rock consisting of quartz and any kind of feldspar with biotite and/or amphibole and/or pyroxene. *(Johannsen,*

1911, p.319; Tomkeieff p.230)

GRANIDE. A revised spelling recommended to replace the field term graneid. Now obsolete. *(Johannsen, 1926, p.182; Johannsen v.1, p.56)*

GRANILITE. An obsolete term for a granite containing more than three essential minerals. *(Kirwan, 1794, p.346; Tomkeieff p.231)*

GRANITE. A plutonic rock consisting essentially of quartz, alkali feldspar and sodic plagioclase in variable amounts usually with biotite and/or hornblende. Now defined modally in QAPF field 3 (Fig. 2.4, p.22). See also A-type granite, I-type granite, M-type granite, S-type granite, two-mica granite etc. *(A term of great antiquity usually attributed to Caesalpino, 1596, p.89 – see Johannsen for further discussion; Tröger 856; Johannsen v.2, p.124; Tomkeieff p.231)*

GRANITEL. An obsolete term attributed to Saussure for a crystalline rock consisting of two of the minerals quartz, feldspar or a ferromagnesian mineral. *(Pinkerton, 1811a, p.203)*

GRANITELL. An obsolete term for a quartz tourmaline or quartz hornblende rock, attributed to D'Aubenton. *(Kirwan, 1784, p.343)*

GRANITELLE. An erroneous spelling of granitell. *(Tomkeieff et al., 1983, p.231)*

GRANITELLO. An obsolete Italian name for a fine-grained granite. *(Brückmanns, 1778, p.213; Johannsen v.2, p.302; Tomkeieff p.232)*

GRANITIN. An obsolete term for a fine-grained granite. *(Pinkerton, 1811a, p.201; Tomkeieff p.232)*

GRANITITE. A confusing term redefined several times for various varieties of biotite and hornblende granite. Should not be used. *(Leonhard, 1823a, p.54; Tröger 857; Johannsen v.2, p.226; Tomkeieff p.232)*

GRANITOID. A term originally used for rocks resembling granite but of different composition, it is now commonly used as a synonym for a granitic rock, i.e. any plutonic rock consisting essentially of quartz, alkali feldspar and/or plagioclase. Now proposed for preliminary use in the QAPF "field" classification (Fig. 2.10, p.29) for plutonic rocks tentatively identified as granite, granodiorite or tonalite. *(Pinkerton, 1811a, p.209; Tomkeieff p.233)*

GRANITON. An obsolete term for a coarse-grained granite. *(Pinkerton, 1811a, p.202; Tomkeieff p.233)*

GRANITONE. An obsolete term originally proposed for rapakivi from Finland and mentioned by Ferber in his letters from Italy. Also used in Italy as an old name for gabbro. *(Kirwan, 1784, p.149; Tomkeieff p.233)*

GRANITOPHYRE. An obsolete group name for porphyritic granitic rocks. *(Gümbel, 1888, p.111; Tomkeieff p.233)*

GRANODIORITE. A plutonic rock consisting essentially of quartz, sodic plagioclase and lesser amounts of alkali feldspar with minor amounts of hornblende and biotite. Name first used by Becker on maps of the Gold Belt of the Sierra Nevada. Now defined modally in QAPF field 4 (Fig. 2.4, p.22). *(Lindgren, 1893, p.202; Tröger 105; Johannsen v.2, p.318; Tomkeieff p.233)*

GRANODOLERITE. An obsolete term for a plutonic rock containing labradorite-bytownite and orthoclase. *(Shand, 1917, p.466; Tröger 119)*

GRANOFELSOPHYRE. An obsolete term for a porphyry with a groundmass texture between felsitic and granitic. *(Vogelsang, 1875, p.160; Tomkeieff p.234)*

GRANOGABBRO. An obsolete term for an orthoclase-bearing variety of quartz gabbro analogous to granodiorite. *(Johannsen, 1917, p.89; Tröger 110; Johannsen v.2, p.367; Tomkeieff p.234)*

GRANOLIPARITE. An obsolete name for a recent granitic rock containing vitreous feldspar. *(Lapparent, 1893, p.620; Tomkeieff p.234)*

GRANOLITE. An obsolete general term first suggested by Pirsson for all granular igneous rocks. *(Turner, 1899, p.141; Tröger 858; Tomkeieff p.234)*

GRANOMASANITE. An obsolete local term for a porphyritic granite with plagioclase phenocrysts. Effusive equivalent of eutectofelsite. *(Kotô, 1909, p.186; Ku-ryong, Masanpho, South Korea; Tröger 89)*

GRANOPHYRE. A term originally used for a granite porphyry with a microcrystalline groundmass. Rosenbusch (1877) later re-defined it as it is used today, as a porphyritic rock of granite composition in which the groundmass alkali feldspar and quartz are in micrographic intergrowth. *(Vogelsang, 1872, p.534; Tröger 859; Johannsen v.2, p.288; Tomkeieff p.234)*

GRANOPHYRITE. An obsolete term for non-porphyritic rocks with a microcrystalline texture. *(Vogelsang, 1872, p.534)*

GRANULITE. Apart from metamorphic usage, this term has also been used for fine-grained muscovite or two mica leuco-granites but is now obsolete in this sense. *(Michel-Lévy, 1874, p.177; Tröger 860; Tomkeieff p.235)*

GRANULOPHYRE. A quartz porphyry in which the groundmass has a microgranitic texture. *(Lapparent, 1885, p.602; Morvan, France; Tomkeieff p.236)*

GRAPHIC GRANITE. A name originally suggested by Haüy for a variety of granite in which the quartz and feldspar are intergrown in such a way that it has the appearance of cuneiform or runic writing. *(Brongniart, 1813, p.32; Johannsen v.2, p.72; Tomkeieff p.236)*

GRAPHIPHYRE (GRAPHOPHYRE). Obsolete terms originally suggested to replace the textural term granophyre – graphophyre having a megacrystalline and graphiphyre a microcrystalline groundmass. Tröger (1938) uses graphopyre as a rock name, although he suggests it is better used as an adjective. Tomkeieff et al. (1983) suggest both are rock names. *(Cross et al., 1906, p.704; Tröger 861;*

Tomkeieff p.237)

GRAZINITE. A variety of phonolitic nephelinite containing analcime but no olivine. *(Almeida, 1961, p.135; Mt Grazinas, Trindade Island, South Atlantic Ocean)*

GREENHALGHITE. A variety of quartz latite consisting of phenocrysts of oligoclase-andesine and biotite in a matrix of quartz and K-feldspar, corresponding to the yosemititic magma-type of Niggli. *(Niggli, 1923, p.113; Mt Greenhalgh, Silverton, South Carolina, USA; Tröger 97; Tomkeieff p.238)*

GREENLANDITE. See grönlandite.

GREISEN. An old Saxon mining term originally used for a variety of granitic rocks containing tin ore and virtually no feldspar. Later used for rocks which appear to be pneumatolytically altered granite and consisting essentially of quartz and mica, with or without topaz, tourmaline, fluorite and Sn, W, Be, Li, Nb, Ta etc. mineralization. *(Leonhard, 1823a, p.59; from the German Greisstein = ash-coloured stone; Tröger 6; Johannsen v.2, p.5; Tomkeieff p.239)*

GRENNAITE. A variety of agpaitic nepheline syenite with phenocrysts of catapleiite and/or eudialyte in a fine-grained groundmass rich in aegirine and showing gneissose textures. The rock is most probably a deformed and partly recrystallized lujavrite. *(Adamson, 1944, p.123; Grenna, now Gränna, Norra Kärr complex, Sweden; Tomkeieff p.239)*

GRIMMAITE. A variety of charnockite porphyry to quartz mangerite porphyry of QAPF fields 3b to 8* (Fig. 2.4, p.22), respectively. It is suggested (Streckeisen, 1974, p.358) that this term should be abandoned. *(Ebert, 1968, p.1044; Grimma, Saxony, Germany)*

GRIQUAITE. A variety of garnet clinopyroxenite occurring as inclusions in kimberlites, composed essentially of pyrope garnet, with diopside, and minor phlogopite, olivine, spinel and ore. *(Beck, 1907, p.301; named after Griqualand, South Africa; Tröger 716;*

Tomkeieff p.240)

GRÖBAITE. An obsolete name for a leucocratic variety of monzodiorite consisting of major amounts of sodic plagioclase, with lesser amounts of orthoclase, augite, hornblende and biotite. *(Grahmann, 1927, p.33; Gröba, Saxony, Germany; Tröger 277; Tomkeieff p.240)*

GRÖNLANDITE (GREENLANDITE). An obsolete name for a variety of orthopyroxene hornblendite consisting essentially of hornblende and hypersthene (= enstatite). *(Machatschki, 1927, p.175; named after Greenland; from Island of Upernivik; Tröger 705; Johannsen v.4, p.447; Tomkeieff p.238)*

GRORUDITE. A local name for a variety of peralkaline microgranite containing aegirine. *(Brögger, 1890, p.66; Grorud, Oslo district, Norway; Tröger 62; Johannsen v.2, p.102; Tomkeieff p.240)*

GUARDIAITE. A variety of tephritic phonolite or nepheline-bearing latite containing plagioclase phenocrysts mantled with alkali feldspar in a groundmass of essential nepheline, augite, biotite and glass. *(Narici, 1932, p.234; Cape Guardia, Ponza Island, Italy; Tröger 531; Tomkeieff p.242)*

HAKUTOITE. A local name, attributed to F. Yamanari, for a variety of peralkaline trachyte that is characterized by sodic pyroxenes and amphiboles and small amounts of quartz. *(Lacroix, 1927a, p.1410; Hakuto San, North Korea; Tröger 77; Tomkeieff p.244)*

HAMRÅNGITE, HAMRONGITE. An obsolete local name for a variety of kersantite (a lamprophyre) containing phenocrysts of biotite in a groundmass of biotite, andesine and a little quartz. *(Eckermann, 1928b, p.13; supposedly from Hamrångefjarden, Gävle, Sweden, but attempts to relocate this rock have failed; Tröger 146; Johannsen v.2, p.405; Tomkeieff p.245)*

HAPLITE. A name suggested as the etymologically correct spelling of aplite.

(Johannsen, 1932, p.91; Tomkeieff p.246)

HAPLO-. A prefix suggested to denote a pure artifical mixture of various feldspars and diopside used in experimental work, e.g. haplodiorite (diopside + sodic plagioclase), haplogabbro (diopside + calcic plagioclase), haplosyenite (diopside + albite). Haplogranite and haplograndiorite are now commonly used by experimental petrologists for mixtures of quartz and synthectic feldspars with no mafic minerals. Cf. aplo-. *(Bowen, 1915, p.161; from the Greek haploos = simple)*

HAPLO-PITCHSTONE. A term suggested for an experimental composition approximating a natural pitchstone. *(Reynolds, 1958, p.388; Tomkeieff p.246)*

HAPLODIORITE. See haplo-.

HAPLOGABBRO. See haplo-.

HAPLOGRANITE. See haplo-.

HAPLOGRANODIORITE. See haplo-.

HAPLOPHYRE. An obsolete term for a granitic rock with a texture intermediate between porphyritic and equigranular. *(Stache & John, 1877, p.189; Tomkeieff p.246)*

HAPLOSYENITE. See haplo-.

HARRISITE. A variety of troctolite in which the large black lustrous olivines have a branching habit of growth and are orientated perpendicular to the layering. The matrix to the olivines is composed of highly calcic plagioclase and minor pyroxene. *(Harker, 1908, p.71; Glen Harris, Island of Rhum, Scotland, UK; Tröger 401; Johannsen v.3, p.349; Tomkeieff p.246)*

HARTUNGITE. A local name for a variety of melteigite consisting of nepheline, pyroxene and wollastonite with or without alkali feldspar. *(Eckermann, 1942, p.402; Hartung, Alnö Island, Västernorrland, Sweden; Tomkeieff p.247)*

HARZBURGITE. An ultramafic plutonic rock composed essentially of olivine and orthopyroxene. Now defined modally in the ultramafic rock classification (Fig. 2.9, p.28).

(Rosenbusch, 1887, p.269; Harzburg, Harz Mts, Lower Saxony, Germany; Tröger 732; Johannsen v.4, p.438; Tomkeieff p.247)

HATHERLITE. A local term for a variety of alkali feldspar syenite containing abundant anorthoclase, with biotite and hornblende. The rock was later called leeuwfonteinite. *(Henderson, 1898, p.46; Hatherly gun powder factory, Magaliesberg Range, South Africa; Tröger 182; Johannsen v.3, p.10; Tomkeieff p.247)*

HAÜYNE BASALT. A general term originally used for a rock in which the place of feldspar is taken by haüyne. Later redefined as a rock of basaltic appearance consisting of haüyne, olivine and clinopyroxene, often with minor leucite and nepheline. The name should not be used as the term basalt is now restricted to a rock containing essential plagioclase. As the rock is a variety of foidite it should be given the appropriate name, e.g. olivine haüynite. *(Trimmer, 1841, p.172; Tröger 650; Johannsen v.4, p.346; Tomkeieff p.247)*

HAÜYNE BASANITE. Now defined in QAPF field 14 (Fig. 2.11, p.31) as a variety of basanite in which haüyne is the most abundant foid. *(Tröger, 1935, p.246; Tröger 598; Johannsen v.4, p.238)*

HAÜYNE PHONOLITE. Now defined in QAPF field 11 (Fig. 2.11, p.31) as a variety of phonolite in which haüyne is an important foid. *(Bořický, 1873, p.16; Tröger 473; Johannsen v.4, p.131)*

HAÜYNFELS. An obsolete name based on the incorrect identification of haüyne in a rock that was later called ditroite. *(Haidinger, 1861, p.64; Ditro (now Ditrau), Transylvania, Romania; Johannsen v.4, p.98; Tomkeieff p.247)*

HAÜYNITE. A term originally used for a basaltic rock with haüyne as the only foid. Now defined as a variety of foidite of field 15c of the QAPF diagram (Fig. 2.11, p.31). *(Reinisch, 1917, p.68; Morgenberg, Blatt Wiesental, Saxony, Germany; Tröger 650;*

Tomkeieff p.247)

HAÜYNITITE. An obsolete term proposed for a volcanic rock consisting of haüyne and mafic minerals without olivine. *(Johannsen, 1938, p.336; Tröger 650; Tomkeieff p.248)*

HAÜYNOLITH. An obsolete term proposed for a monomineralic volcanic rock consisting of haüyne. *(Johannsen, 1938, p.337; Tröger 633; Tomkeieff p.248)*

HAÜYNOPHYRE. A volcanic rock name equivalent to leucitophyre or nephelinite for rocks that are rich in haüyne and nosean. The matrix may contain glass, pyroxene, haüyne and other foids. *(Abich, 1839, p.337; Tröger 537; Johannsen v.4, p.362; Tomkeieff p.248)*

HAWAIITE. A term originally defined as a variety of olivine-bearing basalt in which the normative plagioclase is oligoclase or andesine. Now defined chemically as the sodic variety of trachybasalt in TAS field S1 (Fig. 2.14, p.35). *(Iddings, 1913, p.198; named after Hawaii, USA; Tröger 327; Johannsen v.3, p.171; Tomkeieff p.248)*

HEDRUMITE. A local name for a porphyritic fine-grained variety of alkali feldspar syenite with a trachytic or foyaitic texture consisting essentially of microcline-microperthite. The groundmass is rich in biotite and contains minor sodic amphibole, aegirine and nepheline. Cf. pulaskite. *(Brögger, 1898, p.183; Hedrum, Oslo district, Norway; Tröger 190; Johannsen v.4, p.25; Tomkeieff p.248)*

HELSINKITE. A local name for an equigranular dyke rock consisting essentially of albite and epidote, which was originally thought to be primary. *(Laitakari, 1918, p.3; named after Helsinki; Tröger 199; Johannsen v.3, p.142; Tomkeieff p.249)*

HEPTORITE. A variety of lamprophyre consisting of phenocrysts of titanian augite, barkevikite (see p.44), olivine, and haüyne in a groundmass of glass and labradorite laths. *(Busz, 1904, p.86; from the Greek hepta = seven; Siebengebirge, near Bonn, Germany;*

Tröger 561; Johannsen v.4, p.244; Tomkeieff p.249)

HERONITE. A variety of analcime monzosyenite which contains spheroidal aggregates of orthoclase in a matrix of analcime, labradorite and aegirine. *(Coleman, 1899, p.436; Heron Bay, Lake Superior, Ontario, Canada; Tröger 516; Johannsen v.4, p.287; Tomkeieff p.250)*

HEUMITE. A local name for a fine-grained variety of nepheline-bearing syenite consisting essentially of alkali feldspar, barkevikite (see p.44) and lepidomelane with minor amounts of plagioclase and nepheline. *(Brögger, 1898, p.90; Heum, Larvik, Oslo Igneous Province, Norway; Tröger 497; Johannsen v.4, p.170; Tomkeieff p.251)*

HIGH-ALUMINA BASALT. A collective term for basalts from the tholeiitic, calc-alkaline and alkaline associations in which Al_2O_3 is generally greater than 16%. They usually, but not invariably, have phenocrysts of plagioclase. Tholeiitic basalts are commonly and calc-alkaline basalts are almost exclusively of this type. *(Kuno, 1960, p.122)*

HIGH-K. A chemical term originally applied to andesites with $K_2O > 2.5\%$ but later applied to volcanic rocks which plot above a line on the SiO_2–K_2O diagram, i.e. for their SiO_2 values they are higher than usual in K_2O. The field is now defined as an optional qualifier for certain rock names in the TAS classification (Fig. 2.17, p.37). *(Taylor, 1969, p.45)*

HIGH-MG ANDESITE. A term proposed as an alternative to boninite to avoid the equivalence between some ophiolite basalts and high-Mg andesites implied by some authors for the term boninite. *(Jenner, 1981, p.307)*

HIGHWOODITE. A variety of nepheline-bearing monzonite consisting essentially of anorthoclase, labradorite, augite and biotite. The light minerals only just exceed the dark minerals in abundance. *(Johannsen, 1938, p.40; Highwood Mts, Montana, USA; Tröger(38) 237½; Tomkeieff p.252)*

HILAIRITE. A local name for a coarse-grained variety of sodalite nepheline syenite consisting of nepheline, alkali feldspar, sodalite and aegirine with a little eudialyte. *(Johannsen, 1938, p.289; Mt St Hilaire, Quebec, Canada; Tröger(38) 415½; Tomkeieff p.252)*

HIRNANTITE. An obsolete name for an intrusive rock consisting essentially of albite with minor chlorite, iron ore and secondary quartz and calcite. Possibly an albite keratophyre or an albitized tholeiite. *(Travis, 1915, p.79; Craigddu, Hirnant, Berwyn Hills, Wales, UK; Tröger 219; Johannsen v.3, p.143; Tomkeieff p.253)*

HOBIQUANDORTHITE. An unwieldy name constructed by Johannsen to illustrate some of the possibilties of the mnemonic classification of Belyankin (1929) for a granitic rock consisting of *ho*rnblende, *bi*otite, *qu*artz, *and*esine and *orth*oclase. Cf. biquahororthandite and topatourbiolilepiquorthite. *(Johannsen, 1931, p.125)*

HOLLAITE. A local name for a coarse-grained variety of calcite melteigite consisting of pyroxene (55%), calcite and nepheline. *(Brögger, 1921, p.217; Holla Church, Fen Complex, Telemark, Norway; Tröger 699; Tomkeieff p.253)*

HOLMITE. An obsolete term suggested by Johannsen for a lamprophyre in which melilite was thought to occur. Apparently unknown to Johannsen, it had already been shown (Flett, 1935, p.185) that the mineral originally identified as melilite was, in fact, apatite. The rock is actually a monchiquite. *(Johannsen, 1938, p.378; Holm Island, Orkney Islands, Scotland, UK; Tröger(38) 746½; Tomkeieff p.253)*

HOLOLEUCOCRATIC. Now defined (see Table 2.2, p.5) as a rock whose colour index ranges from 0 to 10. *(Streckeisen, 1967; from the Greek holos = whole, leukos = white, krateein = to predominate)*

HOLOMELANOCRATIC. Now defined (see Table 2.2, p.5) as a rock whose colour index ranges from 90 to 100. *(Streckeisen, 1967; from the Greek holos = whole, melas = dark, krateein = to predominate)*

HOLYOKEITE. A local name for an albitized dolerite with an ophitic texture. The principal minerals are albite, some zoisite, calcite, chlorite and opaques. *(Emerson, 1902, p.508; Holyoke, Massachusetts, USA; Tröger 197; Johannsen v.3, p.139; Tomkeieff p.254)*

HONGITE. A name given to a hypothetical set of data used to illustrate the problems of the statistical interpretation of rock compositions. *(Aitchison, 1984, p.534)*

HOOIBERGITE. A local name for a melanocratic variety of what, according to the Subcommission classification, is a granite just inside QAPF field 3b (Fig. 2.4, p.22) consisting essentially of green hornblende with minor amounts of labradorite, orthoclase and quartz. Traditionally the rock has been called a variety of gabbro. *(Westermann, 1932, p.46; Hooiberg, Aruba Island, Lesser Antilles; Tröger 281; Tomkeieff p.256)*

HÖRMANNSITE. An obsolete name for a variety of monzodiorite containing sodic plagioclase and lesser amounts of orthoclase with biotite and calcite. Thought to have formed by the assimilation of marble. *(Ostadal, 1935, p.122; Hörmanns, N.W. Waldvirtel, Austria; Tröger(38) 863½; Tomkeieff p.256)*

HORNBERG. An old German name for greisen. *(Brückmanns, 1778, p.214; Johannsen v.2, p.18; Tomkeieff p.256)*

HORNBLENDE GABBRO. A variety of gabbro in which primary hornblende occurs. Now defined modally in the gabbroic rock classification (Fig. 2.6, p.25). *(Streng & Kloos, 1877, p.113; Tröger 349; Johannsen v.3, p.227; Tomkeieff p.256)*

HORNBLENDE PERIDOTITE. An ultramafic plutonic rock consisting essentially of olivine with up to 50% amphibole. Now defined modally in the ultramafic rock classification (Fig. 2.9, p.28). *(Wyllie, 1967, p.2)*

HORNBLENDE PYROXENITE. A collective term for pyroxenites containing up to 50% of amphibole. Now defined modally in the ultramafic rock classification (Fig. 2.9, p.28). *(Wyllie, 1967, p.2)*

HORNBLENDITE. An ultramafic plutonic rock composed almost entirely of hornblende. Now defined modally in the ultramafic rock classification (Fig. 2.9, p.28). *(Phillips, 1848, p.40; Tröger 701; Johannsen v.4, p.442; Tomkeieff p.257)*

HORTITE. An obsolete term for a melanocratic variety of monzosyenite consisting of alkali feldspar, plagioclase and minor calcite with abundant pyroxene and hornblende. Probably formed from gabbro by the assimilation of limestone. *(Vogt, 1915, p.5; Hortavaer, N.W. Leka, Trondheim, Norway; Tröger 261; Johannsen v.3, p.66; Tomkeieff p.258)*

HORTONOLITITE. A variety of dunite composed essentially of hortonolite. *(Polkanov & Eliseev, 1941, p.28; Tomkeieff p.258)*

HOVLANDITE. A local name for a variety of monzogabbro composed of large crystals of bytownite, biotite, hypersthene (= enstatite) and olivine. The rock is a hybrid type. *(Barth, 1944, p.68; Hovland Farm, Modum, Oslo district, Norway; Tomkeieff p.258)*

HRAFTINNA. An Icelandic name for obsidian. *(Johannsen, 1932, p.279; Tomkeieff p.258)*

HUDSONITE. A term given to a variety of pyroxene olivine hornblendite which was later replaced by the name cortlandtite as hudsonite (= hastingsite) was already in use as a mineral name. *(Cohen, 1885, p.242; Hudson River, New York, USA; Tröger 866; Johannsen v.4, p.425; Tomkeieff p.259)*

HUNGARITE. A local term for a variety of andesite containing hornblende. *(Lang, 1877, p.196; named after Hungary; Tröger 867; Johannsen v.3, p.169; Tomkeieff p.260)*

HURUMITE. A local name for a medium-grained variety of monzonite containing andesine and orthoclase, with biotite, clinopyroxene and quartz. *(Brögger, 1931, p.72; Hurum, Oslo district, Norway; Tröger 92; Tomkeieff p.260)*

HUSEBYITE. A local name for a medium-grained variety of plagioclase-bearing nepheline syenite containing aegirine-augite and alkali amphibole. The term is synonymous with essexite-foyaite. *(Brögger, 1933, p.35; Husebyås, Akerhus, Oslo Igneous Province, Norway; Tröger 508; Johannsen v.4, p.165; Tomkeieff p.261)*

HYALO-. Suggested as an optional prefix (p.5) that can be used to indicate the presence of glass in a rock that has been named using the TAS classification (section 2.12.2, p.33). *(Le Maitre (Editor) et al., 1989, p.5; from the Greek hyalos = glass)*

HYALOCLASTITE. A consolidated pyroclastic rock composed of angular fragments of glass which may or may not be devitrifed. *(Rittmann, 1962, p.72)*

HYALOMELANE. An obsolete term for a basaltic glass insoluble in acids. *(Hausmann, 1847, p.545; from the Greek hyalos = glass, melas = dark; Tröger 869; Johannsen v.3, p.290; Tomkeieff p.261)*

HYALOMICTE. An obsolete term for a greisen consisting of hyaline quartz and mica. *(Brongniart, 1813, p.34; Tomkeieff p.261)*

HYALOPSITE. An obsolete name for obsidian. *(Gümbel, 1888, p.41; Johannsen v.2, p.279; Tomkeieff p.261)*

HYDROTACHYLYTE. A term originally used for a glass formed from a partially melted sandstone, but later used for tachylytes with high water contents (up to 13%). *(Petersen, 1869, p.32; from the Greek hydor = water, tachys = rapid, lytos = soluble; Tröger 870; Johannsen v.3, p.290; Tomkeieff p.265)*

HYPABYSSAL. Pertaining to rocks whose type of emplacement is intermediate between plutonic and volcanic. Often applied to rocks from minor intrusions, e.g. sills and dykes. *(Brögger, 1894, p.97; from the Greek hypo = under or from, abyssos = bottomless or unfathomable; Johannsen v.1, p.176; Tomkeieff p.266)*

HYPERACIDITE. A term proposed for highly acid rocks. *(Loewinson-Lessing, 1896, p.175; Tomkeieff p.266)*

HYPERITE. An old name originally used for a variety of norite consisting of hypersthene (= enstatite), plagioclase and augite, but redefined with various other meanings, e.g. diabase, gabbro-norite and for noritic rocks with hypersthene coronas around olivine. *(Naumann, 1850, p.594; from an old Swedish name; Tröger 354; Johannsen v.3, p.238; Tomkeieff p.266)*

HYPERITITE. An obsolete term for a bronzite (= enstatite) diabase. *(Törnebohm, 1877a, p.42; Tomkeieff p.266)*

HYPERSTHENITE. A variety of orthopyroxenite composed almost entirely of hypersthene (= enstatite). *(Naumann, 1850, p.595; Tröger 676; Johannsen v.4, p.458; Tomkeieff p.267)*

HYPOBASITE. An obsolete term for an ultrabasic rock. *(Loewinson-Lessing, 1898, p.42; Tomkeieff p.267)*

HYSTEROBASE. A variety of diabase in which the augite is replaced by brown hornblende. *(Lossen, 1886, p.925; Tröger 321; Tomkeieff p.268)*

I-TYPE GRANITE. A general term for a range of metaluminous calc-alkali granitic rocks, mainly tonalites to granodiorites and granites, characterized by essential quartz, variable amounts of plagioclase and alkali feldspar, hornblende and biotite. Muscovite is absent. The prefix I implies that the source rocks have igneous compositions. *(Chappell & White, 1974, p.173)*

ICELANDITE. A variety of intermediate volcanic rock containing phenocrysts of andesine, clinopyroxene and/or orthopyroxene and/or pigeonite and less commonly olivine in a

groundmass of the same minerals. It differs chemically from a typical orogenic andesite in being poorer in alumina and richer in iron. *(Carmichael, 1964, p.442; named after Iceland)*

IGNEOUS ROCK. A rock that has solidified from a molten state either within or on the surface of the Earth or extraterrestrial bodies. *(Kirwan, 1794, p.455; from the Latin ignis = fire; Johannsen v.1, p.177; Tomkeieff p.271)*

IGNIMBRITE. An indurated tuff consisting of crystal and rock fragments in a matrix of glass shards which are usually welded together. In some cases the welding is so extreme that the original texture shown by the glass shards is lost. The composition is usually acid to intermediate. *(Marshall, 1932, p.200; from the Latin ignis = fire, imber = shower; Tröger(38) 871½; Tomkeieff p.271)*

IJOLITE. A plutonic rock consisting of pyroxene with 30%–70% nepheline. Now defined modally as a variety of foidolite in QAPF field 15 (Fig. 2.8, p.27). *(Ramsay & Berghell, 1891, p.304; Iijoki, now Iivaara, Kuusamo, Finland; Tröger 607; Johannsen v.4, p.313; Tomkeieff p.271)*

IJUSSITE. A local name for a variety of teschenite without olivine and with kaersutite and titanian augite as the mafic phases. *(Rachkovsky, 1911, p.257; Ijuss River, Upper Yenisey River, Siberia, Russian Federation; Tröger 567; Tomkeieff p.272)*

ILMEN-GRANITE. A biotite nepheline syenite later called miaskite. *(Tomkeieff et al., 1983, p.272; Ilmen Mts, Urals, Russian Federation)*

ILMENITITE. A rock consisting almost entirely of ilmenite. *(Kolderup, 1896, p.14; Tröger 768; Johannsen v.4, p.470; Tomkeieff p.272)*

ILVAITE. Alleged to be a variety of granite from the Island of Ilva, now Elba, Italy but the reference cited (Fournet, 1845) does not contain the name. However, the term should not be used as a rock name as it has been used since 1811 for a silicate of Fe and Ca. *(Tomkeieff et al., 1983, p.272)*

ILZITE. An obsolete local name for an aplite consisting of sodic plagioclase, quartz and biotite with minor alkali feldspar. *(Frentzel, 1911, p.176; Ilzgebirge, Passau, Bavaria, Germany; Tröger(38) 278½; Tomkeieff p.272)*

IMANDRITE. A local name for a variety of granite consisting of graphically intergrown quartz and albite with chloritized biotite. A hybrid rock formed by the interaction of nepheline syenite magma with greywacke. *(Ramsay & Hackman, 1894, p.46; Imandra, Kola Peninsula, Russian Federation; Tröger 57; Tomkeieff p.272)*

INNINMORITE. A local name for an andesitic to dacitic rock composed of phenocrysts of calcic plagioclase and pigeonite in a fine-grained to glassy groundmass. *(Thomas & Bailey, 1915, p.209; Inninmore Bay, Morvern, Scotland, UK; Tröger 126; Tomkeieff p.276)*

INTERMEDIATE. A commonly used chemical term now defined in the TAS classification (Fig. 2.14, p.35) as a rock containing more than 52% and less than 63% SiO_2, i.e. lying between acid and basic rocks. *(Judd, 1886a, p.51; Johannsen v.1, p.181; Tomkeieff p.277)*

INTRITE. An obsolete term for porphyritic rocks. *(Pinkerton, 1811a, p.75; Tomkeieff p.279)*

INTRODACITE. A term for a normal dacite that was later replaced by phanerodacite. *(Belyankin, 1923, p.99; Tomkeieff p.279)*

IOPHYRE. An obsolete term for blue to chocolate-brown porphyritic rocks containing amphiboles. *(Rozière, 1826, p.310; Tomkeieff p.281)*

ISENITE. A local name proposed for a variety of trachyandesite thought to contain hornblende, biotite and nosean. As the "nosean" was later shown to be apatite (Dannenberg, 1897), the name is not justified. *(Bertels, 1874, p.175; River Eis (Latin name Isena), Nassau, Ger-*

many; Tröger 872; Tomkeieff p.281)

ISOPHYRE. An obsolete name for obsidian. *(Tomkeieff et al., 1983, p.283)*

ISSITE. A local name for a fine-grained variety of hornblendite consisting essentially of hornblende, with cores of augite and a little plagioclase. *(Duparc & Pamphil, 1910, p.1136; River Iss, Urals, Russian Federation; Tröger 712; Johannsen v.3, p.336; Tomkeieff p.283)*

ITALITE. A rock consisting almost entirely of leucite held together by small amounts of glass. Now defined modally as a leucocratic variety of foidolite in QAPF field 15 (Fig. 2.8, p.27) in which the foid is predominantly leucite. *(Washington, 1920, p.33; Alban Hills, near Rome, Italy; Tröger 627; Johannsen v.4, p.311; Tomkeieff p.284)*

ITSINDRITE. A local name for a fine-grained variety of nepheline syenite consisting essentially of microcline and nepheline in granophyric intergrowth, with aegirine, biotite and melanite. *(Lacroix, 1922, p.388; Itsindra Valley, Madagascar; Tröger 417; Johannsen v.4, p.145; Tomkeieff p.284)*

IVERNITE. An obsolete name for a granite-like rock consisting of phenocrysts of orthoclase and plagioclase in a groundmass of euhedral feldspars with minor hornblende, mica and quartz. *(McHenry & Watt, 1895, p.93; Iverness, County Limerick, Ireland; Tröger 873; Tomkeieff p.284)*

IVOIRITE. A variety of clinopyroxene norite of QAPF field 10 (Fig. 2.4, p.22) belonging to the charnockitic rock series. It is suggested (Streckeisen, 1974, p.358) that this term should be abandoned. *(Lacroix, 1910; Mt Marny, Ivory Coast)*

IVREITE. A name suggested for a group of pyroxene quartz diorites. *(Marcet Riba, 1925, p.293; Ivrea, Piedmont, Italy; Tomkeieff p.284)*

JACUPIRANGITE. A variety of alkali pyroxenite consisting essentially of titanian augite with minor amounts of titanomagnetite, nepheline, apatite, perovskite and melanite garnet. *(Derby, 1891, p.314; Jacupiranga, São Paulo, Brazil; Tröger 687; Johannsen v.4, p.463; Tomkeieff p.285)*

JADEOLITE. A term for a green chrome-bearing variety of syenite which has the appearance of jade. *(Kunz, 1908, p.810; Tomkeieff p.285)*

JERSEYITE. A local name for a quartz-bearing variety of minette. *(Lacroix, 1933, p.187; Jersey, Channel Islands, UK; Tröger 68; Tomkeieff p.286)*

JERVISITE. An oceanic volcanic rock of the icelandite group, in which groundmass plagioclase is oligoclase. Cf. duncanite. *(Stewart & Thornton, 1975, p.568; named after Jervis Island, Galapagos Islands, Pacific Ocean)*

JOSEFITE. An obsolete name for a poorly defined ultramafic rock composed of titanian augite and olivine with secondary calcite and chlorite. *(Szádeczky, 1899, p.215; "Big Quarry", 2 km S.E. of Aswân, Egypt; Tröger 697; Johannsen v.4, p.438; Tomkeieff p.286)*

JOTUN-NORITE. The term is synonymous with jotunite. *(Goldschmidt, 1916, p.34; Jotunheim, Norway)*

JOTUNITE. A member of the charnockitic rock series equivalent to orthopyroxene monzonorite in QAPF field 9 (Table 2.10, p.20). The term is synonymous with jotunnorite. *(Hødal, 1945, p.140; Jotunheim, Norway)*

JUMILLITE. Originally described as a variety of olivine phonolitic leucitite composed essentially of leucite and diopside, and with subordinate olivine, alkali feldspar and phlogopite. Now regarded as an olivine-diopside-richterite madupitic lamproite (Table 2.7, p.17). *(Osann, 1906, p.306; Jumilla, Murcia, Spain; Tröger 505; Johannsen v.4, p.263; Tomkeieff p.286)*

JUVITE. A local name for a coarse-grained variety of nepheline syenite in which K-feldspar is more abundant than Na-feldspar; sometimes used as a general term for potassic nepheline syenites. *(Brögger, 1921, p.93; Juvet, Fen Complex, Telemark, Norway; Tröger 413; Johannsen v.4, p.104; Tomkeieff p.286)*

KAGIARITE. See khagiarite.

KAHUSITE. A rock described as a silica-rich magnetite rhyolite consisting essentially of quartz, lesser amounts of magnetite and minor tourmaline, biotite and graphite. Possibly a re-melted iron-rich quartzite. *(Sorotchinsky, 1934, p.192; Kahusi Volcano, S.W. of Lake Kivu, Democratic Republic of Congo; Tröger(38) 873½; Tomkeieff p.287)*

KAIWEKITE. A local name for a variety of trachyte that usually carries phenocrysts of anorthoclase and titanian augite. *(Marshall, 1906, p.400; Kaiweke (= Long Beach), Kaikorai Valley, Otago, New Zealand; Tröger 216; Johannsen v.4, p.38; Tomkeieff p.287)*

KAJANITE. An obsolete term for a melanocratic variety of leucitite largely composed of biotite and clinopyroxene and with subordinate leucite. *(Lacroix, 1926, p.600; Oele Kajan, Kalimantan, Indonesia; Tröger 648; Johannsen v.4, p.371; Tomkeieff p.287)*

KAKORTOKITE. A local name for a variety of agpaitic nepheline syenite displaying pronounced cumulate textures and igneous layering with a repetition of layers enriched in alkali feldspar, eudialyte and arfvedsonite. *(Ussing, 1912, p.43; Kakortok (now Qaqortoq), Ilímaussaq, Greenland; Tröger 878; Johannsen v.4, p.118; Tomkeieff p.287)*

KALIOPLETE. An obsolete term for an igneous rock rich in potassium. *(Brögger, 1898, p.266; Tomkeieff p.288)*

KALMAFITE. A mnemonic name for a mixture of kalsilite plus mafic minerals. *(Hatch et al., 1961, p.358)*

KALSILITITE. Now defined in the kalsilite-bearing rock classification (section 2.5, p.12) as a rock containing kalsilite, but no leucite or melilite. Cf. mafurite. *(Woolley et al., 1996, p.177)*

KALSILITOLITE. A medium-grained holocrystalline rock mainly composed of kalsilite and clinopyroxene with small amounts of leucite, melanite, biotite and melilite occurring as ejected blocks in pyroclastic deposits. *(Federico, 1976, p.5; Colle Cimino, Alban Hills, near Rome, Italy)*

KAMAFUGITE. A name for the distinctive series of consanguineous rocks katungite, mafurite and ugandite (section 2.5, p.12). *(Sahama, 1974, p.96; Toro-Ankole, Uganda)*

KAMCHATITE. A local name for a porphyritic igneous rock consisting of orthoclase, bluish hornblende, pyroxene and epidote. Although similar in chemical composition to monzonite, it contains no plagioclase. *(Morozov, 1938, p.19; Sredinnyi Range, Kamchatka, Russian Federation; Tomkeieff p.288)*

KAMMGRANITE. An old term reused by Niggli for a dark coloured K-rich amphibole biotite granite. *(Groth, 1877, p.396; from the German Kamm = divide; granite of Vosges Divide, France; Tröger 54; Tomkeieff p.289)*

KAMMSTEIN. An old German (Saxon) term for serpentinite. *(Tomkeieff et al., 1983, p.289)*

KAMPERITE. A local name for a medium- to fine-grained highly potassic dyke rock composed of almost equal amounts of orthoclase and biotite with minor oligoclase. The biotite is a late crystallization product. *(Brögger, 1921, p.104; Kamperhoug, Fen Complex, Telemark, Norway; Tröger 248; Johannsen v.3, p.87; Tomkeieff p.289)*

KANZIBITE. A local name for a K-rich variety of rhyolite in which the orthoclase phenocrysts are coloured black by graphite inclusions. *(Sorotchinsky, 1934, p.190; Kanzibi Lake, Kivu, Democratic Republic of Congo; Tröger(38) 878½; Tomkeieff p.289)*

KARITE. A term tentatively suggested for a quartz-rich variety of grorudite or peralkaline microgranite. *(Karpinskii, 1903, p.31; Kara*

River, Transbaikal region, Russian Federation; Tröger 63; Tomkeieff p.289)*

KARJALITE. A local name for a rock consisting essentially of albite with variable amounts of quartz, carbonate, amphibole, chlorite and iron ore. *(Väyrynen, 1938, p.74; from the Finnish name for Karelia; Tomkeieff p.289)*

KARLSTEINITE. A local name for a variety of peralkaline granite containing abundant microcline and some alkali amphiboles. *(Hackl & Waldmann, 1935, p.263; Karlstein, near Raabs, Lower Austria, Austria; Tröger(38) 56½; Tomkeieff p.290)*

KÄRNÄITE. A glassy rock with the composition of dacite, containing phenocrysts of feldspar and abundant inclusions of tuff-like material. It has been identified as an impactite. *(Saksela, 1948, p.20; Kärnä Island, Lappajärvi, Finland; Tomkeieff p.290)*

KASANSKITE. See kazanskite.

KÅSENITE (KOSENITE). A local name for a variety of calcite carbonatite consisting of sodic pyroxene, a little nepheline and 50%–60% calcite. *(Brögger, 1921, p.222; Kåsene, Fen Complex, Telemark, Norway; Tröger 756; Tomkeieff p.290)*

KASSAITE. A porphyritic dyke rock with phenocrysts of haüyne, labradorite with oligoclase rims, barkevikite (see p.44) and augite in a groundmass of hastingsite and andesine rimmed by oligoclase and orthoclase. *(Lacroix, 1918, p.542; Kassa Island, Îles de Los, Conakry, Guinea; Tröger 519; Johannsen v.4, p.190; Tomkeieff p.290)*

KATABUGITE. A variety of jotunite or norite of QAPF fields 9-10 (Fig. 2.4, p.22). It is a member of the bugite series with 50% to 58% SiO_2, 15% to 40% hypersthene (= enstatite), oligoclase-andesine antiperthite and quartz. It is suggested (Streckeisen, 1974, p.358) that this term should be abandoned. *(Bezborod'ko, 1931, p.145; Bug River, Podolia, Ukraine; Tröger(38) 308⅓; Tomkeieff p.290)*

KATNOSITE. A local name for a variety of

nordmarkite (a quartz-bearing syenite) containing biotite or aegirine. *(Brögger, 1933, p.86; Lake Katnosa, Nordmarka, Oslo Igneous Province, Norway; Tomkeieff p.291)*

KATUNGITE. A potassic melanocratic variety of olivine melilitite composed essentially of olivine and melilite with subordinate leucite, kalsilite and nepheline in a glassy matrix. A kamafugitic rock and now regarded as a kalsilite-leucite-olivine melilitite (Tables 2.5 and 2.6, p.12). *(Holmes, 1937, p.201; Katunga, Uganda; Tröger(38) 672½; Johannsen v.4, p.362; Tomkeieff p.291)*

KATZENBUCKELITE. A porphyritic variety of haüyne nepheline microsyenite containing phenocrysts of nepheline, nosean or analcime, biotite and olivine in a matrix of nepheline, analcime (leucite), alkali feldspar and sodic pyroxene and amphibole. *(Osann, 1902, p.403; Katzenbuckel, Odenwald, Germany; Tröger 446; Johannsen v.4, p.273; Tomkeieff p.292)*

KAUAIITE. A variety of monzodiorite with titanian augite, a little olivine and feldspar zoned from labradorite to oligoclase and mantled by lime anorthoclase. *(Iddings, 1913, p.173; Kauai, Hawaiian Islands, USA; Tröger 284; Johannsen v.3, p.118; Tomkeieff p.292)*

KAUKASITE. See caucasite.

KAULAITE. A variety of olivine basalt containing anemousite instead of plagioclase, corresponding to the kaulaitic magma-type of Niggli (1936, p.366) This is the same as the olivine pacificite of Barth (1930). *(Tröger, 1938, p.67; Kaula Gorge, Ookala, Mauna Kea, Hawaii, USA; Tröger(38) 385; Tomkeieff p.292)*

KAXTORPITE. A local name for a coarse-grained, sometimes schistose, variety of nepheline syenite consisting of alkali feldspar, nepheline, alkali amphibole (eckermannite) with or without pectolite and aegirine. *(Adamson, 1944, p.188; Kaxtorp, Norra Kärr complex, Sweden; Tomkeieff p.292)*

KAZANSKITE (KASANSKITE). A local term for a mafic to ultramafic dyke rock consisting essentially of olivine with magnetite and small amounts of bytownite. It is a variety of plagioclase-bearing dunite or melanocratic troctolite. *(Duparc & Grosset, 1916, p.106; Kazansky, Central Urals, Russian Federation; Tröger 402; Johannsen v.3, p.336; Tomkeieff p.292)*

KEDABEKITE. A local term for a gabbroic rock composed of bytownite and approximately equal amounts of hedenbergite and andradite. Probably a hybrid or skarn. *(Fedorov, 1901, p.135; Kedabek, Azerbaijan; Tröger 365; Johannsen v.3, p.241; Tomkeieff p.292)*

KEMAHLITE. A local name for a fine-grained variety of monzonite containing pseudoleucite. *(Lacroix, 1933, p.190; Kemahl, Turkey; Tröger 520; Tomkeieff p.293)*

KENNINGITE. An obsolete name given to a rock which was thought to be a leucocratic variety of basalt consisting mainly of labradorite with minor augite and serpentinized olivine, and representing the volcanic equivalent of anorthosite. However, Lundqvist (1975) showed that the rock was similar to other dolerite dykes in the area and has no resemblance to anorthosite. *(Eckermann, 1938a, p.277; Känningen Island, Rödö Archipelago, Sweden; Tomkeieff p.293)*

KENTALLENITE. A local term for a melanocratic variety of monzonite composed of olivine, augite, zoned plagioclase, brown and green micas and interstitial alkali feldspar. *(Hill & Kynaston, 1900, p.532; Kentallen Quarry, Ballachulish, Scotland, UK; Tröger 260; Johannsen v.3, p.99; Tomkeieff p.293)*

KENYTE (KENYITE). A variety of phonolitic trachyte characterized by rhomb-shaped phenocrysts of anorthoclase with or without augite and olivine set in a glassy *ne*-normative groundmass. *(Gregory, 1900, p.211; Mt Kenya, Kenya; Tröger 467; Johannsen v.4,*

p.130; Tomkeieff p.294)

KERAMIKITE. A group name for a series of cordierite-bearing pumices and obsidians of rhyolitic composition which grade into cordierite-bearing microtinites (an obsolete term for trachyte). *(Kotô, 1916a, p.197; Sakura-jima Volcano, Ryukyu Island, Japan; Tröger(38) 40⅓; Tomkeieff p.294)*

KERATOPHYRE. A term originally used for a quartz-bearing orthoclase-plagioclase rock with a dense groundmass, but later used for albitized felsic extrusive rocks consisting essentially of albite with minor mafic minerals, often altered to chlorite. Potassic keratophyres are also recognized in which the feldspar is orthoclase. Commonly associated with spilites. *(Gümbel, 1874, p.43; from the Greek keras = horn; Tröger 213; Johannsen v.3, p.47; Tomkeieff p.294)*

KERATOPHYRITE. A keratophyre with the composition of adamellite. *(Loewinson-Lessing, 1928, p.142; Tröger(38) 880½; Tomkeieff p.294)*

KERSANTITE. A variety of lamprophyre consisting of phenocrysts of Mg-biotite, with or without hornblende, olivine or pyroxene in a groundmass of the same minerals plus plagioclase and occasional alkali feldspar. Now defined in the lamprophyre classification (Table 2.9, p.19). *(Delesse, 1851, p.164; named after its similarity to the rock called kersanton from Kersanton, Brittany, France; Tröger 317; Johannsen v.3, p.187; Tomkeieff p.295)*

KERSANTON. An obsolete local name for a rock that was later named kersantite. *(Rivière, 1844, p.537; Kersanton, Brittany, France; Tröger 881; Johannsen v.3, p.187; Tomkeieff p.295)*

KHAGIARITE (KAGIARITE). An obsolete local name for a glassy pantellerite with distinct flow texture. *(Washington, 1913, p.708; Khagiar, Pantelleria Island, Italy; Tröger 73; Tomkeieff p.295)*

KHEWRAITE. A K-rich effusive rock composed of K-feldspar and enstatite, with minor ilmenite, hematite and apatite. The enstatite is altered to antigorite, vermiculite and quartz. *(Mosebach, 1956, p.200; Khewra Gorge, Pakistan)*

KHIBINITE (CHIBINITE). A variety of eudialyte-bearing nepheline syenite with aegirine, alkali amphibole and many accessory minerals, particularly those containing Ti and Zr. Although originally spelt chibinite, the preferred spelling is now khibinite. *(Ramsay & Hackman, 1894, p.80; Khibina complex, Kola Peninsula, Russian Federation; Tröger 418; Johannsen v.4, p.107; Tomkeieff p.98)*

KIIRUNAVAARITE (KIRUNAVAARITE). A local name for an ultramafic rock consisting almost entirely of magnetite. Cf. shishimskite. *(Rinne, 1921, p.182; Kiirunavaara, Lapland, Sweden; Tröger 761; Johannsen v.4, p.466; Tomkeieff p.296)*

KILAUEITE. An obsolete term for an aphanitic basaltic rock rich in magnetite. *(Silvestri, 1888, p.186; Kilauea, Hawaii, USA; Tröger(38) 881½; Tomkeieff p.296)*

KIMBERLITE. An ultramafic rock consisting of major amounts of serpentinized olivine with variable amounts of phlogopite, orthopyroxene, clinopyroxene, carbonate and chromite. Characteristic accessory minerals include pyrope garnet, monticellite, rutile and perovskite. It cannot be defined but is characterized by the mineralogical criteria given in section 2.6, p.13. *(Lewis, 1888, p.130; Kimberley, South Africa; Tröger 742; Johannsen v.4, p.413; Tomkeieff p.296)*

KIVITE. A local name for a variety of leucite tephrite largely composed of clinopyroxene and plagioclase with subordinate leucite and minor olivine and biotite. *(Lacroix, 1923, p.265; Lake Kivu, Democratic Republic of Congo; Tröger 584; Johannsen v.4, p.241; Tomkeieff p.297)*

KJELSÅSITE. A local name given to a plagioclase-rich larvikite which is a variety of augite syenite or monzonite. *(Brögger, 1933, p.45; Kjelsås, Sørkedal, Oslo district, Norway; Tröger 275; Johannsen v.3, p.115; Tomkeieff p.297)*

KLAUSENITE. A local term used as a group name for two different types of rocks: (1) for dioritic norites and quartz gabbros from the Tyrol, (2) for a series of lamprophyric rocks ranging from diorite to tonalite in composition but with a graphic groundmass. *(Cathrein, 1898, p.275; Klausen, near Bressanone, Alto Adige, Italy; Tröger 338; Tomkeieff p.298)*

KLEPTOLITH. An obsolete term suggested for basic lamprophyric dyke rocks which intrude many Scandinavian granites. *(Sederholm, 1934, p.17; Tomkeieff p.298)*

KLINGHARDTITE. A variety of nepheline phonolite containing phenocrysts of sanidine. *(Kaiser-Gießen, 1913, p.597; Klinghardt Mts, S.E. of Lüderitz, Namibia; Tröger(38) 881¾; Tomkeieff p.298)*

KLINGSTEIN (CLINKSTONE). An old term, used before the mineral composition of rocks was known, for rocks which ring when hit with a hammer. Later replaced by the name phonolite. *(Werner, 1787, p.11; from the German klingen = to sound; Johannsen v.4, p.121)*

KLOTDIORITE. A Swedish term for an orbicular diorite. *(Holst & Eichstädt, 1884, p.137; Tomkeieff p.298)*

KLOTGRANITE. A Swedish term for an orbicular granite. *(Holst & Eichstädt, 1884, p.137; Tomkeieff p.298)*

KODURITE. A local name suggested by L.L. Fermor for an intrusive rock consisting of K-feldspar, manganese garnet and apatite. *(Holland, 1907, p.22; Kodur Mines, Vizagapatam, now Vishakhapatnam, Andhra Pradesh, India; Tröger 224; Tomkeieff p.299)*

KOHALAITE. A general term proposed for intermediate volcanic rocks in which the normative feldspar is oligoclase; they some-times contain normative or modal olivine. *(Iddings, 1913, p.193; Kohala, Waimea, Hawaii, USA; Tröger 289; Johannsen v.3, p.169; Tomkeieff p.299)*

KOHLIPHYRE. An obsolete term proposed for igneous rocks intruding coal formations. *(Ebray, 1875, p.291)*

KOKKITE (COCCITE). An obsolete group name for crystalline non-schistose igneous and sedimentary rocks. *(Gümbel, 1888, p.85; Tomkeieff p.108)*

KOLDERUPITE. A name suggested for a group of pyroxene tonalites. *(Marcet Riba, 1925, p.293; named after C.F. Kolderup)*

KOMATIITE. A variety of ultramafic lavas that crystallize from high temperature magmas with 18% to 32% MgO. They often form pillows and have chilled flow-tops and usually display well-developed spinifex textures with intergrown skeletal and bladed olivine and pyroxene crystals set in abundant glass. The more highly magnesian varieties are often termed peridotitic komatiite. Now defined chemically in the TAS classification (Fig. 2.13, p.34). *(Viljoen & Viljoen, 1969, p.83; Komati River, Barberton, Transvaal, South Africa)*

KOMATIITIC BASALT. See basaltic komatiite.

KONGITE. A name given to a hypothetical set of data used to illustrate the problems of the statistical interpretation of rock compositions. *(Aitchison, 1984, p.534)*

KOSENITE. See kåsenite.

KOSWITE. A local term for a variety of olivine clinopyroxenite composed of clinopyroxene, olivine and magnetite. *(Duparc & Pearce, 1901, p.892; Koswinski Mts, Urals, Russian Federation; Tröger 683; Johannsen v.4, p.440; Tomkeieff p.300)*

KOTUITE. A local name for a dyke rock consisting of augite, nepheline, biotite and ore minerals with minor aegirine, apatite, perovskite and phlogopite. *(Butakova, 1956, p.213; Kotui River, S. of Khatanga, N. Siberia,*

Russian Federation)

KOVDITE. A local name for a variety of pyroxene hornblendite composed of green amphibole and orthopyroxene, with small amounts of biotite and plagioclase and occasional garnet. It may be a metamorphic rock. The description given in Tomkeieff *et al.* (1983) appears to be incorrect. *(Fedorov, 1903, p.215; Kovda, White Sea, Russian Federation; Tomkeieff p.300)*

KOVDORITE. A local term for a variety of turjaite (a melilitolite) containing olivine. *(Zlatkind, 1945, p.92; Lake Kovdor, Kola Peninsula, Russian Federation; Tomkeieff p.300)*

KRABLITE. An obsolete term for a rhyolitic crystal tuff. *(Preyer & Zirkel, 1862, p.317; Krabla, Iceland; Tröger 882; Tomkeieff p.300)*

KRAGERITE. An alternative spelling of krageröite. *(Watson, 1912, p.509; Tröger 313; Tomkeieff p.300)*

KRAGERÖITE. A local name for an aplite consisting essentially of sodic plagioclase and a considerable amount of rutile with minor quartz and orthoclase. *(Brögger, 1904, p.30; Kragerö, Arendal, Norway; Johannsen v.3, p.124; Tomkeieff p.301)*

KRISTIANITE (CHRISTIANITE). A local name for a biotite granite of the Oslo Igneous Province. *(Brögger, 1921, p.371; Kristiania, now Oslo, Norway; Tröger 883; Tomkeieff p.100)*

KUGDITE. A variety of olivine melilitolite consisting of melilite, olivine, pyroxene, nepheline and titanomagnetite. May be used as an optional term in the melilitic rocks classification if olivine > 10% (section 2.4, p.11). Synonomous with olivine melilitolite. *(Egorov, 1969, p.29)*

KUGELDIORITE. An orbicular variety of gabbro composed essentially of calcic plagioclase with hornblende, minor hypersthene (= enstatite) and quartz. This rock has also been called corsite, miagite and napoleonite.

(Leonhard, 1823a, p.107; Tröger 361; Johannsen v.3, p.320)

KULAITE. An obsolete term originally defined as a subgroup of basalts in which hornblende is more abundant than augite. Redefined (Washington, 1900) as a basic alkaline volcanic rock in which orthoclase and plagioclase are present in about equal amounts together with up to 25% nepheline. *(Washington, 1894, p.115; Kula, E. of Izmir, Turkey; Tröger 578; Johannsen v.4, p.198; Tomkeieff p.301)*

KULLAITE. A local name for porphyritic variety of monzodiorite composed of phenocrysts of andesine and Na-orthoclase in an ophitic groundmass of oligoclase-andesine, chlorite and epidote after augite, quartz and magnetite. *(Hennig, 1899, p.19; Kullagården, Kullen, Sweden; Tröger 288; Johannsen v.3, p.121; Tomkeieff p.303)*

KUSELITE. See cuselite.

KUSKITE. A discredited name for a light-coloured dyke rock originally thought to consist essentially of phenocrysts of quartz and scapolite in a groundmass of quartz, orthoclase and muscovite. The scapolite was later shown to be quartz and the rock a granite porphyry. *(Spurr, 1900b, p.315; Kuskokwim River, Holiknuk, Alaska, USA; Tröger 885; Johannsen v.2, p.300; Tomkeieff p.302)*

KUZEEVITE (KUSEEVITE). A poorly defined term used for a banded variety of the charnockitic rock series containing quartz, plagioclase, hypersthene (= enstatite), biotite, hornblende and garnet. Tomkeieff transliterates the name as kuseevite. *(Ainberg, 1955, p.111; Kuzeeva River, tributary of Yenisey River, Siberia, Russian Federation; Tomkeieff p.302)*

KVELLITE. A local name for an ultramafic dyke rock containing abundant phenocrysts of lepidomelane, olivine and barkevikite (see p.44) in a sparse groundmass of anorthoclase laths and some nepheline. *(Brögger, 1906, p.128; Kvelle, Larvik, Oslo Igneous Province, Norway; Tröger 715; Johannsen v.4,*

p.158; Tomkeieff p.302)

KYLITE. A local term for a variety of nepheline-bearing gabbro containing abundant olivine predominating over titanian augite, labradorite and minor nepheline. *(Tyrrell, 1912, p.121; Kyle district, near Dalmellington, Scotland, UK; Tröger 554; Johannsen v.4, p.226; Tomkeieff p.303)*

KYSCHTYMITE. A local name for a corundum-rich variety of anorthosite composed of 45%–50% corundum, calcic plagioclase and minor poikilitic biotite, spinel, zircon and apatite. *(Morozewicz, 1899, p.202; Kyschtym, source of River Borsowska, Urals, Russian Federation; Tröger 301; Johannsen v.3, p.327; Tomkeieff p.303)*

LAACHITE. A local term for a variety of sanidinite containing anorthoclase and biotite occurring as ejecta in volcanic tuff. *(Kalb, 1936, p.190; Laacher See, near Koblenz, Germany; Tröger(38) 226½; Tomkeieff p.304)*

LABRADITE. An obsolete term proposed for a coarse-grained rock consisting essentially of labradorite. *(Turner, 1900, p.110; Tröger 296; Johannsen v.3, p.198; Tomkeieff p.304)*

LABRADOPHYRE. An obsolete general term for rocks consisting of phenocrysts of labradorite in a groundmass of labradorite and pyroxene. *(Coquand, 1857, p.87; Tröger 887; Johannsen v.3, p.307; Tomkeieff p.304)*

LABRADORITE. An old name once used as a group name for gabbros, basalts etc. Also used by the Russians and French for labradorite-phyric basalts. *(Senft, 1857, Table II; Tröger 890; Tomkeieff p.304)*

LABRADORITITE. A name suggested for a variety of anorthosite composed almost entirely of labradorite. Johannsen (p.198) states that the name comes from an earlier 1920 publication, but this is in error. *(Johannsen, 1937, p.196; Tröger 296; Tomkeieff p.304)*

LADOGALITE. A melanocratic variety of alkali feldspar syenite, composed of clinopyroxene, alkali feldspar, hornblende, biotite, and apa-tite, with minor titanite and allanite. *(Khazov, 1983, p.1200)*

LADOGITE. A melanocratic apatite-rich variety of alkali feldspar syenite, composed of clinopyroxene, alkali feldspar, and apatite, with minor biotite, hornblende, titanite and allanite. *(Khazov, 1983, p.1200; Lake Ladoga, near St Petersburg, Russian Federation)*

LAHNPORPHYRY. An obsolete term for a rock variously described as a keratophyre and as an anchimetamorphic aegirine trachyte. The reference cited does not contain the name, although porphyries from the area are described. *(Koch, 1858, p.97; Lahntal, Hessen, Germany; Tröger 891; Tomkeieff p.306)*

LAKARPITE. A local name for a coarse-grained variety of nepheline syenite consisting of alkali feldspar, altered nepheline, abundant arfvedsonite, aegirine and a little pectolite. *(Törnebohm, 1906, p.18; Lakarp, Norra Kärr complex, Sweden; Tröger 422; Johannsen v.4, p.161; Tomkeieff p.306)*

LAMPROITE. A comprehensive term originally used for lamprophyric extrusive rocks rich in potassium and magnesium, corresponding to the lamproitic magma-type of Niggli. Lamproite is no longer regarded as a lamprophyre. Now part of the IUGS classification and although it cannot be defined it is characterized by the mineralogical and chemical criteria given in section 2.7, p.16. *(Niggli, 1923, p.184; Tröger 892; Tomkeieff p.307)*

LAMPROPHYRE. A name for a distinctive group of rocks which are strongly porphyritic in mafic minerals, typically biotite, amphiboles and pyroxenes, with any feldspars being confined to the groundmass. They commonly occur as dykes or small intrusions and often show signs of hydrothermal alteration. Further details of the subdivision of these rocks are given in section 2.9 and Table 2.9, p.19. *(Gümbel, 1874, p.36; from the Greek lampros = glistening; Tröger 893; Johannsen v.3, p.32; Tomkeieff p.307)*

LAMPROSYENITE. A comprehensive term used for mesocratic biotite syenites corresponding to the lamprosyenitic magma-type of Niggli. *(Niggli, 1923, p.182; Tröger 894; Tomkeieff p.307)*

LANGITE. An obsolete name suggested for a group of pyroxene diorites. *(Marcet Riba, 1925, p.293; named after H.A. Lang)*

LAPIDITE. A variety of ignimbrite consisting of angular fragments of cognate rock in a fine welded matrix of glass shards. *(Marshall, 1935, p.358; Tröger(38) 871½; Tomkeieff p.308)*

LAPILLI. Now defined in the pyroclastic classification (section 2.2.1, p.7) as a pyroclast of any shape with a mean diameter between 2 mm and 64 mm. *(Lyell, 1835, p.391; from the Latin lapillus = little stone; Johannsen v.1, p.179; Tomkeieff p.308)*

LAPILLI TUFF. Now defined in the pyroclastic classification (section 2.2.2, p.8 and Fig. 2.1, p.8) as a pyroclastic rock in which bombs and/or blocks < 25%, and both lapilli and ash < 75%.

LAPILLISTONE. Now defined in the pyroclastic classification (section 2.2.2, p.8 and Fig. 2.1, p.8) as a pyroclastic rock in which lapilli > 75%.

LARDALITE (LAURDALITE). A local name for a coarse-grained variety of nepheline syenite characterized by rhomb-shaped alkali or ternary feldspar crystals and large crystals of nepheline. *(Brögger, 1890, p.32; Lardal, Oslo, Norway; Tröger 419; Johannsen v.4, p.102; Tomkeieff p.308)*

LARVIKITE (LAURVIKITE). A variety of augite syenite or monzonite consisting of rhomb-shaped ternary feldspars (with a distinctive schiller), barkevikite (see p.44), titanian augite and lepidomelane. Minor nepheline, iron-rich olivine or quartz may be present. Common ornamental stone. *(Brögger, 1890, p.30; Larvik, Oslo Igneous Province, Norway; Tröger 183; Johannsen v.4, p.9; Tomkeieff*

p.308)

LASSENITE. A local name for a fresh glass of trachyte composition. *(Wadsworth, 1893, p.97; Lassen Peak, California, USA; Tröger 895; Johannsen v.3, p.77; Tomkeieff p.309)*

LATHUS PORPHYRY. An obsolete local name for fine-grained acid rocks with flow structure and K_2O much greater than Na_2O. *(Holtedahl, 1943, p.32; Lathusås, Bærum, Oslo Igneous Province, Norway)*

LATIANDESITE. A synonym for latite andesite. *(Rittmann, 1973, p.133)*

LATIBASALT. A synonym for latite basalt. *(Rittmann, 1973, p.133)*

LATITE. A term originally proposed for a rock which chemically lies between trachyte and andesite, but later used as a volcanic rock composed of approximately equal amounts of alkali feldspar and sodic plagioclase, i.e. the volcanic equivalent of monzonite. Now defined modally in QAPF field 8 (Fig. 2.11, p.31) and, if modes are not available, chemically as the potassic variety of trachyandesite in TAS field S3 (Fig. 2.14, p.35). *(Ransome, 1898, p.355; from Latium an Italian region; Tröger 270; Johannsen v.3, p.100; Tomkeieff p.311)*

LATITE ANDESITE. A name originally proposed as a comprehensive term for volcanic rocks in QAPF field 9, intermediate between latite and andesite and having a colour index less than 40, to cover such rock types as mugearite, shoshonite, benmoreite and hawaiite. The term was not widely accepted by English-speaking petrologists and was dropped from later versions of the volcanic QAPF classification. The term is synonymous with andelatite. *(Streckeisen, 1967, p.161)*

LATITE BASALT. A name orignally proposed as a comprehensive term for volcanic rocks in QAPF field 9, intermediate between latite and basalt and having a colour index higher than 40. The term was not widely accepted by English-speaking petrologists and was dropped from later versions of the volcanic QAPF classification. The term is synonymous with basalatite. *(Streckeisen, 1967, p.161)*

LAUGENITE. An obsolete name for a leucocratic rock originally described as a variety of diorite in which the plagioclase is oligoclase rather than the more common andesine. However, as the rock was said to be like an akerite and also contains alkali feldspar it is better described as a variety of monzodiorite of QAPF field 9 (Fig. 2.4, p.22). *(Iddings, 1913, p.164; Tuft, Laugendal, Lardal, Oslo district, Norway; Tröger 896; Johannsen v.3, p.118; Tomkeieff p.311)*

LAURDALITE. See lardalite.

LAURVIKITE. See larvikite.

LECKSTONE. A name given to teschenite used for lining the bottoms of ovens in Scotland. *(Tomkeieff et al., 1983, p.315)*

LEDMORITE. A local name for a coarse-grained variety of nepheline syenite consisting of alkali feldspar, altered nepheline, melanite, pyroxene and biotite. *(Shand, 1910, p.384; Ledmore River, Borralan complex, Assynt, Scotland, UK; Tröger 486; Johannsen v.4, p.117; Tomkeieff p.315)*

LEEUWFONTEINITE. A local term for a variety of alkali feldspar syenite containing abundant anorthoclase, with biotite and hornblende. The rock had previously been called hatherlite. *(Brouwer, 1917, p.775; Leeuwfontein, Bushveld, South Africa; Tröger 238; Johannsen v.3, p.11; Tomkeieff p.315)*

LEHMANITE. An obsolete term for a feldspar quartz rock. *(Pinkerton, 1811a, p.206; named after J.G. Lehman; Johannsen v.2, p.46; Tomkeieff p.315)*

LEIDLEITE. A local name for an andesitic rock with microlites of plagioclase, augite and iron ore in a fine-grained to glassy groundmass. *(Thomas & Bailey, 1915, p.207; Glen Leidle, Island of Mull, Scotland, UK; Tröger 124; Tomkeieff p.315)*

LENGAITE (LENGAIITE). A carbonatite lava

extruded in 1960 composed of Na-Ca-K carbonate minerals including nyerereite and gregoryite. The earlier name of natrocarbonatite is preferred. *(Dawson & Gale, 1970, p.222; Oldoinyo Lengai, Tanzania)*

LENNEPORPHYRY. An obsolete term for a quartz-rich keratophyre. Although the rock was originally described by Dechen (1845) he did not name the rock. *(Mügge, 1893, p.535; Tröger 11; Tomkeieff p.316)*

LENTICULITE. A variety of ignimbrite which contains elongated lenticles of glass in a welded matrix. *(Marshall, 1935, p.358; Tröger(38) 871½; Tomkeieff p.316)*

LEOPARD ROCK. A local name for spotted rocks of several types, e.g. a syenite from Ontario, a gabbro from Quebec, Canada. The reference states the name had been in use for some time. *(Gordon, 1896, p.99; Johannsen v.3, p.86; Tomkeieff p.316)*

LEOPARDITE. A name suggested for a spotted rock. *(Hunter, 1853, p.377; Mecklenburg County, near Charlotte, New York, USA; Tröger 897; Tomkeieff p.316)*

LEPAIGITE. A colourless to pale brown or blue volcanic glass of rhyolite composition containing small euhedral phenocrysts of cristobalite, cordierite and sillimanite. It occurs as lapilli of globular habit thought to have been ejected during a fissure eruption of ignimbrites. *(Mueller, 1964, p.374; named after Padre G. le Paige; rock from San Pedro de Atacama, Chile)*

LESTIWARITE. A local name for a variety of microsyenite composed almost entirely of microperthite. *(Rosenbusch, 1896, p.464; Lestiware, Kola Peninsula, Russian Federation; Tröger 170; Johannsen v.3, p.25; Tomkeieff p.317)*

LEUCILITE. An obsolete name for leucitophyre. *(Naumann, 1850, p.655; Johannsen v.4, p.351; Tomkeieff p.317)*

LEUCITE BASALT. A term used for a volcanic

rock consisting essentially of leucite, olivine and clinopyroxene. The name should not be used as the term basalt is now restricted to a rock containing essential plagioclase. As the rock is a variety of foidite it should be given the appropriate name, e.g. olivine leucitite. *(Zirkel, 1870, p.108; Tröger 900; Johannsen v.4, p.351; Tomkeieff p.317)*

LEUCITE BASANITE. Now defined modally in the leucite-bearing rock classification (section 2.8, p.18) as a rock falling into QAPF field 14 and consisting essentially of leucite, clinopyroxene, plagioclase and olivine > 10%. *(Rosenbusch, 1887, p.760; Tröger 595; Tomkeieff p.318)*

LEUCITE PHONOLITE. A term originally used for a rock consisting essentially of sanidine and leucite without nepheline. Zirkel (1894a) redefined it as a phonolite (sanidine + nepheline) in which leucite is an important foid. Now defined modally in the leucite-bearing rock classification (section 2.8, p.18) in the original sense as a rock falling into QAPF field 11 and consisting essentially of leucite, clinopyroxene and sanidine. *(Rosenbusch, 1877, p.234; Tröger 471; Johannsen v.4, p.132)*

LEUCITE TEPHRITE. Now defined modally in the leucite-bearing rock classification (section 2.8, p.18) as a rock falling into QAPF field 14 and consisting essentially of leucite, clinopyroxene, plagioclase and olivine < 10%. *(Rosenbusch, 1877, p.492; Tröger 581; Johannsen v.4, p.235)*

LEUCITE TRACHYTE. A term originally used for a volcanic rock consisting of alkali feldspar, leucite and minor mafic minerals. As such a rock falls in the QAPF field 11 it would now be called leucite phonolite. *(Rath, 1868, p.297; Tröger 477; Johannsen v.4, p.133; Tomkeieff p.318)*

LEUCITE-BASITE. A term originally used for a leucite basalt as originally defined, i.e. a rock composed of leucite, olivine and clinopyroxene. Later used for all leucite-bearing ultrabasic rocks. Obsolete. *(Vogelsang, 1872, p.542; Tomkeieff p.318)*

LEUCITITE. Now defined modally in the leucite-bearing rock classification (section 2.8, p.18) in two senses: (1) *sensu lato:* as a group name for all leucite-bearing rocks falling into QAPF field 15, and (2) *sensu stricto:* as a rock falling into QAPF field 15c and consisting essentially of leucite, clinopyroxene and olivine > 10%. *(Senft, 1857, p.287; Tröger 643; Johannsen v.4, p.351; Tomkeieff p.318)*

LEUCITITE BASANITE. A variety of leucite basanite in which the leucite exceeds the plagioclase. *(Johannsen, 1938, p.301; Tomkeieff p.318)*

LEUCITITE TEPHRITE. A variety of leucite tephrite in which the leucite exceeds the plagioclase. *(Johannsen, 1938, p.301)*

LEUCITOID-BASALT. An obsolete name for a basalt which does not contain phenocrysts of leucite, but may contain leucite in the groundmass. *(Bořický, 1874, p.41; between Turtsch and Duppau (now Turee and Doupov), Bohemia, Czech Republic; Tomkeieff p.319)*

LEUCITOLITH. An obsolete term proposed for a monomineralic volcanic rock consisting of leucite. *(Johannsen, 1938, p.336; Tomkeieff p.319)*

LEUCITONEPHELINITE. A general term for feldspar-free volcanic rocks that contain leucite and nepheline in almost equal amounts. *(Jung & Brousse, 1959, p.88)*

LEUCITOPHYRE. A variety of leucitite characterized by leucite phenocrysts and essentially composed of leucite, nepheline and clinopyroxene. *(Humboldt, 1837, p.257; Eifel district, near Koblenz, Germany; Tröger 641; Johannsen v.4, p.359; Tomkeieff p.319)*

LEUCO-. Originally used as a prefix for rocks with more than 95% felsic minerals, but may now be used in the modal QAPF classification as indicating that the rock has considerably more felsic minerals than would be regarded as normal for that rock type (Fig. 2.7, p.26 and Fig. 2.8, p.27). *(Johannsen, 1920a, p.48; Johannsen v.1, p.153; Tomkeieff p.319)*

LEUCO-APHANEID. An obsolete alternative term for felseid. *(Johannsen, 1911, p.321)*

LEUCOCRATE. A general term proposed for rocks rich in light-coloured minerals. *(Brögger, 1898, p.264; Johannsen v.1, p.183)*

LEUCOCRATIC. Now defined (see Table 2.2, p.5) as a rock whose colour index ranges from 10 to 35. *(Brögger, 1898, p.264; from the Greek leukos = white, krateein = to predominate; Johannsen v.1, p.183; Tomkeieff p.319)*

LEUCOGRANITE. A term originally defined as the leucocratic variety of granite (Fig. 2.7, p.26). Now frequently used as a synonym for two-mica granite. *(Lameyre, 1966, p.14)*

LEUCOLITE. An obsolete term for a leucocratic igneous rock. *(Loewinson-Lessing, 1901, p.118; Tomkeieff p.319)*

LEUCOPHYRE. An obsolete term for light-coloured altered diabases and also a variety of serpentinized peridotite. *(Gümbel, 1874, p.33; from the Greek leukos = white; Tröger 899; Johannsen v.3, p.317; Tomkeieff p.320)*

LEUCOPHYREID. An obsolete alternative term for felseid porphyry. *(Johannsen, 1911, p.321)*

LEUCOPHYRIDE. A revised spelling recommended to replace the field term leucophyreid. Now obsolete. *(Johannsen, 1926, p.182; from the Greek leukos = white; Johannsen v.1, p.58; Tomkeieff p.320)*

LEUCOPTOCHE. An obsolete adjectival term for igneous rocks poor in light-coloured minerals. *(Loewinson-Lessing, 1901, p.118; Tomkeieff p.320)*

LEUCOSTITE. An obsolete group name for porphyritic trachtyes, phonolites etc. *(Cordier, 1868, p.92; Tomkeieff p.320)*

LEUMAFITE. A mnemonic name for rocks consisting of *leu*cite and *maf*ic minerals. *(Hatch et al., 1961, p.385)*

LHERCOULITE. Alleged to be another name for lherzolite; however, the reference cited (Cordier, 1842) does not appear to contain the name. *(Tomkeieff et al., 1983, p.320; Lhercoul, near Lhers, Ariège, Pyrénées, France)*

LHERZITE. A variety of hornblendite which occurs as dykes and consists essentially of hornblende with minor biotite; similar in composition to theralite. *(Lacroix, 1917c, p.385; Lac de Lherz, now Lhers, Ariège, Pyrénées, France; Tröger 704; Johannsen v.4, p.444; Tomkeieff p.320)*

LHERZOLINE. An obsolete name for a fine-grained lherzolite. *(Cordier, 1868, p.128; Lac de Lherz, now Lhers, Ariège, Pyrénées, France; Tomkeieff p.320)*

LHERZOLITE. An ultramafic plutonic rock composed of olivine with subordinate orthopyroxene and clinopyroxene. Now defined modally in the ultramafic rock classification (Fig. 2.9, p.28). *(Delamétherie, 1795, p.454; Lac de Lherz, now Lhers, Ariège, Pyrénées, France; Tröger 735; Johannsen v.4, p.422; Tomkeieff p.320)*

LIMBURGITE. A basic volcanic rock containing phenocrysts of pyroxene, olivine and opaques in a glassy groundmass containing the same minerals. No feldspars are present. May be used as synonym for a hyalo-nepheline basanite (p.5). *(Rosenbusch, 1872, p.53; Limburg, Kaiserstuhl, Baden, Germany; Tröger 593; Tomkeieff p.321)*

LINDINOSITE. A local name for a variety of peralkaline granite containing nearly 60% riebeckite. *(Lacroix, 1922, p.580; Lindinosa, Corsica, France; Tröger 59; Tomkeieff p.322)*

LINDÖITE. A local name for a dyke rock that is a leucocratic variety of trachyte or rhyolite containing minor amounts of arfvedsonite. *(Brögger, 1894, p.131; Lindö, Oslo district, Norway; Tröger 33; Johannsen v.2, p.100;*

Tomkeieff p.322)

LINOPHYRE. A porphyritic rock in which the phenocrysts occur in lines and streaks. *(Iddings, 1909, p.224; Tomkeieff p.324)*

LINOSAITE. An obsolete local name for a variety of alkali basalt containing small amounts of nepheline. *(Johannsen, 1938, p.68; Linosa Island, Italy; Tröger(38) 381½; Tomkeieff p.324)*

LIPARITE. A name given independently to the same rocks that were called rhyolites by Richthofen in the previous year, 1860. Although the term was also used in a broader sense later it has not been widely used. May be used as a synonym for rhyolite (p.30). *(Roth, 1861, p.xxxiv; Lipari Island, Italy; Tröger 40; Johannsen v.2, p.265; Tomkeieff p.324)*

LITCHFIELDITE. A coarse-grained, somewhat foliated variety of nepheline syenite consisting of K-feldspar, albite, nepheline, cancrinite, sodalite and lepidomelane. *(Bayley, 1892, p.243; Litchfield, Maine, USA; Tröger 415; Johannsen v.4, p.181; Tomkeieff p.325)*

LITHIC TUFF. Now defined in the pyroclastic classification (section 2.2.2, p.8) as a variety of tuff in which lithic fragments predominate. *(Pirsson, 1915, p.193; Tomkeieff p.326)*

LITHOIDITE. A non-porphyritic rhyolite. *(Richthofen, 1860, p.183; Tomkeieff p.327)*

LLANITE. A local term for a granite porphyry containing abundant K-feldspar and quartz phenocrysts with groundmass albite. Although the name is attributed by Johannsen to Iddings, the cited reference only contains a description of the rock from Llano and not the name. *(Iddings, 1904; Llano, Texas, USA; Tröger 60; Johannsen v.2, p.117; Tomkeieff p.329)*

LOSERO. See lozero.

LOSSENITE. A name suggested for a group of quartz gabbros *sensu stricto*. *(Marcet Riba, 1925, p.293; named after K.A. Lossen)*

LOW-K. A chemical term originally applied to andesites with $K_2O < 0.7\%$ but later applied to volcanic rocks which plot below a line on the SiO_2–K_2O diagram, i.e. for their SiO_2 values they are lower than usual in K_2O. The field is now defined as an optional qualifier for certain rock names in the TAS classification (Fig. 2.17, p.37). *(Taylor, 1969, p.45)*

LOZERO (LOSERO). A local Mexican name for a bedded tuff used for tiles. *(Humboldt, 1823, p.217; from the Spanish lozas = thin plates of rock; Tomkeieff p.332)*

LUCIITE. An obsolete term for a coarse-grained variety of the lamprophyric rock malchite consisting essentially of intermediate plagioclase and hornblende with minor quartz, orthoclase and biotite. Cf. orbite. *(Chelius, 1892, p.3; Luciberg, Melibocus, Odenwald, Germany; Tröger 335; Tomkeieff p.332)*

LUGARITE. A local name for a variety of teschenite containing prominent phenocrysts of titanian augite and kaersutite; labradorite is minor and analcime abundant. *(Tyrrell, 1912, p.121; Lugar, Scotland, UK; Tröger 564; Johannsen v.4, p.304; Tomkeieff p.332)*

LUHITE. A local name for a rock consisting of phenocrysts of olivine and pyroxene in a groundmass of pyroxene, melilite, haüyne, perovskite and biotite cemented by nepheline and calcite. Considered to be a nepheline-rich alnöite. *(Scheumann, 1922, p.523; Luhov, S. of Mimoň, N. Bohemia, Czech Republic; Tröger 666; Johannsen v.4, p.387)*

LUJAVRITE (LUIJAURITE, LUJAUVRITE). A term, originally spelt luijaurite, for a melanocratic agpaitic variety of nepheline syenite rich in eudialyte, arfvedsonite and aegirine with perthitic alkali feldspar or separate microcline and albite. A pronounced igneous lamination is characteristic, as is the abundance in minerals rich in incompatible elements such as the REE, U, Th, Li etc. *(Brögger, 1890, p.204; Luijaur, now Lujavr-Urt, Lovozero complex, Kola Peninsula, Russian Federation; Tröger 421; Johannsen v.4, p.106;*

Tomkeieff p.332)

LUJAVRITITE. A variety of ijolite composed mainly of aegirine-augite and nepheline with much titanite and some apatite and characterized by the presence of small amounts of K-feldspar. *(Antonov, 1934, p.25; Khibina complex, Kola Peninsula, Russian Federation; Tröger(38) 607⅔; Tomkeieff p.332)*

LUNDYITE. A local name for a rock which is close to the boundary of alkali feldspar syenite and alkali feldspar granite and contains small amounts of katophorite. *(Hall, 1915, p.53; Lundy Island, Bristol Channel, England, UK; Tröger 75)*

LUPATITE. An incompletely described rock named as a nepheline feldspar porphyry. *(Mennell, 1929, p.536; Lupata Gorge, Zambezi River, Mozambique; Tröger(38) 906½; Tomkeieff p.333)*

LUSCLADITE. A variety of theralite poor in nepheline with the mode dominated by titanian augite and labradorite. *(Lacroix, 1920, p.21; Ravin de Lusclade, Mont Dore, Auvergne, France; Tröger 544; Johannsen v.4, p.197; Tomkeieff p.333)*

LUSITANITE. A variety of peralkaline alkali feldspar syenite containing riebeckite and aegirine. Also used as family name for rocks consisting essentially of alkali feldspar. *(Lacroix, 1916c, p.283; from Lusitania, the Roman name for Portugal; Tröger 222; Johannsen v.3, p.46; Tomkeieff p.333)*

LUTALITE. An obsolete name for a variety of leucitite composed essentially of clinopyroxene and leucite and with subordinate olivine, nepheline and plagioclase. The value of Na_2O/K_2O is higher than in ordinary leucitites. *(Holmes & Harwood, 1937, p.10; Lutale, Birunga volcanic field, Uganda; Tröger(38) 642⅓; Tomkeieff p.333)*

LUXULIANITE (LUXULLIANITE, LUXULYANITE). A local name for a variety of porphyritic granite containing abundant tourmaline replacing various minerals. It was spelt luxuliane in the

original reference. *(Pisani, 1864, p.913; Luxulion, now Luxulyan, near Lostwithiel, Cornwall, England, UK; Tröger 907; Johannsen v.2, p.58; Tomkeieff p.333)*

M-CHARNOCKITE, M-ENDERBITE ETC. The prefix m- was suggested for use with members of the charnockitic rock series which contain mesoperthite (p.20). *(Tobi, 1971, p.201)*

M-TYPE GRANITE. A general term for granitic rocks occurring in some continental margins and having the chemical and isotopic compositions of island arc volcanic rocks. The prefix M implies a mantle origin as they are assumed to have formed by partial melting of the subducted oceanic crust. *(White, 1979, p.539)*

MACEDONITE. A local name for a fine-grained variety of basaltic trachyandesite with trachytic texture consisting of plagioclase and alkali feldspar with minor biotite, hornblende, olivine and pyroxene. *(Skeats, 1910, p.205; Mt Macedon, Victoria, Australia; Tröger 217; Johannsen v.3, p.120; Tomkeieff p.335)*

MACUSANITE. An acid volcanic glass, with phenocrysts of andalusite and sillimanite and more rarely staurolite and cordierite, chemically characterized by high alumina (16%-20%) and fluorine. Originally thought to be a tectite. *(Martin & de Sitter-Koomans, 1955, p.152; Macusani, Puno, Peru)*

MADEIRITE. An obsolete name for a melanocratic picrite consisting of phenocrysts of titanian augite and partially serpentinized olivine in a fine-grained groundmass of plagioclase, augite and iron ore with some secondary calcite. *(Gagel, 1913, p.382; Ribeira de Massapez, Madeira, North Atlantic Ocean; Tröger 404; Johannsen v.4, p.73; Tomkeieff p.336)*

MADUPITE. Originally described as a melanocratic variety of leucitite essentially composed of diopside and phlogopite in a glassy matrix, which has the composition of nepheline and leucite. Wyomingite is a more felsic variety. Now regarded as a diopside-madupitic lamproite (Table 2.7, p.17). *(Cross, 1897, p.129; from the Shoshone Indian madúpa = Sweetwater, Wyoming, USA; Tröger 645; Johannsen v.4, p.370; Tomkeieff p.336)*

MAENAITE. A local name for a variety of trachyte which occurs as dykes and consists essentially of albite and orthoclase. *(Brögger, 1898, p.206; Lake Maena, Gran, Oslo Igneous Province, Norway; Tröger 193; Johannsen v.3, p.123; Tomkeieff p.336)*

MAFIC. A collective term for modal ferromagnesian minerals, such as olivine, pyroxene etc., which was introduced to stop the normative term femic incorrectly being used for that purpose. See also felsic and salic. *(Cross et al., 1912, p.561; Johannsen v.1, p.180; Tomkeieff p.336)*

MAFITITE. A term originally proposed for rocks with a colour index (M) of 90–100, but later withdrawn in favour of ultramafitite, now replaced by ultramafic rock. *(Streckeisen, 1967, p.163)*

MAFRAITE. A variety of monzogabbro in which kaersutite exceeds titanian augite. *(Lacroix, 1920, p.22; Tifão de Mafra and Rio Touro, Cintra, Portugal; Tröger 285; Johannsen v.4, p.55; Tomkeieff p.337)*

MAFURITE. An ultrabasic rock consisting of phenocrysts of olivine and minor pyroxene in a groundmass of diopside and kalsilite with small amounts of perovskite, olivine and biotite. It is a kamafugitic rock and now regarded as an olivine-pyroxene kalsilitite (Tables 2.5 and 2.6, p.12). *(Holmes, 1942, p.199; Mafuru craters, Uganda; Tomkeieff p.337)*

MAGMABASALT. An obsolete name for a basaltic rock consisting of augite, magnetite and glass and without visible feldspar. Cf. limburgite. *(Bořický, 1874, p.40; Tröger 908;*

Tomkeieff p.337)

MAGNESIOCARBONATITE. A chemically defined variety of carbonatite in which wt % $MgO > (FeO+Fe_2O_3+MnO)$ and $CaO / (CaO+MgO+FeO+Fe_2O_3+MnO) < 0.8$ (Fig. 2.2, p.10). See also dolomite-carbonatite and rauhaugite. *(Woolley, 1982, p.16)*

MAGNETITE HÖGBOMITITE. An obsolete name for a variety of magnetitite containing up to 15% of högbomite (a mineral of the corundum-hematite group). On account of the small amount of högbomite present högbomite magnetitite would have been a more logical name (Tröger, 1935, Johannsen, 1938). *(Gavelin, 1916, p.306; Routivare, Norrbotten, Lapland, Sweden; Tröger 764; Johannsen v.4, p.469; Tomkeieff p.339)*

MAGNETITE SPINELLITE. A rock consisting of titanomagnetite and ilmenite, with abundant spinel, and minor olivine and hypersthene (= enstatite). *(Sjögren, 1893, p.63; Routivare, Norrbotten, Lapland, Sweden; Johannsen v.4, p.469; Tomkeieff p.540)*

MAGNETITITE. An ultramafic rock consisting essentially of magnetite. *(Johannsen, 1920c, p.225; Tröger 761; Johannsen v.4, p.466; Tomkeieff p.339)*

MALCHITE. A dyke rock consisting of small rare phenocrysts of hornblende and labradorite and occasionally biotite, in a groundmass of hornblende, andesine and quartz. Cf. gabbrophyre or odinite. *(Osann, 1892, p.386; Malchen or Melibokus, Odenwald, Germany; Tröger 334; Johannsen v.2, p.402; Tomkeieff p.339)*

MALIGNITE. A mesocratic nepheline syenite containing abundant aegirine-augite and some orthoclase and nepheline in equal amounts. Many other mafic minerals, such as amphibole, garnet, biotite, may be present. Now defined modally as a mesocratic variety of foid syenite in QAPF field 11 (Fig. 2.8, p.27). *(Lawson, 1896, p.337; Maligne River, Ontario, Canada; Tröger 487; Johannsen v.4,*

p.115; Tomkeieff p.340)

MAMILITE. Originally described as a rock consisting essentially of leucite and magnophorite (= titanian potassian richterite) with subordinate phlogopite in a glassy matrix. Now regarded as a leucite-richterite lamproite (Table 2.7, p.17). *(Wade & Prider, 1940, p.68; Mamilu Hill, Kimberley district, West Australia, Australia; Tomkeieff p.340)*

MANDCHURITE (MANCHURITE, MANDCHOURITE, MANDSCHURITE). A glassy variety of basanite in which plagioclase is not present as a mineral phase. *(Lacroix, 1928a, p.47; named after Manchuria, China; Tröger 592; Tomkeieff p.340)*

MANDELSTEIN. An old German name for an amygdaloidal rock. *(Werner, 1787, p.13; from the German Mandel = almond, Stein = rock; Tröger 910; Tomkeieff p.340)*

MANDSCHURITE. See mandchurite.

MANGERITE. An intermediate member of the charnockitic rock series equivalent to orthopyroxene monzonite of QAPF field 8 (Table 2.10, p.20) and frequently containing mesoperthite. *(Kolderup, 1903, p.109; Kalsaas, near Manger, Rado, Bergen district, Norway; Tröger 278; Johannsen v.3, p.63; Tomkeieff p.341)*

MAPPAMONTE. A local name for a loosely textured grey tuff-like material found near Naples. *(Lorenzo, 1904, p.309; Tomkeieff p.341)*

MARCHITE. An obsolete name for a variety of websterite composed of diopside and enstatite. *(Kretschmer, 1917, p.149; March River (now Morava), Moravia, Czech Republic; Tröger 686; Johannsen v.4, p.461; Tomkeieff p.341)*

MAREKANITE. A term given to the more or less rounded glass fragments, often showing concave indentations, that are formed when perlite fractures. *(Judd, 1886b, p.241; Marekanka River, Ochotsk Sea, Siberia, Russian Federation; Tröger 911; Johannsen v.2, p.284; Tomkeieff p.341)*

MAREUGITE. A variety of foid-bearing gabbro dominated by titanian augite and bytownite and also containing haüyne. *(Lacroix, 1917a, p.587; Mareuges, Mont Dore, Auvergne, France; Tröger 553; Johannsen v.4, p.222; Tomkeieff p.341)*

MARIANITE. A variety of boninite consisting of clinoenstatite, bronzite (= enstatite) and augite phenocrysts and microlites in a glassy matrix with shaped pseudomorphs after olivine. The new name was proposed on the grounds that the type boninite was richer in olivine. *(Sharaskin et al., 1980, p.473; Mariana Trench, N.W. Pacific Ocean)*

MARIENBERGITE. A variety of natrolite phonolite with phenocrysts of Na-sanidine, andesine, and augite in a matrix of Na-sanidine, natrolite and sodalite. *(Johannsen, 1938, p.169; Mt Marienberg, near Ústí nad Labem, N. Bohemia, Czech Republic; Tröger(38) 526; Tomkeieff p.342)*

MARIUPOLITE. A leucocratic variety of nepheline syenite characterized by the absence of K-feldspar and the presence of albite and aegirine. *(Morozewicz, 1902, p.244; Mariupol, Sea of Azov, Ukraine; Tröger 416; Johannsen v.4, p.211; Tomkeieff p.342)*

MARKFIELDITE. A coarse-grained dioritic rock with idiomorphic andesine, augite or hornblende and micrographic intergrowths of quartz and orthoclase in the groundmass. The term was later used for lamprophyric rocks of similar mineralogy. *(Hatch, 1909, p.219; Markfield, Charnwood Forest, Leicester, England, UK; Tröger(38) 116; Johannsen v.2, p.295; Tomkeieff p.342)*

MARLOESITE. A local name for an altered variety of trachyte containing olivine, hornblende and augite. *(Thomas, 1911, p.198; Marloes Bay, Dyfed, Wales, UK; Tröger 214; Johannsen v.3, p.175; Tomkeieff p.343)*

MAROSITE. An obsolete local name for a melanocratic variety of monzogabbro rich in biotite and augite with equal but small amounts of sanidine and bytownite and minor nepheline. *(Iddings, 1913, p.246; Pic de Maros, Sulawesi, Indonesia; Tröger 262; Johannsen v.4, p.41; Tomkeieff p.344)*

MARSCOITE. A local name for a hybrid rock formed by the mixing of ferrodiorite with rhyolite. Variable in texture and composition the rock contains partially resorbed xenocrysts of quartz and andesine. *(Harker, 1904, p.175; Marsco, Island of Skye, Scotland, UK; Tröger 332; Tomkeieff p.344)*

MARTINITE. An obsolete name for a variety of phonolitic tephrite composed essentially of plagioclase, leucite and alkali feldspar and with accessory clinopyroxene. *(Johannsen, 1938, p.200; Croce di San Martino, Vico Volcano, near Viterbo, Italy; Tröger(38) 569½; Tomkeieff p.344)*

MARUNDITE. A local mnemonic name for a *mar*garite cor*und*ite pegmatite. *(Hall, 1922, p.43; Tröger 776; Tomkeieff p.344)*

MASAFUERITE. A variety of picrite basalt composed of 50% olivine phenocrysts in a groundmass of augite, calcic plagioclase and ores. *(Johannsen, 1937, p.334; Masafuera, Juan Fernandez Islands, Pacific Ocean; Tröger(38) 410; Tomkeieff p.344)*

MASANITE. A variety of granite porphyry containing phenocrysts of zoned oligoclase in a micropegmatitic groundmass of quartz, orthoclase and minor biotite. *(Kotô, 1909, p.190; Ku-ryong Copper Mine, Ma-san-pho, South Korea; Tröger 89; Tomkeieff p.345)*

MASANOPHYRE. An obsolete name for a variety of masanite containing quartz and oligoclase phenocrysts in a micropegmatitic groundmass. *(Kotô, 1909, p.192; Tröger 89; Tomkeieff p.345)*

MASEGNA. A local name for a trachyte from the Euganean Hills, near Padova, Italy. *(Rio, 1822, p.350; Tomkeieff p.345)*

MATRAITE. An obsolete term for a variety of augite andesite in which the feldspar is anorthite. *(Judd, 1876, p.302; Matra district, Hungary; Tomkeieff p.345)*

MEDIUM-K. A chemical term, now defined as an optional qualifier for certain rock names in the TAS classification (Fig. 2.17, p.37), for rocks that lie between the high-K and low-K fields in the SiO_2–K_2O diagram.

MEIMECHITE (MEYMECHITE). An ultramafic volcanic rock composed of olivine phenocrysts in a groundmass of olivine, clinopyroxene, magnetite and glass. Now defined chemically in the TAS classification (Fig. 2.13, p.34). *(Moor & Sheinman, 1946, p.141; River Meimecha, tributary of River Kheta, N. Siberia, Russian Federation; Tomkeieff p.348)*

MELA-. Originally used as a prefix for rocks with between 5% and 50% felsic minerals, but may now be used in the modal QAPF classification as indicating that the rock has considerably more mafic minerals than would be regarded as normal for that rock type (Fig. 2.7, p.26 and Fig. 2.8, p.27). *(Johannsen, 1920a, p.48; Johannsen v.1, p.153; Tomkeieff p.348)*

MELANEPHELINITE. Originally defined as a volcanic rock consisting of abundant pyroxene and some nepheline but with no olivine. Now commonly used for undersaturated basic and ultrabasic volcanic rocks consisting of abundant augite phenocrysts in a groundmass of indeterminate mineralogy and defined chemically as a rock falling in TAS fields U1 or F in which normative *ne* and *ab* are both present but with *ne* < 20% and *ab* < 5% (p. 36). *(Johannsen, 1938, p.363; Johannsen v.4, p.363)*

MELANIDE. An obsolete term suggested to replace names, such as greenstone, in which the ferromagnesian minerals cannot be distinguished in hand specimen. *(Milch, 1927, p.63; Johannsen v.1, p.57; Tomkeieff p.348)*

MELANO-APHANEID. An obsolete alternative term for anameseid. *(Johannsen, 1911, p.321)*

MELANOCRATE. A general term proposed for

rocks rich in dark-coloured minerals. *(Brögger, 1898, p.263; from the Greek melas = dark; Johannsen v.1, p.184)*

MELANOCRATIC. Now defined (see Table 2.2, p.5) as a rock whose colour index ranges from 65 to 90. *(Brögger, 1898, p.263; from the Greek melas = dark, krateein = to predominate; Johannsen v.1, p.184; Tomkeieff p.348)*

MELANOLITE. An obsolete term for a melanocratic igneous rock. *(Loewinson-Lessing, 1901, p.118)*

MELANOPHYREID. An obsolete alternative term for anameseid porphyry. *(Johannsen, 1911, p.321; Tomkeieff p.349)*

MELANOPHYRIDE. A revised spelling recommended to replace the field term melanophyreid. Now obsolete. *(Johannsen, 1926, p.182; from the Greek melas = dark; Johannsen v.1, p.58; Tomkeieff p.349)*

MELANOPTOCHE. An obsolete adjectival term for igneous rocks poor in dark-coloured minerals. *(Loewinson-Lessing, 1901, p.118; Tomkeieff p.349)*

MELAPHYRE. A very old name replacing trapporphyr and originally used for amygdaloidal rocks composed of plagioclase and augite but later used for Upper Palaeozoic rocks, usually basaltic. *(Brongniart, 1813, p.40; from the Greek melas = dark; Tröger 388; Johannsen v.3, p.296; Tomkeieff p.349)*

MELAPORPHYRE. An obsolete name for a dark-coloured labradorite porphyry. *(Senft, 1857, p.271; Tomkeieff p.349)*

MELFITE. An obsolete local name for a haüynophyre. *(Lacroix, 1933, p.199; Melfi, Vulture, Italy; Tröger 649; Tomkeieff p.349)*

MELILITE BASALT. A term used for a volcanic rock consisting of phenocrysts of augite and olivine in a groundmass essentially of melilite, augite, olivine and occasional nepheline. The name should not be used as the term basalt is now restricted to a rock containing essential plagioclase. As the rock is a variety of foidite

it should be given the appropriate name, e.g. olivine melilitite. *(Stelzner, 1882, p.229; Hochbohl, near Owen, Württemberg, Germany; Tröger 914; Johannsen v.4, p.347; Tomkeieff p.349)*

MELILITE LEUCITITE. An old varietal term now defined as a rock falling in field F of the TAS classification in which normative cs (larnite) is present but is < 10% and the dominant feldspathoid mineral is leucite (p.38). *(Woolley et al., 1996, p.176; Tröger 671)*

MELILITE NEPHELINITE. An old varietal term now defined as a rock falling in field F of the TAS classification in which normative cs (larnite) is present but is < 10% and the dominant feldspathoid mineral is nepheline (p.38). *(Woolley et al., 1996, p.176; Tröger 668)*

MELILITHITE. An obsolete term proposed for a monomineralic volcanic rock consisting of melilite. The term is synonymous with melilitholith. *(Loewinson-Lessing, 1901, p.114; Tröger 749; Johannsen v.4, p.346; Tomkeieff p.349)*

MELILITHOLITH. A term proposed to replace melilithite, for a monomineralic volcanic rock consisting of melilite. Now obsolete. *(Johannsen, 1938, p.337)*

MELILITITE. An ultramafic volcanic rock consisting essentially of melilite and pyroxene. Perovskite is also commonly present. Now defined as a general term for volcanic rocks in the melilite-bearing rocks classification (section 2.4.2, p.11) or chemically in field F of the TAS classification as a rock which does not contain kalsilite but has normative cs (larnite) > 10% and $K_2O < Na_2O$ (p.38). If modal olivine > 10% the rock should be called an olivine melilitite. *(Lacroix, 1893, p.627; Johannsen v.4, p.346; Tomkeieff p.349)*

MELILITOLITE. An ultramafic plutonic rock consisting essentially of melilite, pyroxene and olivine. Now defined as a general term for plutonic rocks in the melilite-bearing rocks classification (section 2.4.1, p.11). *(Lacroix, 1933, p.197; Tröger 744; Johannsen v.4, p.337; Tomkeieff p.349)*

MELMAFITE. A mnemonic name for a mixture of *mel*ilite and *maf*ic minerals. *(Hatch et al., 1961, p.358)*

MELTEIGITE. The melanocratic member of the ijolite series, containing 10%–30% of nepheline. Now defined modally as a melanocratic variety of foidite in QAPF field 15 (Fig. 2.8, p.27). *(Brögger, 1921, p.18; Melteig, Fen Complex, Telemark, Norway; Tröger 609; Johannsen v.4, p.327; Tomkeieff p.350)*

MESITE. An obsolete name for a rock of intermediate composition. *(Loewinson-Lessing, 1898, p.39; Tomkeieff p.351)*

MESO-. Originally used as a prefix for rocks with between 95% and 50% felsic minerals. *(Johannsen, 1920a, p.48; Johannsen v.1, p.153; Tomkeieff p.351)*

MESOBUGITE. A member of the bugite series with 53%–66% SiO_2, 5%–20% hypersthene (= enstatite), oligoclase antiperthite and quartz. As all other members of the bugite series have been recommended as terms to be abandoned (Streckeisen, 1974) it is suggested that this term should also be abandoned. *(Bezborod'ko, 1931, p.142; Bug River, Podolia, Ukraine; Tomkeieff p.80)*

MESOCRATIC. Now defined (see Table 2.2, p.5) as a rock whose colour index ranges from 35 to 65. *(From the Greek mesos = between, krateein = to predominate)*

MESTIGMERITE. A variety of melanocratic nepheline syenite or malignite with abundant aegirine-augite, nepheline and orthoclase with some titanite and apatite. *(Duparc, 1926, p.b120; Mestigmer, W. Oujda, Morocco; Tröger(38) 487½; Tomkeieff p.353)*

METABOLITE. An obsolete term for an altered trachyte glass. *(Wadsworth, 1893, p.97;*

Johannsen v.3, p.77; Tomkeieff p.353)

METALUMINOUS. A chemical term used for rocks in which molecular $(Na_2O + K_2O) < Al_2O_3 < (CaO + Na_2O + K_2O)$. This produces prominent anorthite *(an)* in the CIPW norm and typically such Al-bearing minerals as hornblende, biotite and melilite in the mode. *(Shand, 1927, p.128)*

MEYMECHITE. See meimechite.

MIAGITE. An orbicular variety of gabbro composed essentially of calcic plagioclase with hornblende, minor hypersthene (= enstatite) and quartz. This rock has also been called corsite, kugeldiorite and napoleonite. *(Pinkerton, 1811b, p.63; Glacier Miage, Mont Blanc, France; Tröger(38) 916½; Johannsen v.3, p.232; Tomkeieff p.359)*

MIAROLITE. A textural variety of fine-grained granite containing many irregular drusy cavities. *(Fournet, 1845, p.495; from the Italian name of the rock = miarolo; Johannsen v.2, p.130; Tomkeieff p.360)*

MIASKITE (MIASCITE). A leucocratic variety of biotite nepheline monzosyenite with oligoclase and perthitic orthoclase. May be used as a special term for oligoclase nepheline monzosyenite of QAPF field 12 (p.24). *(Rose, 1839, p.375; Miask, Ilmen Mts, Urals, Russian Federation; Tröger 509; Tomkeieff p.360)*

MIASKITIC (MIASCITIC). A general term for nepheline syenites in which the molecular ratio of $(Na_2O + K_2O) / Al_2O_3 < 1$. Cf. agpaitic. *(Fersman, 1929, p.63; Tomkeieff p.360)*

MICROCLINITE. A variety of alkali feldspar syenite composed almost entirely of microcline. *(Loewinson-Lessing, 1901, p.114; Tröger 165; Johannsen v.3, p.5; Tomkeieff p.361)*

MICROTINITE. An obsolete term for plagioclase-bearing trachytes, but later used by Lacroix (1900) for recrystallized ejected blocks of dioritic composition that contain glassy plagioclase (microtine). *(Wolf, 1866, p.33;*

Tröger 307; Tomkeieff p.364)

MID-OCEAN RIDGE BASALT (MORB). A variety of low-K tholeiitic basalt, low in Ti, erupted at mid-ocean ridges consisting of Mg-olivine, Ca-rich clinopyroxene, plagioclase, titanomagnetite and variable amounts of pale brown glass. Pigeonite and ilmenite occur occasionally. *(Sun et al., 1979, p.119)*

MIENITE. An obsolete name for a variety of rhyolite glass containing crystals of labradorite and minor hypersthene (= enstatite). *(Scheumann, 1925, p.283; Lake Mien, Småland, Sweden; Tröger 96; Tomkeieff p.365)*

MIHARAITE. An obsolete name for a variety of basalt with phenocrysts of bytownite and hypersthene (= enstatite) in a groundmass of labradorite, augite, opaques and glass. *(Tsuboi, 1918, p.47; Volcano Mihara, Oshima Island, Japan; Tröger 162; Johannsen v.3, p.284; Tomkeieff p.366)*

MIJAKITE. An obsolete term for a Mn-rich variety of basalt consisting of phenocrysts of bytownite, augite and hypersthene (= enstatite), in a groundmass of andesine, magnetite and pyroxene, which is probably Mn-rich. *(Petersen, 1890a, p.47; Mijakeshima, Bonin Islands, now Ogaswara-gunto Islands, Japan; Tröger 161; Johannsen v.3, p.281; Tomkeieff p.366)*

MIKENITE. An obsolete term for a variety of leucitite essentially composed of leucite, nepheline and clinopyroxene and in which the Na_2O/K_2O ratio is higher than usual for rocks of this type. *(Lacroix, 1933, p.196; Mikeno Volcano, Birunga volcanic field, Rwanda; Tröger 642; Tomkeieff p.366)*

MIMESITE. This name does not exist in the reference given (Cordier, 1868) but matches the description given for mimosite. *(Tröger, 1935, p.325; Tröger 918)*

MIMOPHYRE. An obsolete general term for volcanic ash, tuff, agglomerate etc. *(Brongniart, 1813, p.46; from the Greek mimos = imitation; Tomkeieff p.366)*

MIMOSE. An obsolete name originally suggested by Haüy for dolerite. *(Brongniart, 1813, p.32; from the Greek mimos = imitation; Johannsen v.3, p.303; Tomkeieff p.366)*

MIMOSITE. An obsolete term originally used for melanocratic basaltic rocks, but redefined by Macdonald (1949) for highly undersaturated picrobasalts with >15% normative foids but no recognizable nepheline. *(Cordier, 1842, vol.8, p.227; from the Greek mimos = imitation; Tröger 919; Johannsen v.3, p.303; Tomkeieff p.366)*

MINETTE. Originally an old miners' term for oolitic ironstones. Later used for a variety of lamprophyre consisting of phenocrysts of phlogopite-biotite and occasionally amphiboles in a groundmass of the same minerals plus orthoclase and minor plagioclase. Mg-olivine and diopsidic pyroxene may also be present. Now defined in the lamprophyre classification (Table 2.9, p.19). *(Elie de Beaumont, 1822, p.524; Minkette Valley, Vosges, France; Tröger 247; Johannsen v.3, p.33; Tomkeieff p.368)*

MINETTEFELS. An obsolete term suggested to replace the term minette, when used as a lamprophyre dyke rock name, because oolitic ironstones of Lorraine were also called minette. *(Kretschmer, 1917, p.21; Tröger 920; Tomkeieff p.368)*

MINIMITE. A term suggested for experimental compositions that crystallize at the minimum temperature for a given pressure. *(Reynolds, 1958, p.390; Tomkeieff p.368)*

MINVERITE. A local name for an albite dolerite with spilitic affinities that carries a primary brown hornblende. *(Dewey, 1910, p.46; St Minver, Cornwall, England, UK; Tröger 235; Johannsen v.3, p.141; Tomkeieff p.369)*

MISSOURITE. A melanocratic intrusive rock composed of clinopyroxene and subordinate leucite and olivine. Now defined modally as a melanocratic variety of foidite in QAPF field 15 (Fig. 2.8, p.27). *(Weed & Pirsson, 1896, p.323; Missouri River, Highwood Mts, Montana, USA; Tröger 631; Johannsen v.4, p.334; Tomkeieff p.369)*

MODLIBOVITE. A local term for a variety of polzenite containing olivine and biotite phenocrysts in a groundmass of melilite, lazurite, biotite, augite and nepheline. *(Scheumann, 1922, p.496; Modlibov, N. Bohemia, Czech Republic; Tröger 664; Johannsen v.4, p.388; Tomkeieff p.370)*

MODUMITE. A local name for an anorthositic facies of the Oslo alkali gabbro. *(Brögger, 1933, p.35; Modum, Oslo district, Norway; Tröger 298; Johannsen v.4, p.66; Tomkeieff p.370)*

MONCHIQUITE. A variety of lamprophyre similar to camptonite except that the groundmass is feldspar-free and composed of combinations of glass and feldspathoids, especially analcime. Now defined in the lamprophyre classification (Table 2.9, p.19). *(Hunter & Rosenbusch, 1890, p.447; Caldas de Monchique, Algarve, Portugal; Tröger 374; Johannsen v.4, p.375; Tomkeieff p.371)*

MONDHALDEITE. An obsolete local name for a "camptonite-like" dyke rock largely composed of equal proportions of plagioclase and alkali feldspar, with subordinate clinopyroxene, amphibole and leucite, in a glassy matrix. *(Gruss, 1900, p.89; Mondhalde, Kaiserstuhl, Baden, Germany; Tröger 264; Johannsen v.4, p.55; Tomkeieff p.371)*

MONMOUTHITE. A variety of urtite composed essentially of nepheline and some hastingsite with minor albite and calcite. *(Adams & Barlow, 1910, p.277; Monmouth Township, Ontario, Canada; Tröger 606; Johannsen v.4, p.317; Tomkeieff p.372)*

MONNOIRITE. A poorly defined local name for a coarse-grained porphyritic rock transitional between essexite and pulaskite. *(Osborne & Wilson, 1934, p.181; Monnoir, now Mt Johnson, Quebec, Canada; Tomkeieff p.372)*

MONTREALITE. A highly melanocratic variety of alkali gabbro, dominated by olivine, titanian augite and kaersutite. Labradorite can be in small amounts and minor nepheline may occur. *(Adams, 1913, p.38; Mount Royal, Montreal, Quebec, Canada; Tröger 397; Johannsen v.4, p.430; Tomkeieff p.374)*

MONZODIORITE. A term suggested to replace syenodiorite for a plutonic rock intermediate between monzonite and diorite. Now defined modally in QAPF field 9 (Fig. 2.4, p.22). *(Johannsen, 1920b, p.174; Tröger 994; Tomkeieff p.374)*

MONZOGABBRO. A term suggested to replace syenogabbro for a plutonic rock of gabbroic aspect that contains minor but essential orthoclase as well as calcic plagioclase. Now defined modally in QAPF field 9 (Fig. 2.4, p.22). *(Johannsen, 1920c, p.212; Tröger 995; Johannsen v.3, p.126; Tomkeieff p.374)*

MONZOGRANITE. An optional term for a variety of granite in QAPF field 3b (Fig. 2.4, p.22) having roughly equal amounts of alkali feldspar and plagioclase. Lacroix (1933, p.188) used the similar term granite monzonitique. *(Streckeisen, 1967, p.166)*

MONZONITE. There has been considerable divergence in the use of the term due to the variety of rock types found in the Monzoni district but it is now commonly used for a plutonic rock containing almost equal amounts of plagioclase and alkali feldspar with minor amphibole and/or pyroxene. Now defined modally in QAPF field 8 (Fig. 2.4, p.22). *(Lapparent, 1864, p.260; Mt Monzoni, Alto Adige, Italy; Tröger 259; Johannsen v.3, p.94; Tomkeieff p.374)*

MONZONORITE. A plutonic rock of gabbroic aspect enriched in hypersthene (= enstatite) that contains plagioclase (oligoclase to labradorite) with minor but essential orthoclase. *(Johannsen, 1920c, p.212; Tomkeieff p.374)*

MONZOSYENITE. A collective name originally

given to acid rocks of variable composition from the Monzoni region that contained both alkali feldspar and plagioclase. The same rocks were later called monzonite. *(Buch, 1824, p.344; Johannsen v.3, p.95; Tomkeieff p.374)*

MORB. See mid-ocean ridge basalt.

MOYITE. A term proposed for a variety of biotite granite in which quartz exceeds orthoclase. *(Johannsen, 1920b, p.158; Moyie Sill, Purcell Range, British Columbia, Canada; Tröger 55; Johannsen v.2, p.28; Tomkeieff p.376)*

MUGEARITE. A volcanic rock, often exhibiting flow texture, containing small phenocrysts of olivine, augite and magnetite in a matrix of oligoclase, augite and magnetite with interstitial alkali feldspar. Now defined chemically as the sodic variety of basaltic trachyandesite in TAS field S2 (Fig. 2.14, p.35). *(Harker, 1904, p.264; Mugeary, Island of Skye, Scotland, UK; Tröger 290; Johannsen v.3, p.118; Tomkeieff p.377)*

MUGODZHARITE. A local term for a plutonic rock consisting essentially of quartz and alkali felspar with minor epidote. Cf. alaskite. *(Chumakov, 1946, p.295; Mudgodzhar Hills, between the Ural Mts and Aral Sea, Kazakhstan; Tomkeieff p.378)*

MULATOPORPHYRY (MULATTOPHYRE). An obsolete name for a variety of quartz porphyry. *(Klipstein, 1843, p.78; Mt Mulato, Predazzo, Alto Adige, Italy; Tomkeieff p.378)*

MULDAKAITE. An ultramafic rock consisting of uralite (= actinolite pseudomorph after pyroxene), augite and hornblende. *(Karpinskii, 1869, p.231; Muldakaeva, Urals, Russian Federation; Tomkeieff p.378)*

MUNIONGITE. A variety of phonolite with phenocrysts of nepheline in a groundmass of sanidine, aegirine-augite and glass. *(David et al., 1901, p.377; from the Aboriginal name muniong = Kosciusko Plateau, New South Wales, Australia; Tröger 456; Johannsen v.4,*

p.268; Tomkeieff p.378)

MURAMBITE. A melanocratic variety of leucite basanite composed of plagioclase, clinopyroxene, olivine and with subordinate leucite. *(Holmes & Harwood, 1937, p.136; Murambe Volcano, Bufumbira, Uganda; Tröger(38) 595¾; Tomkeieff p.379)*

MURAMBITOID. A term used for a melanocratic variety of murambite containing very little leucite. *(Holmes & Harwood, 1937, p.145; Murambe Volcano, Bufumbira, Uganda; Tomkeieff p.379)*

MURITE. A variety of phonolite containing phenocrysts of fayalite and titanian augite (rimmed by aegirine-augite) in a groundmass of nepheline, sanidine and aegirine-augite. *(Lacroix, 1927b, p.32; Cape Muri, Rarotonga, Cook Islands, Pacific Ocean; Tröger 501; Johannsen v.4, p.262; Tomkeieff p.379)*

NADELDIORITE. An obsolete local name for a diorite porphyrite containing needle-shaped hornblende. *(Gümbel, 1868, p.348; Rohrbach, Regen, Bavaria; Germany; Tröger 316)*

NAPOLEONITE. An orbicular variety of gabbro composed essentially of calcic plagioclase with hornblende, minor hypersthene (= enstatite) and quartz. This rock has also been called corsite, kugeldiorite and miagite. *(Cotta, 1866, p.155; named after Napoleon I; Tröger 361; Johannsen v.3, p.232; Tomkeieff p.381)*

NATRIOPLETE. An obsolete term for a leucocratic igneous rock rich in sodium. *(Brögger, 1898, p.266; Tomkeieff p.381)*

NATROCARBONATITE. A rare variety of carbonatite lava, currently only known from one locality, consisting essentially of the Na-Ca-K carbonate minerals, nyerereite and gregoryite. It has also been called lengaite. *(Du Bois et al., 1963, p.446; Oldoinyo Lengai, Tanzania; Tomkeieff p.381)*

NAUJAITE. A local name for an agpaitic variety of nepheline sodalite syenite characterized by a poikilitic texture with small crystals of sodalite enclosed in large grains of alkali feldspar, arfvedsonite, aegirine and eudialyte. The content of sodalite may exceed 50% of the rock. *(Ussing, 1912, p.32; Naujakasik (now Naajakasik), Ilímaussaq, Greenland; Tröger 635; Johannsen v.4, p.250; Tomkeieff p.382)*

NAVITE. An obsolete name for a basaltic rock composed of phenocrysts of andesine, augite, enstatite and iddingsite in a groundmass of these minerals, opaques and glass. *(Rosenbusch, 1887, p.512; Nave, now Nahe district, Rheinland-Pfalz, Germany; Tröger 346; Johannsen v.3, p.298; Tomkeieff p.382)*

NAXITE. A corundum, phlogopite, plagioclase rock containing patches of blue tourmaline. Not found *in situ* but possibly from the contact zone between a muscovite granite and a peridotite. *(Papastamatiou, 1939, p.2089; Island of Naxos, Greece; Tomkeieff p.382)*

NEAPITE. A mnemonic name from *nep*heline and *apa*tite, used for alkaline intrusive rocks consisting mainly of nepheline and apatite with minor aegirine and biotite. *(Vlodavets, 1930, p.34; Khibina complex, Kola Peninsula, Russian Federation; Tröger(38) 608⅔; Tomkeieff p.382)*

NECROLITE. A local name for a variety of vesicular biotite latite occurring in the Viterbo and Tolfa regions in Tuscany; used for making Etruscan sarcophagi. *(Brocchi, 1817, p.156; Tomkeieff p.382)*

NEIVITE. A local name for a melanocratic variety of alkali feldspar syenite consisting of major amounts of hornblende with lesser amounts of albite and minor magnetite, pyrite, apatite, titanite and calcite. *(Sobolev, 1959, p.115; River Neiva, Urals, Russian Federation; Tomkeieff p.383)*

NELSONITE. A granular dyke rock consisting essentially of ilmenite and apatite with or without rutile. Several varieties are distinguished according to mineral prefixes. *(Watson, 1907, p.300, states that the name was proposed "elsewhere" but gives no reference; Nelson County, Virginia, USA; Tröger 770; Johannsen v.4, p.471; Tomkeieff p.383)*

NEMAFITE. A mnemonic name for a mixture of *ne*pheline and *maf*ic minerals. *(Hatch et al., 1961, p.358)*

NEMITE. A local name for a melanocratic variety of leucitite. *(Lacroix, 1933, p.199; Lake Nemi, Alban Hills, near Rome, Italy; Tröger 646; Tomkeieff p.383)*

NEPHELINE ANDESITE. A term suggested for volcanic rocks with sodic plagioclase and some nepheline but without olivine. They differ from basanite and tephrite which usually contain calcic plagioclase. *(Johannsen, 1938, p.215; Tröger 574; Tomkeieff p.385)*

NEPHELINE BASALT. A term used for a volcanic rock composed essentially of pyroxene, nepheline and olivine. The name should not be used as the term basalt is now restricted to a rock containing essential plagioclase. As the rock is a variety of foidite it should be given the appropriate name, e.g. olivine nephelinite. *(Naumann, 1850, p.650; Wickenstein, Silesia, Poland; Tröger 923; Johannsen v.4, p.338; Tomkeieff p.385)*

NEPHELINE BASANITE. Now defined in QAPF field 14 (Fig. 2.11, p.31) as a variety of basanite in which nepheline is the most abundant foid. As this type of basanite is the commonest variety it is often known simply as basanite. *(Rosenbusch, 1887, p.763; Böhmisches Mittelgebirge (now České Středohoří), N. Bohemia, Czech Republic; Tröger 591; Johannsen v.4, p.231; Tomkeieff p.385)*

NEPHELINE BASITE. An obsolete term used in two senses: (1) for basanite and (2) for all nepheline-bearing ultrabasic rocks. *(Vogelsang, 1872, p.542; Tomkeieff p.385)*

NEPHELINE BENMOREITE. A variety of benmoreite in which nepheline is present in small amounts. *(Coombs & Wilkinson, 1969, p.493)*

NEPHELINE DIORITE. Now defined in

QAPF field 14 (Fig. 2.4, p.22) as a variety of foid diorite in which nepheline is the most abundant foid. *(Johannsen, 1920b, p.177; Tröger 547; Johannsen v.4, p.214)*

NEPHELINE GABBRO. Now defined in QAPF field 14 (Fig. 2.4, p.22) as a variety of foid gabbro in which nepheline is the most abundant foid. The special term theralite may be used as an alternative. *(Lacroix, 1902, p.191; Tröger 552; Tomkeieff p.385)*

NEPHELINE HAWAIITE. A variety of hawaiite in which nepheline is present in small amounts. *(Coombs & Wilkinson, 1969, p.493)*

NEPHELINE LATITE. A term proposed in a classification for a volcanic rock in which foids > 10% of the total rock and alkali feldspar is 40%–60% of the total feldspar. *(Nockolds, 1954, p.1008)*

NEPHELINE MONZODIORITE. Now defined in QAPF field 13 (Fig. 2.4, p.22) as a variety of foid monzodiorite in which nepheline is the most abundant foid. The special term essexite may be used as an alternative. *(Johannsen, 1920b, p.177)*

NEPHELINE MONZOGABBRO. Now defined in QAPF field 13 (Fig. 2.4, p.22) as a variety of foid monzogabbro in which nepheline is the most abundant foid. The special term essexite may be used as an alternative. *(Johannsen, 1920c, p.216; Tröger 542)*

NEPHELINE MONZONITE. An alkaline plutonic rock with essential nepheline and roughly equal amounts of alkali feldspar and plagioclase. *(Lacroix, 1902, p.33; Tröger 510)*

NEPHELINE MONZOSYENITE. Now defined in QAPF field 12 (Fig. 2.4, p.22) as a variety of foid monzosyenite in which nepheline is the most abundant foid. The term is synonymous with nepheline plagisyenite.

NEPHELINE MUGEARITE. A variety of mugearite in which nepheline is present in small amounts. *(Coombs & Wilkinson, 1969, p.493)*

NEPHELINE PLAGISYENITE. Now defined in QAPF field 12 (Fig. 2.4, p.22) as a variety of foid plagisyenite in which nepheline is the most abundant foid. The term is synonymous with nepheline monzosyenite.

NEPHELINE SYENITE. Now defined in QAPF field 11 (Fig. 2.4, p.22) as a variety of foid syenite in which nepheline is the most abundant foid. *(Rosenbusch, 1877, p.203; Tröger 412; Johannsen v.4, p.77; Tomkeieff p.385)*

NEPHELINE TEPHRITE. Now defined in QAPF field 14 (Fig. 2.11, p.31) as a variety of tephrite in which nepheline is the most abundant foid. As this is the common variety it is often simply called tephrite. *(Rosenbusch, 1877, p.492; Tröger 576; Johannsen v.4, p.231)*

NEPHELINE TRACHYANDESITE. A variety of trachyandesite in which nepheline is present in small amounts. *(Coombs & Wilkinson, 1969, p.493; Tröger 571)*

NEPHELINE TRACHYBASALT. A variety of trachybasalt in which nepheline is present in small amounts. *(Coombs & Wilkinson, 1969, p.493)*

NEPHELINE TRISTANITE. A variety of tristanite in which nepheline is present in small amounts. *(Coombs & Wilkinson, 1969, p.493)*

NEPHELINITE. A term originally used for a nepheline-bearing basaltic rock but now used for rocks consisting essentially of nepheline and clinopyroxene. Now defined modally as a variety of foidite of QAPF field 15c (Fig. 2.11, p.31) and, if modes are not available, chemically as a rock falling in TAS fields U1 or F in which normative $ne > 20\%$ (p. 36). *(Cordier, 1842, vol.8, p.618; Tröger 615; Johannsen v.4, p.338; Tomkeieff p.385)*

NEPHELINITE BASANITE. A variety of nepheline basanite in which nepheline exceeds plagioclase. *(Johannsen, 1938, p.301; Tomkeieff p.386)*

NEPHELINITE TEPHRITE. A variety of nepheline tephrite in which nepheline exceeds plagioclase. *(Johannsen, 1938, p.301)*

NEPHELINITOID. An obsolete term for a variety of nephelinite in which the interstitial nepheline cannot be determined optically, but may be inferred chemically. *(Bořický, 1874, p.41; Hasenberg, Bohemia, Czech Republic; Johannsen v.4, p.344; Tomkeieff p.386)*

NEPHELINOLITE. A plutonic rock now defined as a variety of foidolite of QAPF field 15c (Fig. 2.4, p.22) in which nepheline is the most abundant foid. The term is subdivided into urtite, ijolite and melteigite on a basis of the mafic mineral content (Fig. 2.8, p.27). The rock is the plutonic equivalent of the volcanic rock nephelinite. *(Streckeisen, 1976, p.21; Tomkeieff p.386)*

NEPHELINOLITH. An obsolete term proposed for a monomineralic volcanic rock consisting of nepheline. *(Loewinson-Lessing, 1901, p.114; Tröger 603; Johannsen v.4, p.336; Tomkeieff p.386)*

NEVADITE. A local name for a porphyritic variety of rhyolite containing abundant phenocrysts of quartz, sanidine and plagioclase with minor biotite and hornblende. *(Richthofen, 1868, p.16; named after Nevada, USA; Tröger 42; Johannsen v.2, p.273; Tomkeieff p.387)*

NEVOITE. A melanocratic, apatite-rich ultrabasic rock, composed of biotite, hornblende, apatite, alkali feldspar and augite with minor titanite. *(Khazov, 1983, p.1200; Lake Nevo, now Ladoga, near St Petersburg, Russian Federation)*

NEWLANDITE. An obsolete name for a variety of garnet websterite composed of garnet, Cr-diopside and enstatite. Occurs as inclusions in kimberlites. *(Bonney, 1899, p.315; Newlands diamond pipe, South Africa; Tröger 717; Tomkeieff p.387)*

NGURUMANITE. A medium-grained variety of melteigite composed of pyroxene, altered nepheline in an iron-rich mesostasis with vugs of zeolites and calcite. *(Saggerson & Williams, 1963, p.479; Nguruman Escarpment, Kenya)*

NIKLESITE. An obsolete name for a variety of websterite composed of diallage (= altered diopside), enstatite and diopside, with lamellae of enstatite and diallage. *(Kretschmer, 1917, p.164; Nikles (now Raškov), Moravia, Czech Republic; Tröger 928; Johannsen v.4, p.461; Tomkeieff p.387)*

NILIGONGITE. A medium-grained variety of melilite-bearing leucite ijolite with alkali pyroxenes, and with equal proportions of leucite and nepheline. *(Lacroix, 1933, p.198; Niligongo Volcano, Birunga, Kivu, Democratic Republic of Congo; Tröger 630; Johannsen v.1 (2nd Edn), p.269; Tomkeieff p.387)*

NIOLITE. An obsolete term for a variety of felsite containing spherules of radiating feldspar. *(Pinkerton, 1811b, p.74; Chain of Niolo, Corsica, France; Tomkeieff p.388)*

NONESITE. An obsolete local name for a variety of basalt composed of phenocrysts of labradorite, augite and olivine in a groundmass of andesine, augite, orthoclase, opaques and glass. *(Lepsius, 1878, p.163; Mendola Pass, Nonsberg, Alto Adige, Italy; Tröger 345; Johannsen v.3, p.285; Tomkeieff p.389)*

NORDMARKITE. A variety of quartz-bearing alkali feldspar syenite composed mainly of microperthite with minor biotite, alkali amphibole or pyroxene. *(Brögger, 1890, p.54; Nordmarka, Oslo Igneous Province, Norway; Tröger 185; Johannsen v.3, p.6; Tomkeieff p.389)*

NORDSJÖITE. A local name for a coarse-grained variety of calcite nepheline syenite in which nepheline exceeds alkali feldspar. Aegirine-augite is usually present, sometimes with melanite. *(Johannsen, 1938, p.247; Söve, Nordsjö, Fen Complex, Telemark, Norway; Tröger(38) 421⅓; Tomkeieff p.389)*

NORITE. A plutonic rock composed essentially of bytownite, labradorite or andesine and orthopyroxene. Now defined modally in the gabbroic rock classification (Fig. 2.6, p.25). *(Esmark, 1823, p.207; Tröger 355; Johannsen v.3, p.233; Tomkeieff p.390)*

NORTHFIELDITE. A variety of quartzolite occurring as a border phase in a granite and resembling vein quartz or greisen composed essentially of quartz with minor muscovite and traces of tourmaline and tremolite or actinolite. Cf. esmeraldite. *(Emerson, 1915, p.212; Crag Mtn., Northfield, Massachusetts, USA; Tröger 4; Johannsen v.2, p.18; Tomkeieff p.390)*

NOSEAN BASANITE. Now defined in QAPF field 14 (Fig. 2.11, p.31) as a variety of basanite in which nosean is the most abundant foid. *(Tröger, 1935, p.246; Tröger 597; Johannsen v.4, p.238)*

NOSEANITE. A variety of nephelinite containing considerable amounts of nosean and amphibole. Now defined as a variety of foidite of QAPF field 15c (Fig. 2.11, p.31). *(Bořický, 1874, p.41; St Georgenberg (now Oip), Říp, Bohemia, Czech Republic; Tröger 651; Johannsen v.4, p.345; Tomkeieff p.390)*

NOSEANOLITH. An obsolete term proposed for a monomineralic volcanic rock consisting of nosean. *(Johannsen, 1938, p.337; Tröger 634; Tomkeieff p.391)*

NOSELITITE. An obsolete term proposed for a variety of noseanite consisting mainly of nosean, pyroxene and amphibole. The term is synonymous with noseanite. *(Johannsen, 1938, p.345; Tomkeieff p.391)*

NOSYKOMBITE. A local name for a variety of nepheline monzosyenite consisting of sanidine, nepheline, and kaersutite with small amounts of plagioclase; corresponds to the nosykombitic magma-type of Niggli. *(Niggli, 1923, p.157; Nosy-Komba Island, Madagascar; Tröger 507; Tomkeieff p.391)*

NOTITE. An obsolete name for a variety of

porphyry containing quartz, feldspar and mica. *(Jurine, 1806, p.376; Mont Blanc, France; Tomkeieff p.391)*

OBSIDIAN. A common term for a volcanic glass, usually with a water content < 1%, often dark in colour, massive and with a conchoidal fracture. *(A term of great antiquity usually attributed to Theophrastus, 320 BC – see Johannsen for further discussion; Tröger 930; Johannsen v.2, p.276; Tomkeieff p.392)*

OCEANIC PLAGIOGRANITE. A term proposed for a series of plutonic rocks consisting of plagioclase, ranging in composition from oligoclase to anorthite, quartz and minor amounts of hornblende and pyroxene. They frequently show the effects of low-grade metamorphism developing epidote, chlorite, actinolite and albite. *(Coleman & Peterman, 1975, p.1099)*

OCEANITE. A variety of melanocratic picritic basalt consisting of abundant phenocrysts of olivine and lesser amounts of augite in a groundmass of augite, olivine and plagioclase. *(Lacroix, 1923, p.49; Piton de la Fournaise, Réunion, Indian Ocean; Tröger 409; Johannsen v.3, p.306; Tomkeieff p.392)*

ODINITE. A dyke rock of basaltic composition consisting of labradorite and rare augite phenocrysts in a felted groundmass of hornblende needles with occasional quartz. Cf. malchite. Gabbrophyre was suggested as an alternative term. *(Chelius, 1892, p.3; Frankenstein, Odenwald, Hessen, Germany; Tröger 319; Johannsen v.3, p.326; Tomkeieff p.393)*

OKAITE. A local name for a variety of haüyne melilitolite consisting mainly of melilite and haüyne with some biotite and perovskite. Mineralogically the rock is similar to turjaite with haüyne taking the place of nepheline. May be used as an optional term in the melilitic rocks classification if haüyne > 10% (section 2.4, p.11). *(Stansfield, 1923a, p.440; Oka*

Hills, Montreal, Quebec, Canada; Tröger 661; Johannsen v.4, p.324; Tomkeieff p.394)

OKAWAITE. A local name for a glassy aegirine-augite rhyolite. *(Nemoto, 1934, p.300; Okawa River, Tokati, Hokkaido, Japan; Tröger(38) 47½; Tomkeieff p.394)*

OLIGOCLASE ANDESITE. A term originally given to a variety of andesite in which the normative plagioclase is oligoclase, but later (Washington & Keyes, 1928) applied to a volcanic rock consisting essentially of oligoclase with minor pyroxene. *(Iddings, 1913, p.193; Tröger 324; Tomkeieff p.394)*

OLIGOCLASE BASALT. A misnomer for a rock in which andesine was probably incorrectly identified as oligoclase (Johannsen, 1937). *(Bořický, 1874, p.43; Johannsen v.3, p.289)*

OLIGOCLASITE. An obsolete term originally and inappropriately used for a plutonic rock consisting of orthoclase and lesser oligoclase, later replaced by the name cavalorite. Also used for both a plutonic rock consisting entirely of oligoclase (Kolderup, 1898) and a fine-grained rock of trachytic texture composed essentially of oligoclase (Washington, 1923). *(Bombicci, 1868, p.79; Mt Cavaloro, near Bologna, Italy; Tröger 292; Johannsen v.3, p.145; Tomkeieff p.394)*

OLIGOPHYRE. An obsolete general term for rocks consisting of phenocrysts of oligoclase in a groundmass of the same mineral. *(Coquand, 1857, p.96; Tröger 932; Johannsen v.3, p.183; Tomkeieff p.395)*

OLIGOSITE. An obsolete term proposed for a coarse-grained rock consisting essentially of oligoclase. *(Turner, 1900, p.110; Tröger 934; Johannsen v.3, p.145; Tomkeieff p.395)*

OLIVINE BASALT. A commonly used term for a basalt containing olivine as an essential constituent. The reference cited actually contains the term olivine-bearing basalt. *(Rosenbusch, 1896, p.1018; Tröger 379; Johannsen v.3, p.281; Tomkeieff p.395)*

OLIVINE CLINOPYROXENITE. An ultramafic plutonic rock consisting essentially of clinopyroxene and up to 50% olivine. Now defined modally in the ultramafic rock classification (Fig. 2.9, p.28). *(Streckeisen, 1973, p.26)*

OLIVINE GABBRO. A commonly used name for a gabbro containing essential olivine. Now defined modally in the gabbroic rock classification (Fig. 2.6, p.25) as a variety of gabbro in which olivine is between 5% and 85%. *(Lasaulx, 1875, p.310; Tröger 351; Johannsen v.3, p.224; Tomkeieff p.396)*

OLIVINE GABBRONORITE. A collective term for plutonic rocks consisting of 10%–90% calcic plagioclase and accompanied by olivine, orthopyroxene and clinopyroxene in various amounts. Now defined modally in the gabbroic rock classification (Fig. 2.6, p.25) as a variety of gabbronorite in which olivine is between 5% and 85%. *(Streckeisen, 1973, p.27)*

OLIVINE HORNBLENDE PYROXENITE. An ultramafic plutonic rock consisting of more than 30% pyroxene accompanied by amphibole and olivine in various amounts. Now defined modally in the ultramafic rock classification (Fig. 2.9, p.28). *(Streckeisen, 1973, p.26)*

OLIVINE HORNBLENDITE. An ultramafic plutonic rock consisting essentially of amphibole and up to 50% olivine. Now defined modally in the ultramafic rock classification (Fig. 2.9, p.28). *(Streckeisen, 1973, p.26; Tröger 707)*

OLIVINE MELILITITE. A collective term for ultramafic volcanic rocks consisting of melilite, clinopyroxene and olivine in various amounts. Now defined modally in the melilite-bearing rocks classification (Fig. 2.3, p.11) or chemically in field F of the TAS classification as a rock which does not contain kalsilite but has normative *cs* (larnite) > 10% and $K_2O < Na_2O$ (p.38). If modal olivine < 10% the rock should be called a melilitite. *(Streckeisen, 1978, p.13)*

OLIVINE MELILITOLITE. Originally defined as an ultramafic plutonic rock consisting essentially of melilite and olivine with minor clinopyroxene (see 1st Edition, Fig. B.3, p.12), but not required by the new melilite-bearing rocks classification (section 2.4.1, p.11). *(Streckeisen, 1978, p.13)*

OLIVINE NORITE. An old term for a norite containing essential olivine. Now defined modally in the gabbroic rock classification (Fig. 2.6, p.25) as a variety of norite in which olivine is between 5% and 85%.

OLIVINE ORTHOPYROXENITE. An ultramafic plutonic rock consisting essentially of orthopyroxene and up to 50% olivine. Now defined modally in the ultramafic rock classification (Fig. 2.9, p.28). *(Streckeisen, 1973, p.26)*

OLIVINE PACIFICITE. A term for an anemousite olivine basalt with the chemistry of nepheline basanite. Cf. pacificite. Niggli (1936, p.366) used the chemistry of this rock for the kaulaitic magma-type and Tröger (1938, p.67) renamed it kaulaite. *(Barth, 1930, p.65; Kaula Gorge, Ookala, Mauna Kea, Hawaii, USA; Tröger 385)*

OLIVINE PYROXENE HORNBLENDITE. An ultramafic plutonic rock consisting of more than 30% amphibole accompanied by pyroxene and olivine in various amounts. Now defined modally in the ultramafic rock classification (Fig. 2.9, p.28). *(Streckeisen, 1973, p.26)*

OLIVINE PYROXENE MELILITOLITE. Originally defined as an ultramafic plutonic rock consisting essentially of melilite, clinopyroxene and lesser amounts of olivine (see 1st Edition, Fig. B.3, p.12), but not required by the new melilite-bearing rocks classification (section 2.4.1, p.11). *(Streckeisen, 1978, p.13)*

OLIVINE PYROXENITE. An ultramafic plutonic rock consisting essentially of

pyroxene and up to 50% olivine. Now defined modally in the ultramafic rock classification (Fig. 2.9, p.28). *(Streckeisen, 1973, p.26)*

OLIVINE THOLEIITE. Chemically defined as an olivine-hypersthene normative basalt. It is an exceedingly abundant rock and contains phenocrysts of olivine and/or plagioclase and pyroxenes in a groundmass of Ca-poor pyroxene, labradorite, opaques and sometimes glass. Now regarded as a variety of subalkali basalt. *(Rosenbusch, 1887, p.515; Tröger 344)*

OLIVINE UNCOMPAHGRITE. Originally proposed as a special term in the melilitic rocks classification synonomous with olivine pyroxene melilitolite, but no longer necessary. *(Streckeisen, 1978, p.14)*

OLIVINE WEBSTERITE. An ultramafic plutonic rock consisting of 10%–40% olivine with various amounts of clinopyroxene and orthopyroxene. Now defined modally in the ultramafic rock classification (Fig. 2.9, p.28). *(Streckeisen, 1973, p.26)*

OLIVINITE. A term originally used for ore-bearing olivine rocks and later for plutonic rocks composed of olivine with pyroxene and/or amphibole. In the Russian Republic the term is used for olivine rocks with accessory magnetite to distinguish them from dunite, which contains accessory chromite. *(Sjögren, 1876, p.58; Tröger 937; Johannsen v.4, p.402; Tomkeieff p.396)*

ONGONITE. A name proposed for quartz keratophyres containing topaz which are the subvolcanic equivalents of REE-, Li-, F-rich granites. See also topaz rhyolite. *(Kovalenko et al., 1971, p.430; Ongon-Khairkhan, Mongolia)*

ONKILONITE. A variety of nephelinite containing olivine, augite, nepheline, leucite and perovskite in two generations. *(Backlund, 1915, p.307; Onkilone tribe, Vilkitzky Island, New Siberia Islands, Russian Federa-*

tion; Tröger 618; Johannsen v.4, p.368; Tomkeieff p.396)

OPDALITE. A member of the charnockitic rock series equivalent to orthopyroxene granodiorite of QAPF field 4 (Table 2.10, p.20), consisting of zoned plagioclase, microcline, quartz, biotite, hypersthene (= enstatite) and diopside. The term is synonymous with charno-enberbite. *(Goldschmidt, 1916, p.70; Opdal, Trondheim, Norway; Tröger 108; Johannsen v.2, p.347; Tomkeieff p.399)*

OPHIGRANITONE. A variety of gabbro containing scaly serpentine, saussurite and diallage (= altered diopside). *(Mazzuoli & Issel, 1881, p.326; Tomkeieff p.399)*

OPHIOLITE. A term originally applied to rocks consisting mainly of serpentine, but later extended to the rock suite of Alpine-type peridotites, gabbros, dolerites, spilites and keratophyres. Now used to define the association of basic to ultrabasic rocks thought to represent oceanic crust. *(Brongniart, 1813, p.37; from the Greek ophis = snake, on account of the appearence; Tomkeieff p.399)*

OPHITE. A term originally used by Pliny for a greenish mottled ornamental marble, but later applied to doleritic rocks from the Pyrénées, many of which were uralitized. *(A term of great antiquity usually attributed to Pliny, AD 77 – see Johannsen for further discussion; from the Greek ophites = like a serpent; Tröger 938; Johannsen v.3, p.319; Tomkeieff p.399)*

OPHITONE. An obsolete term for a greenish diabase. *(Cordier, 1842, vol.8, p.135; Tomkeieff p.399)*

ORANGEITE. A name synonymous with micaceous kimberlite and Group II kimberlites. For further details see p.14. *(Mitchell, 1995)*

ORANGITE. A name originally suggested to replace micaceous kimberlite and now replaced by orangeite. *(Wagner, 1928, p.140; Lion Hill Dyke, Orange Free State, South Africa)*

ORBICULITE. A group name for plutonic rocks with orbicular structure. *(Sederholm, 1928, p.72; Tröger 939; Tomkeieff p.400)*

ORBITE. A coarse-grained lamprophyric dyke rock composed of phenocrysts of hornblende in a plagioclase-rich groundmass. Cf. luciite. *(Chelius, 1892, p.3; Orbishöhe, near Zwingenberg, Odenwald, Germany; Tröger 336; Johannsen v.3, p.309; Tomkeieff p.400)*

ORDANCHITE. A local name for a volcanic rock, variously described as a variety of trachyandesite and tephrite, consisting essentially of andesine and haüyne with resorbed hornblende, augite and minor olivine. *(Lacroix, 1917a, p.582; Banne d'Ordanche, Auvergne, France; Tröger 572; Johannsen v.4, p.216; Tomkeieff p.400)*

ORDOSITE. A melanocratic variety of alkali feldspar syenite containing abundant aegirine, microcline and a little phlogopite. *(Lacroix, 1925, p.482; Ordos Province, Inner Mongolia, China; Tröger 223; Johannsen v.4, p.17; Tomkeieff p.401)*

ORENDITE. Originally described as a variety of leucite phonolite essentially composed of leucite and alkali feldspar with subordinate clinopyroxene, mica and amphibole. Now regarded as a diopside-sanidine-phlogopite lamproite (Table 2.7, p.17). *(Cross, 1897, p.123; Orenda Butte, Leucite Hills, Wyoming, USA; Tröger 478; Johannsen v.4, p.260; Tomkeieff p.401)*

ORNÖITE. A local name for a variety of diorite consisting essentially of sodic plagioclase with minor hornblende, microcline, biotite and quartz. *(Cederström, 1893, p.107; Ornö Island, Stockholm, Sweden; Tröger 305; Johannsen v.3, p.115; Tomkeieff p.402)*

OROGENIC ANDESITE. A term for a variety of andesite defined chemically as having SiO_2 between 53% and 63%, *hy* in the norm, TiO_2 < 1.75% and $K_2O < (0.145 \times SiO_2 - 5.135)$. *(Gill, 1981, p.2)*

OROTVITE. A melanocratic to mesocratic variety of nepheline-bearing diorite rich in kaersutite and with mafic minerals more abundant than andesine-oligoclase. Nepheline and alkali feldspar can be present in minor amounts. *(Streckeisen, 1938, p.159; Orotva Valley, Ditrau, Transylvania, Romania; Tröger(38) 308⅔; Johannsen v.4, p.215; Tomkeieff p.402)*

ORTHO-. A prefix which has been applied to rock names in several senses. For example, to indicate the presence of orthopyroxene (e.g. orthoandesite, orthobasalt) or the abundance of orthoclase and lack of plagioclase (e.g. orthogranite) and later for rocks that are neither oversaturated nor undersaturated with respect to silica (e.g. orthosyenite, orthogabbro). *(Johannsen, 1917; Tröger 940; Tomkeieff p.403)*

ORTHOANDESITE. An obsolete group name for an orthopyroxene-bearing andesite. The term is attributed to Oebbeke who supposedly used it for rocks from the Phillipines, but no actual reference is given. The only known publication is Oebbeke (1881), but it does not contain the name. *(Kotô, 1916b, p.116; Johannsen v.3, p.173)*

ORTHOBASALT. An obsolete term proposed for an olivine-bearing bronzite (= enstatite) basalt. *(Kotô, 1916b, p.119)*

ORTHOBASE. According to Tomkeieff *et al.* (1983) the rock is a diabase or porphyrite containing orthoclase. However, the term does not appear in the reference cited. The same incorrect reference is also cited by Loewinson-Lessing *et al.* (1932). *(Belyankin, 1911, p.363; Tomkeieff p.403)*

ORTHOCLASITE. A general term for a rock consisting essentially of orthoclase. Now used for fenitic rocks composed of K-rich feldspars (Sutherland, 1965) for which the terms orthoclase-rock and feldspar-rock have been used. If fluorite is present the rock is called borengite. *(Senft, 1857, p.51; Tröger 941;*

Johannsen v.3, p.5; Tomkeieff p.403)

ORTHOFELSITE. An obsolete term synonymous with orthophyre. *(Teall, 1888, p.291; Tröger 256; Tomkeieff p.404)*

ORTHOFOYAITE. An obsolete term proposed for a variety of foyaite in which plagioclase is practically absent. *(Johannsen, 1920b, p.161; Tomkeieff p.404)*

ORTHOGABBRO. A variety of gabbro that contains no quartz or foids, i.e. it closely agrees with the definition of the "ideal" type. *(Hatch et al., 1949, p.279)*

ORTHOGRANITE. An obsolete term proposed for a variety of granite in which plagioclase is practically absent. The term was later withdrawn (Johannsen, 1932) in favour of kaligranite. *(Johannsen, 1920a, p.53; Johannsen v.2, p.51; Tomkeieff p.404)*

ORTHOLITHE (ORTHOLITE). An obsolete name originally used in France for a variety of minette containing orthoclase and mica. *(Lapparent, 1906, p.581; Tomkeieff p.404)*

ORTHOPHONITE. An obsolete term for nepheline syenite, being a mnemonic name from *ortho*clase and *phono*lite. *(Lasaulx, 1875, p.318; Tomkeieff p.405)*

ORTHOPHYRE. An obsolete name used for older trachytes and for rocks consisting of phenocrysts of orthoclase in a groundmass of orthoclase. *(Coquand, 1857, p.65; Tröger 256; Johannsen v.3, p.80; Tomkeieff p.405)*

ORTHOPYROXENE GABBRO. A basic plutonic rock consisting mainly of calcic plagioclase, clinopyroxene and minor orthopyroxene. Now defined modally in the gabbroic rock classification (Fig. 2.6, p.25). *(Johannsen, 1937, p.238)*

ORTHOPYROXENITE. An ultramafic plutonic rock consisting almost entirely of orthopyroxene. Now defined modally in the ultramafic rock classification (Fig. 2.9, p.28). *(Wyllie, 1967, p.2)*

ORTHORHYOLITE. An obsolete term proposed for a variety of rhyolite in which plagioclase is practically absent. The term was later withdrawn (Johannsen, 1932) in favour of kalirhyolite. *(Johannsen, 1920b, p.159; Tomkeieff p.405)*

ORTHOSHONKINITE. An obsolete term originally proposed for a variety of shonkinite containing no plagioclase, but later withdrawn (Johannsen, 1938). *(Johannsen, 1920a, p.51; Johannsen v.4, p.15)*

ORTHOSITE. A term proposed for a coarse-grained variety of alkali feldspar syenite consisting essentially of orthoclase. *(Turner, 1900, p.110; Tröger 163; Johannsen v.3, p.4; Tomkeieff p.406)*

ORTHOSYENITE. A term originally used for a syenite in which the amount of alkali feldspar is more than 95% of the total feldspar, but later withdrawn (Johannsen, 1937) in favour of kalisyenite. The name was later redefined (Hatch *et al.*, 1949, p.232) as a variety of syenite that contains no quartz or foids, i.e. is exactly saturated with respect to silica. *(Johannsen, 1920b, p.160; Johannsen v.3, p.8; Tomkeieff p.406)*

ORTHOTARANTULITE. An obsolete term proposed for rocks consisting of major amounts of quartz with minor orthoclase. The term was later withdrawn (Johannsen, 1932) in favour of arizonite of which Johannsen had been unaware. *(Johannsen, 1920a, p.53; Johannsen v.2, p.32; Tomkeieff p.406)*

ORTHOTRACHYTE. A term suggested to replace alkali trachyte if the alternative name quartz-free liparite is regarded as a contradiction in terms. The name was later redefined (Hatch *et al.*, 1949, p.247) as a variety of trachyte that contains no quartz or foids, i.e. is exactly saturated with respect to silica. *(Rosenbusch, 1908, p.887; Johannsen v.3, p.16)*

ORTLERITE. An altered variety of hornblende andesite or trachyandesite consisting of abundant phenocrysts of hornblende and a few of augite and biotite in a felted groundmass of andesine, chlorite, pyrite, calcite and devitrified glass. *(Stache & John, 1879, p.325; Ortler Alps, Alto Adige, Italy; Tröger 287; Johannsen v.3, p.123; Tomkeieff p.406)*

ORVIETITE. A volcanic rock, close to the boundary between phonolitic tephrite and tephritic phonolite, essentially composed of equal proportions of plagioclase and alkali feldspar, with subordinate leucite and clinopyroxene. It contains less leucite than vicoite. *(Niggli, 1923, p.175; Orvieto, Vulsini district, near Viterbo, Italy; Tröger 540; Tomkeieff p.406)*

OSLO-ESSEXITE. A term suggested to replace the name essexite, which had been incorrectly applied to rocks occurring in the volcanic bosses in the Oslo Igneous Province by Brögger (1933), as they do not contain nepheline. The term includes such types as kauaite, bojite, olivine gabbro, pyroxenite etc. *(Barth, 1944, p.31; Tomkeieff p.407)*

OSLOPORPHYRY. A local name for an oligoclase porphyry. *(Brögger, 1898, p.207; Tröger 944; Tomkeieff p.407)*

OSSIPYTE (OSSIPITE, OSSYPITE). An obsolete local name for a variety of plagioclase-rich olivine gabbro. Dana states the name was suggested by Hitchcock. *(Dana, 1872, p.49; named after Ossipee Indians, New Hampshire, USA; Tröger 352; Johannsen v.3, p.225; Tomkeieff p.407)*

ØSTERN PORPHYRY. A local term for a variety of porphyritic kjelsåsite or plagioclase-rich monzonite. *(Holtedahl, 1943, p.32; Østervann, Oslo district, Norway; Tomkeieff p.407)*

OSTRAITE. An obsolete name for a variety of clinopyroxenite composed of partly uralitized augite and spinel. *(Duparc, 1913, p.18; Ostraïa Sopka, Urals, Russian Federation; Tröger 685; Johannsen v.4, p.464; Tomkeieff p.407)*

OTTAJANITE. A variety of leucite tephrite essentially composed of plagioclase, leucite and clinopyroxene, with subordinate olivine. It is said to have the chemical, but not mineralogical, compostion of sommaite. *(Lacroix,*

1917b, p.208; Ottajanite (now Ottaviano), Mt Somma, Naples, Italy; Tröger 945; Johannsen v.4, p.201; Tomkeieff p.407)

OUACHITITE. An ultramafic lamprophyre containing combinations of olivine, phlogopite, amphibole, and/or clinopyroxene phenocrysts with groundmass feldspathoids as well as carbonates. *(Kemp, 1890, p.393; Ouachita River, Arkansas, USA; Tröger 405; Johannsen v.4, p.391; Tomkeieff p.408)*

OUENITE. A fine-grained variety of basalt which occurs as dykes and consists of anorthite with chrome-diopside, fringed by a little bronzite (= enstatite) and olivine. *(Lacroix, 1911a, p.817; Ouen Island, New Caledonia; Tröger 368; Johannsen v.3, p.349; Tomkeieff p.408)*

OVERSATURATED. A term applied to igneous rocks in which there is an excess of SiO_2 over the other oxides, which gives silica minerals in the mode or quartz in the norm. *(Shand, 1913, p.510; Tomkeieff p.408)*

OWHAROITE. A local name for a strongly welded rhyolitic or dacitic tuff, previously called wilsonite. *(Grange, 1934, p.58; Owharoa, Waihi district, Auckland, New Zealand; Tröger(38) 40⅔; Tomkeieff p.409)*

OXYPHYRE. An obsolete general term for porphyritic acid rocks which were thought to be complementary to lamprophyres. *(Pirsson, 1895, p.118; Tomkeieff p.410)*

OXYPLETE. An obsolete term for a leucocratic igneous rock in which $SiO_2 > 6 R_2O_3$. *(Brögger, 1898, p.266; Tomkeieff p.410)*

PACIFICITE. A tephritic rock composed of phenocrysts of augite, some labradorite and rare olivine in a groundmass of anemousite (plagioclase containing carnegieite), augite and opaques. *(Barth, 1930, p.60; named after Pacific Ocean; Haleakala, Maui, Hawaiian Islands, USA; Tröger 384; Tomkeieff p.411)*

PAISANITE. A local name for a leucocratic variety of alkali feldspar microgranite consisting essentially of anorthoclase and quartz with minor amounts of riebeckite. *(Osann,*

1893, p.131; Paisano Pass, Texas, USA; Tröger 29; Johannsen v.2, p.100; Tomkeieff p.412)

PALAEOPHYRE. An obsolete term for altered porphyrites. *(Gümbel, 1874, p.42; Tröger 946; Johannsen v.3, p.183; Tomkeieff p.412)*

PALAEOPHYRITE. An obsolete term for older post-Cretaceous porphyries. *(Stache & John, 1879, p.352; Tröger 947; Johannsen v.3, p.183; Tomkeieff p.412)*

PALAEOPICRITE. An obsolete term for picrites of Palaeozoic age which were more altered than those of Tertiary age. *(Gümbel, 1874, p.38; Tröger 948; Tomkeieff p.412)*

PALAGONITE TUFF. A term for a tuff containing palagonite which is a yellow or brown completely devitrified basaltic glass. *(Waltershausen, 1846, p.402; Palagonia, Sicily, Italy; Tröger 594; Johannsen v.3, p.302; Tomkeieff p.412)*

PALATINITE. An obsolete term applied to various rocks composed of augite and plagioclase. Rosenbusch redefined it as a bronzite (= enstatite) tholeiite. *(Laspeyres, 1869, p.516; Palatia, now Pfalz, Germany; Tröger 343; Johannsen v.3, p.299; Tomkeieff p.413)*

PALLIOESSEXITE, PALLIOGRANITE. Obsolete terms suggested for varieties of igneous rocks which occur on the margin of an intrusion "a little way within the contact". *(Jevons et al., 1912, p.452; from the Latin pallium = cloak; Johannsen v.4, p.61; Tomkeieff p.414)*

PANTELLERITE. Originally described as a leucocratic variety of alkali rhyolite containing phenocrysts of aegirine-augite, anorthoclase and cossyrite. It is now defined and distinguished from comendite as a variety of peralkaline rhyolite of TAS field R in which $Al_2O_3 < 1.33 \times$ total iron as FeO + 4.4 (Fig. 2.18, p.38). Synonymous with pantelleritic rhyolite. *(Foerstner, 1881, p.537; Pantelleria Island, Italy; Tröger 72; Johannsen v.2, p.64; Tomkeieff p.414)*

PANTELLERITIC RHYOLITE. Now defined as a variety of peralkaline rhyolite of TAS field R in which $Al_2O_3 < 1.33 \times$ total iron as FeO + 4.4 (Fig. 2.18, p.38). Synonymous with pantellerite.

PANTELLERITIC TRACHYTE. Now defined as a variety of peralkaline trachyte of TAS field T in which $Al_2O_3 < 1.33 \times$ total iron as FeO + 4.4 (Fig. 2.18, p.38).

PARAMELAPHYRE. An obsolete term for a variety of mica porphyrite. *(Schmid, 1880, p.67; Tomkeieff p.416)*

PARCHETTITE. An obsolete name for a variety of leucite tephrite composed essentially of clinopyroxene, leucite and plagioclase and with accessory alkali feldspar. *(Johannsen, 1938, p.291; Fosso della Parchetta, Vico Volcano, near Viterbo, Italy; Tröger(38) 539½; Tomkeieff p.417)*

PAROPHITE. An obsolete term for a variety of serpentine. *(Hunt, 1852, p.95; Tomkeieff p.418)*

PATRINITE. An obsolete term for a phonolite. *(Pinkerton, 1811a, p.167; named after E.L.M. Patrin; Tomkeieff p.419)*

PAWDITE. An obsolete name for a dark granular dyke rock composed of plagioclase zoned from bytownite to oligoclase, hornblende, quartz, biotite and epidote. *(Duparc & Grosset, 1916, p.110; Pawdinskaya Datcha, Nikolai Pawda, Urals, Russian Federation; Tröger 333; Johannsen v.3, p.319; Tomkeieff p.419)*

PEARLSTONE. An obsolete term for perlite. The name is attributed to Jameson, but Johannsen gives no reference. *(Johannsen, 1932, p.284; Tomkeieff p.420)*

PECHSTEIN. The German name for pitchstone. *(Schulz & Poetsch, 1759, p.267; Tröger 102; Johannsen v.2, p.280)*

PEDROSITE. A peralkaline variety of hornblendite composed essentially of alkali amphibole (osannite), with some magnetite, and occasionally albite and analcime. *(Osann, 1922, p.260; Alter Pedroso, Alentejo, Portugal; Tröger 703; Johannsen v.4, p.444; Tomkeieff p.421)*

PEGMATITE (PEGMATYTE). A term originally suggested by Haüy as a synonym for graphic granite, but later used for the coarse-grained facies of any type of igneous rock. *(Brongniart, 1813, p.32; from the Greek pegma = bond, framework; Tröger 65; Johannsen v.2, p.72; Tomkeieff p.421)*

PEGMATITOID. An obsolete term proposed for coarse-grained veins, some of which may be late differentiates, occurring in basaltic rocks. *(Lacroix, 1928b, p.322; Tomkeieff p.422)*

PEGMATOID. An term proposed for both feldspathoidal pegmatites and for pegmatites without graphic texture. *(Shand, 1910, p.377; Tröger 951; Tomkeieff p.422)*

PEGMATOPHYRE. An obsolete term, originally proposed to replace granophyre and later for varieties of granophyre without plagioclase. *(Lossen, 1892, p.270; Tröger 952; Johannsen v.2, p.70; Tomkeieff p.422)*

PÉLÉ'S HAIR. A name given to the fine hair-like threads of basaltic glass often formed during lava fountaining. *(Ellis, 1825, p.143; named after Pélé, the Hawaiian goddess of fire; Johannsen v.3, p.291; Tomkeieff p.423)*

PÉLÉ'S TEARS. Tear-drop-shaped lava filaments blown through open channels of lava vents. *(Perret, 1913, p.615; named after Pélé, the Hawaiian goddess of fire; Tomkeieff p.423)*

PELÉEITE. A comprehensive term for andesites and basaltic andesites that correspond to the peléeitic magma-type of Niggli (1923, p.123). *(Tröger, 1935, p.75; Mt Pelée, Martinique, Lesser Antilles; Tröger 157; Tomkeieff p.423)*

PELEITI. An Italian name for Pélé's hair. *(Issel, 1916, p.657; Tomkeieff p.423)*

PENIKKAVAARITE. A local name for a variety of gabbro with 60%–70% mafic minerals. Amphiboles are more abundant than titanian augite and plagioclase. Alkali feldspar may be present in minor amounts. *(Johannsen, 1938, p.52; Penikkavaara, Kuusamo, Finland; Tröger(38) 285½; Tomkeieff p.426)*

PEPERIN-BASALT. An obsolete name for a tuff which forms mud flows and contains large crystals of augite and hornblende. *(Bořický, 1874, p.42; Kostenblatt (now Kostomlaty), Bohemia, Czech Republic; Tomkeieff p.426)*

PEPERINO. A local Italian name for a light-coloured unconsolidated tuff from the Alban Hills containing many dark crystal fragments, giving it a peppery appearance. *(Buch, 1809, p.70; from the Italian pepe = pepper, as it resembles grains of pepper; Tröger 953; Johannsen v.4, p.363; Tomkeieff p.426)*

PÉPÉRITE. A local term for a tuff or breccia formed by the intrusion of magma into wet sediments. Usually consists of fragments of glassy igneous rock and some sedimentary rock. *(Cordier, 1816, p.366; Tomkeieff p.426)*

PERACIDITE. A chemically derived term for an igneous rock consisting almost entirely of quartz. Such rocks should now be called quartzolite. *(Rinne, 1921, p.165; Tröger 1; Johannsen v.2, p.11; Tomkeieff p.426)*

PERALBORANITE. A leucocratic variety of alboranite with less than 12.5% of pyroxene. Obsolete. *(Burri & Parga-Pondal, 1937, p.258; Alboran Island, near Cabo de Gata, Spain; Tröger(38) 304½; Tomkeieff p.426)*

PERALKALINE. A chemical term for alkaline rocks in which the molecular amounts of Na_2O plus K_2O exceeds Al_2O_3. This produces acmite *(ac)* and sometimes sodium metasilicate *(ns)* in the CIPW norm and usually alkali pyroxenes and/or alkali amphiboles in the mode. Cf. agpaitic. *(Winchell, 1913, p.210; Tomkeieff p.426)*

PERALKALINE GRANITE. A term that may be used for a variety of alkali feldspar granite that contains alkali pyroxene and/or amphibole (p.23). This term should be used in preference to alkali granite.

PERALKALINE PHONOLITE. A term that may be used in two ways: (1) for a variety of phonolite of QAPF field 11 that contains alkali pyroxene and/or amphibole (p.32), and (2) for a variety of phonolite of TAS field Ph that has a peralkaline index > 1 (p.38).

PERALKALINE RHYOLITE. A term that may be used in two ways: (1) for a variety of alkali feldspar rhyolite of QAPF field 2 that contains alkali pyroxene and/or amphibole (p.30), and (2) for a variety of rhyolite of TAS field R that has a peralkaline index > 1 (p.37). In both cases this term should be used in preference to alkali rhyolite.

PERALKALINE TRACHYTE. A term that may be used in two ways: (1) for a variety of alkali feldspar trachyte of QAPF field 6 that contains alkali pyroxene and/or amphibole (p.30), and (2) for a variety of trachyte of TAS field T that has a peralkaline index > 1 (p.37). In both cases this term should be used in preference to alkali trachyte.

PERALUMINOUS. A chemical term used for rocks in which molecular $Al_2O_3 > (CaO + Na_2O + K_2O)$. This produces corundum *(C)* in the CIPW norm and typically such minerals as muscovite, corundum, tourmaline, topaz, almandine-spessartine in the mode. *(Shand, 1927, p.128)*

PERIDOTEID. An obsolete field term for a coarse-grained igneous rock consisting of olivine with or without pyroxene, amphibole or biotite. Feldspar is absent. *(Johannsen, 1911, p.321; from the French péridot = olivine; Tomkeieff p.427)*

PERIDOTIDE. A revised spelling recommended to replace the field term peridoteid. Now obsolete. *(Johannsen, 1926, p.182; Johannsen v.1, p.57)*

PERIDOTITE. A collective term for ultramafic rocks consisting essentially of olivine with pyroxene and/or amphibole. Now defined modally in the ultramafic rock classification (Fig. 2.9, p.28). *(Cordier, 1842, vol.9, p.619; from the French péridot = olivine; Tröger 723; Johannsen v.4, p.401; Tomkeieff p.427)*

PERIDOTITIC KOMATIITE. See komatiite.

PERIDOTITOID. An obsolete term proposed for

an eclogitic rock similar to peridotite. *(Holmquist, 1908, p.292; Tomkeieff p.427)*

PERIDOTOID. A group name proposed for rocks composed essentially of olivine, pyroxene and iron ore. *(Gümbel, 1888, p.88; Tomkeieff p.427)*

PERKNIDE. A name recommended to replace the field term pyriboleid. Now obsolete. *(Johannsen, 1926, p.182; Johannsen v.1, p.57; Tomkeieff p.428)*

PERKNITE. A collective name for ultramafic olivine-free rocks composed essentially of amphibole and/or pyroxene or biotite. *(Turner, 1901, p.507; from the Greek perknos = dark especially of fruit or bird; Tröger 954; Johannsen v.4, p.399; Tomkeieff p.429)*

PERLITE. A term used for volcanic glasses which exhibit numerous concentric cracks so that when fragmented the pieces vaguely resemble pearls. Some are high in water content and expand when heated. The term is synonymous with pearlstone and perlstein. *(Beudant, 1822, p.360; from the French perle = pearl; in German Perlstein; Tröger 955; Johannsen v.2, p.284; Tomkeieff p.429)*

PERLSTEIN. The original German term for certain volcanic glasses now called perlite in English. Although the term is attributed to the reference cited, only the term perlartig is used. *(Fichtel, 1791, p.365; from the German Perl = pearl, Stein = rock; Tröger 955; Johannsen v.2, p.284)*

PERTHITOPHYRE. An obsolete name for a variety of anorthosite or leucocratic monzogabbro containing microperthite as an interstitial filling. *(Chrustschoff, 1888, p.476; Horoski, Volhynien (Volyn), Ukraine; Tröger 956; Johannsen v.3, p.126; Tomkeieff p.430)*

PERTHOSITE. A variety of alkali feldspar syenite consisting almost entirely of perthite. *(Phemister, 1926, p.41; Tröger 164; Johannsen v.3, p.43; Tomkeieff p.430)*

PETRISCO. A local term for a variety of leucite trachyte from the Viterbo and Lake Vico areas, Italy. *(Rath, 1868, p.297; Tomkeieff p.431)*

PHANEREID. An obsolete field term for rocks whose different constituents can be seen megascopically. *(Johannsen, 1911, p.318; from the Greek phaneros = distinct, visible, clear; Tomkeieff p.433)*

PHANERIDE. A revised spelling recommended to replace the field term phanereid. Now obsolete. *(Johannsen, 1926, p.182; Tomkeieff p.433)*

PHANERODACITE. A term for a dacite containing the excess silica as quartz and not in glass. Cf. cryptodacite. *(Belyankin, 1923, p.100; Tomkeieff p.433)*

PHENO-. Suggested as an optional prefix (p.5) for giving a provisional name to a volcanic rock based on the visible minerals that can be identified, usually phenocrysts. For example a glassy rock with phenocrysts of quartz and plagioclase could be provisonally called a pheno-dacite. *(Niggli, 1931, p.357)*

PHONOBASANITE. A synonym for phonolitic basanite of QAPF field 13 (Fig. 2.11, p.31). *(Streckeisen, 1978, p.6)*

PHONOFOIDITE. A synonym for phonolitic foidite of QAPF field 15a (Fig. 2.11, p.31). *(Streckeisen, 1978, p.7)*

PHONOLEUCITITE. A synonym for phonolitic leucitite of QAPF field 15a (Fig. 2.11, p.31). *(Rittmann, 1973, p.135)*

PHONOLITE. Now defined in QAPF field 11 (Fig. 2.11, p.31) in the sense of Rosenbusch (1877, p.234) as a volcanic rock consisting essentially of alkali feldspar and any foids. If nepheline is the only foid then the term phonolite may be used by itself but if, for example, leucite is the most abundant foid then the term leucite phonolite should be used etc. If modes are not available phonolite is defined chemically in TAS field Ph (Fig. 2.14, p.35). *(Cordier, 1816, p.151; from the Greek phone = sound, lithos = stone; Tröger 465; Johannsen v.4, p.120; Tomkeieff p.435)*

PHONOLITIC BASANITE. A collective term for alkaline basaltic rocks that are the volcanic equivalent to foid monzodiorites or monzogabbros and consist of plagioclase, feldspathoid, olivine, augite and often minor sanidine. If the amount of olivine is less than 10% it is a phonolitic tephrite. Now defined modally in QAPF field 13 (Fig. 2.11, p.31). *(Streckeisen, 1978, p.6)*

PHONOLITIC FOIDITE. A collective term for alkaline volcanic rocks consisting of foids with some alkali feldspar as defined modally in QAPF field 15a (Fig. 2.11, p.31). If possible the most abundant foid should be used in the name, e.g. phonolitic nephelinite, phonolitic leucitite. *(Streckeisen, 1978, p.7)*

PHONOLITIC LEUCITITE. Now defined modally in the leucite-bearing rock classification (section 2.8, p.18) as a volcanic rock falling into QAPF field 15a and consisting of leucite, clinopyroxene, minor olivine and with plagioclase < sanidine. Other foids may be present in minor amounts. *(Streckeisen, 1978, p.7)*

PHONOLITIC NEPHELINITE. A volcanic rock consisting of nepheline with some alkali feldspar as defined modally in QAPF field 15a (Fig. 2.11, p.31). Other foids may be present in minor amounts. *(Streckeisen, 1978, p.7)*

PHONOLITIC TEPHRITE. A collective term for alkaline volcanic rocks consisting of plagioclase, feldspathoid, augite and often minor olivine and sanidine. If the amount of olivine is greater than 10% it is a phonolitic basanite. Now defined modally in QAPF field 13 (Fig. 2.11, p.31). *(Streckeisen, 1978, p.6)*

PHONOLITOID. A term originally used as a collective name for phonolites and leucitophyres. Later used for a rock with the composition of phonolite but without modal nepheline (Lacroix, 1923). Now proposed for preliminary use in the QAPF "field" classi-

fication (Fig. 2.19, p.39) for volcanic rocks thought to contain essential foids and in which alkali feldspar is thought to be more abundant than plagioclase. *(Gümbel, 1888, p.86; Tomkeieff p.435)*

PHONOLITOID TEPHRITE. A term used for a rock consisting of sandine, nepheline, plagioclase, hornblende and olivine, i.e. between phonolite and tephrite. *(Rosenbusch, 1908, p.1375; Tomkeieff p.435)*

PHONONEPHELINITE. A synonym for phonolitic nephelinite of QAPF field 15a (Fig. 2.11, p.31). *(Rittmann, 1973, p.135)*

PHONOTEPHRITE. A synonym for phonolitic tephrite of QAPF field 13 (Fig. 2.11, p.31), and also defined chemically in TAS field U2 (Fig. 2.14, p.35). *(Rittmann, 1973, p.134)*

PHOSCORITE. A magnetite, olivine, apatite rock usually associated with carbonatites. The name is a mnemonic name from *phos*phate rock around a *co*re of carbonatite. *(Russell et al., 1955, p.199; Tomkeieff p.435)*

PICOTITITE. A rock consisting essentially of picotite (85%) with small amounts of serpentine. *(Tröger, 1935, p.308; Tröger 774; Tomkeieff p.437)*

PICRITE. A term originally used for a variety of dolerite or basalt extremely rich in olivine and pyroxene. Also used as the volcanic equivalent of a feldspar-bearing or alkali peridotite and for olivine-rich varieties of gabbro and teschenite. Now defined chemically in the TAS classification (Fig. 2.13, p.34). *(Tschermak, 1866, p.262; from the Greek pikros = bitter, referring to high MgO; Tröger 743; Johannsen v.4, p.432; Tomkeieff p.437)*

PICRITE BASALT. A melanocratic variety of olivine-rich basalt containing abundant phenocrysts of olivine in a sparse groundmass of augite, labradorite, opaques and interstitial glass. *(Quensel, 1912, p.265; Masafuera, Juan Fernandez Islands, Pacific Ocean; Tröger*

410; Johannsen v.3, p.334; Tomkeieff p.438)

PICROBASALT. A chemical term for volcanic rocks, which will include certain picritic and accumulative rocks, which was introduced for TAS field Pc (Fig. 2.14, p.35). *(Le Maitre, 1984, p.245)*

PICROPHYRE. An obsolete term for a variety of minette (a lamprophyre) containing augite and small amounts of olivine. *(Bořický, 1878, p.494; Libšice, Bohemia, Czech Republic; Tröger 958; Tomkeieff p.438)*

PIENAARITE. A mafic variety of nepheline syenite containing abundant titanite, aegirine-augite and anorthoclase. *(Brouwer, 1909, p.563; Pienaar River, Transvaal, South Africa; Tröger 488; Johannsen v.4, p.118; Tomkeieff p.438)*

PIKEITE. An obsolete name for a variety of phlogopite peridotite consisting of olivine phenocrysts and poikilitic phlogopite with inclusions of olivine and augite. *(Johannsen, 1938, p.427; Pike County, Arkansas, USA; Tröger(38) 721; Tomkeieff p.439)*

PIKROPTOCHE (PYCROPTOCHE). An obsolete adjectival term for igneous rocks poor in lime and magnesia. *(Loewinson-Lessing, 1901, p.118; Tomkeieff p.471)*

PILANDITE. A local term for a porphyritic variety of alkali feldspar syenite called hatherlite. *(Henderson, 1898, p.48; Pilansberg (now Pilanesberg), Bushveld, South Africa; Tröger 202; Johannsen v.3, p.31; Tomkeieff p.439)*

PIPERNO. A local name for a trachytic rock exhibiting eutaxitic texture in the form of light and dark streaks resembling flames. *(Lorenzo, 1904, p.301; Tröger 960; Tomkeieff p.440)*

PISSITE. An obsolete term for a pitchstone with a high melting point. *(Delamétherie, 1795, p.461; Tomkeieff p.441)*

PITCHSTONE. A volcanic glass with a lustre resembling pitch and usually containing a few phenocrysts and a water content between 4% and 10%. This is unlike obsidian which

usually has a water content < 1%. *(Babington, 1799, p.94; from the German Pechstein; Johannsen v.2, p.280; Tomkeieff p.441)*

PLÄDORITE. An obsolete term for a biotite hornblende granite, later renamed plethorite. *(Lang, 1877, p.156; Tomkeieff p.445)*

PLAGIAPLITE. A leucocratic diorite aplite with abundant zoned sodic plagioclase and only minor quartz. With more quartz it passes into gladkaite. *(Duparc & Jerchoff, 1902, p.307; Koswinski Mts, Urals, Russian Federation; Tröger 303; Johannsen v.3, p.184; Tomkeieff p.442)*

PLAGIDACITE. A special term used as a replacement for quartz andesite and now for dacites of QAPF field 5 (Fig. 2.11, p.31), having the composition of tonalites. *(Rittmann, 1973, p.133)*

PLAGIOCLASE GRANITE. An obsolete term for a plutonic rock consisting of oligoclase or andesine, quartz and less than 10% of biotite and hornblende. A synonym for trondhjemite or leuco-tonalite. *(Högbom, 1905, p.221; Tortola & Virgin Gorda, Virgin Islands, Lesser Antilles; Tröger 130; Johannsen v.2, p.382)*

PLAGIOCLASE-BEARING HORNBLENDE PYROXENITE. A term for ultramafic plutonic rocks composed mainly of pyroxene with up to 50% amphibole and minor amounts of plagioclase and often quartz. Now defined modally in the ultramafic rock classification (Fig. 2.9, p.28). *(Streckeisen, 1973, p.27)*

PLAGIOCLASE-BEARING HORNBLENDITE. An ultramafic plutonic rock consisting essentially of amphibole with minor amounts of plagioclase. Now defined modally in the ultramafic rock classification (Fig. 2.9, p.28). *(Streckeisen, 1973, p.27)*

PLAGIOCLASE-BEARING PYROXENE HORNBLENDITE. A term for ultramafic plutonic rocks composed mainly of amphibole with up to 50% pyroxene and minor amounts of plagioclase and often quartz. Now defined modally in the ultramafic rock classification (Fig. 2.9, p.28). *(Streckeisen, 1973, p.27)*

PLAGIOCLASE-BEARING PYROXENITE. An ultramafic plutonic rock consisting essentially of pyroxene with minor amounts of plagioclase. Now defined modally in the ultramafic rock classification (Fig. 2.9, p.28). *(Streckeisen, 1973, p.27)*

PLAGIOCLASITE. A term originally used as a group name for a plagioclase-enriched gabbro. *(Viola, 1892, p.121; Valle del Sinni, Basilicata, Italy; Tröger 961; Tomkeieff p.442)*

PLAGIOCLASOLITE. An obsolete group name for plagioclase rocks ranging from albitite to gabbro. *(Lacroix, 1933, p.191; Tröger 962)*

PLAGIOGRANITE. A contraction of plagioclase granite commonly used in the USSR for a plutonic rock consisting of oligoclase or andesine, quartz and less than 10% of biotite and hornblende. May be used as a synonym for trondhjemite and leucocratic tonalite of QAPF field 5 (p.23). *(Zavaritskii, 1955, p.272)*

PLAGIOLIPARITE. A variety of liparite containing phenocrysts of sodic plagioclase. *(Duparc & Pearce, 1900, p.57; Cape Marsa, Ménerville, Algeria; Tröger 43; Tomkeieff p.442)*

PLAGIOPHYRE. An obsolete group name for fine-grained intrusive rocks consisting of zoned plagioclase with subordinate altered mafic minerals. *(Tyrrell, 1911, p.77; Tröger 120; Tomkeieff p.442)*

PLAGIOPHYRITE. A term used for a porphyritic andesite or microdiorite. *(Szentpétery, 1935, p.26; from a contraction of plagioclase-porphyrite; Tomkeieff p.443)*

PLANOPHYRE. A porphyritic rock in which the phenocrysts occur in layers. *(Iddings, 1909, p.224; Tomkeieff p.443)*

PLATEAU BASALT. A general term for basaltic lavas occurring in continental regions as vast accumulations of subhorizontal flows that have been erupted, probably from fissures, in rapid succession on a regional scale. Many of the lavas are transitional basalts, but tholeiitic basalts are also common. The term is synonymous with flood basalt. *(Geikie, 1903, p.763)*

PLAUENITE. An alternative term that was suggested, but not adopted, for the type syenite from Plauenschen Grund. *(Brögger, 1895, p.59; Plauenscher Grund, Dresden, Saxony, Germany; Tröger 964; Johannsen v.3, p.54; Tomkeieff p.444)*

PLETHORITE. An obsolete name for a biotite hornblende granite, previously called plädorite. *(Zirkel, 1894a, p.34; Tomkeieff p.445)*

PLUMASITE. A plutonic rock consisting largely of oligoclase with lesser amounts of corundum. *(Lawson, 1903, p.219; Spanish Peak, Plumas County, California, USA; Tröger 311; Johannsen v.3, p.185; Tomkeieff p.446)*

PLUTONIC. A loosely defined term pertaining to those igneous processes that occur at considerable depth below the surface of the Earth. Plutonic rocks are usually coarse-grained, but not all coarse-grained rocks are plutonic. *(Kirwan, 1794, p.455; named after Pluto, Greek god of the infernal regions; Johannsen v.1, p.188; Tomkeieff p.446)*

PLUTONITE. A term proposed for deep-seated rocks. *(Scheerer, 1862, p.138; Johannsen v.1 (2nd Edn), p.275; Tomkeieff p.447)*

PLUTOVOLCANITE. An obsolete term for rocks that have been called hypabyssal, i.e. with characteristics between plutonic and volcanic. *(Scheerer, 1864, p.403; Tomkeieff p.447)*

POENEITE. A variety of spilite in which K_2O is much greater than Na_2O. *(Roever, 1940, p.263; River Noil Poene, Moetis region, W. Timor, Indonesia)*

POGONITE. An obsolete term for what is now called Pélé's hair. *(Haüy, 1822, p.580; from the Greek pogon = beard; Tomkeieff p.449)*

POIKILEID. An obsolete field term for a porphyry. *(Johannsen, 1911, p.319; from the Greek poikilos = many-coloured, spotted; Tomkeieff p.449)*

POLLENITE. A local name for a variety of tephritic phonolite containing phenocrysts of sanidine, plagioclase, hornblende and biotite in a groundmass of nepheline, other foids, olivine and glass. Cf. tautirite. *(Lacroix, 1907, p.138; Pollena Valley, Mt Somma, Naples, Italy; Tröger 470; Johannsen v.4, p.167; Tomkeieff p.450)*

POLZENITE. A group name for olivine- and melilite-bearing rocks containing some nepheline but no augite. It includes two varieties: modlibovite and vesecite, characterized by the absence and presence of monticellite, respectively. *(Scheumann, 1913, p.728; Polzen River (now Ploučnice), N. Bohemia, Czech Republic; Tröger 965; Johannsen v.4, p.388; Tomkeieff p.451)*

PONZAITE. A local name for a variety of peralkaline trachyte containing sodic pyroxenes and amphiboles and occasionally nepheline. *(Reinisch, 1912, p.121; Ponza Island, Italy; Johannsen v.3, p.77; Tomkeieff p.452)*

PONZITE. A name given to the nepheline-free varieties of trachyte from Ponza Island. *(Washington, 1913, p.691; Ponza Island, Italy; Tröger 174; Johannsen v.3, p.77; Tomkeieff p.452)*

PORPHYRIN. An obsolete term for a porphyritic rock with feldspar phenocrysts that can only be seen with a hand lens. *(Pinkerton, 1811a, p.87)*

PORPHYRITE. Originally used as a term for a quartz-free porphyry and later as a general term for porphyritic rocks of diorite composition. *(Naumann, 1854, p.664, however, for a detailed discussion of the origin of this term see Johannsen; from the Greek porphyreos = purple; Tröger 325; Johannsen v.3, p.82; Tomkeieff p.453)*

PORPHYRON. An obsolete term for a porphyritic rock with discrete feldspar phenocrysts over

1 inch (2.5 cm) in length. *(Pinkerton, 1811a, p.88; Tomkeieff p.454)*

PORPHYRY. A general term applied to any igneous rock that contains phenocrysts in a finer-grained groundmass. It is often applied to rocks that contain two generations of the same mineral. *(Werner, 1787, p.12, however, for a detailed discussion of the origin of this term see Johannsen; from the Greek porphyreos = purple; Tröger 967; Johannsen v.3, p.81; Tomkeieff p.454)*

POTASSIC MELILITITE. A variety of melilitite now defined chemically in the TAS classification as a rock falling in field F in which normative cs (larnite) > 10%, K_2O > Na_2O and K_2O > 2% (p.38). If modal olivine > 10% the rock should be called a potassic olivine melilitite *(Woolley et al., 1996, p.176)*.

POTASSIC OLIVINE MELILITITE. A variety of olivine melilitite now defined chemically in the TAS classification as a rock falling in field F in which normative cs (larnite) > 10%, K_2O > Na_2O and K_2O > 2% (p.38). If modal olivine < 10% it should be called a potassic melilitite *(Woolley et al., 1996, p.176)*.

POTASSIC TRACHYBASALT. A term introduced for the potassic analogue of hawaiite in TAS field S1 (Fig. 2.14, p.35) to distinguish it from trachybasalt which is the collective name of the field. *(Le Maitre, 1984, p.245)*

POZZOLANA (POSSOLANA, PUZZOLANA). Probably the commonest spelling of an ancient local name for a porous variety of tuff, often pumice-rich and sometimes containing leucite, used for making hydraulic cement. *(Original reference uncertain; Pozzuoli, near Naples, Italy; Johannsen v.3, p.20; Tomkeieff p.455)*

PRASOPHYRE. An obsolete name for a greenish porphyry. *(Le Puillon de Boblaye & Virlet, 1833, p.111; Tomkeieff p.457)*

PRESELITE. A local term proposed by archaeologists for a variety of greenish dolerite containing widely spaced clusters of sodic plagioclase and used for the megaliths of the inner circle of Stonehenge. *(Keiller, 1936, p.221; Presely (now Prescelly) Hills, Dyfed, Wales, UK; Tomkeieff p.457)*

PROPYLITE. A seldom used term for rocks which have suffered propylitization or hydro-thermal alteration. Originally applied to greenstone-like rocks which were considered to be the "precursors of all other volcanic rocks" in the region. *(Richthofen, 1868, p.20; from the Greek propolos = a servant who goes before one; Tröger 970; Johannsen v.3, p.177; Tomkeieff p.461)*

PROTEROBASE. An old term for altered rocks of basaltic composition which contain primary hornblende. *(Gümbel, 1874, p.14; from the Greek proteros = earlier; Tröger 394; Johannsen v.3, p.318; Tomkeieff p.461)*

PROTOGINE. An old term given to the weakly cataclastic granites which occur abundantly in the Alps in the belief that they were part of the original crust of the Earth. Later used for the granites from the Island of Elba, Italy. *(Jurine, 1806, p.372; from the Greek protos = first, gignesthai = to be born; Johannsen v.2, p.242; Tomkeieff p.462)*

PROTOKATUNGITE. A term for a variety of katungite without olivine. *(Holmes, 1942, p.199)*

PROTOPYLITE. An obsolete name for a propylitized porphyritic diorite. *(Stache & John, 1879, p.352; Tomkeieff p.463)*

PROTOTECTITE. An igneous rock which crystallizes directly from a primary magma or its differentiation products. Cf. anatectite and syntectite. *(Loewinson-Lessing, 1934, p.7; Tomkeieff p.463)*

PROWERSITE. A term originally used for a potassic variety of minette (a lamprophyre) consisting of abundant biotite and orthoclase with lesser amounts of augite and altered olivine. *(Rosenbusch, 1908, p.1487; Prowers County, Colorado, USA; Tröger 228; Johannsen v.3, p.41; Tomkeieff p.463)*

PUGLIANITE. A local name for a melanocratic variety of leucite gabbro (or theralite) largely composed of clinopyroxene and subordinate plagioclase, with minor mica and leucite. *(Lacroix, 1917b, p.210; Pugliani, Mt Somma, Naples, Italy; Tröger 546; Johannsen v.4, p.208; Tomkeieff p.469)*

PULASKITE. A variety of nepheline-bearing alkali feldspar syenite containing alkali feldspar and varying amounts of sodic pyroxenes and amphiboles, fayalite, biotite and minor amounts of nepheline. *(Williams, 1891, p.56; Fourche Mt, Pulaski County, Arkansas, USA; Tröger 186; Johannsen v.4, p.5; Tomkeieff p.469)*

PULVERULITE. A variety of ignimbrite containing dust-like shards of glass surrounding the crystal grains. *(Marshall, 1935, p.358; from the Latin pulvis = dust; Tröger(38) 871½; Tomkeieff p.470)*

PUMICE. A textural term applied to extremely vesiculated lavas which resemble froth or foam. *(A term of great antiquity usually attributed to Theophrastus, 320 BC – see Johannsen for further discussion; from the Latin pumex = pumice; Tröger 821; Johannsen v.2, p.282; Tomkeieff p.470)*

PUMICITE. A consolidated pumice. *(Chesterman, 1956, p.5; Tomkeieff p.470)*

PUMITE. An obsolete name for a pumice consisting of feldspathic glass. *(Cordier, 1816, p.372; Tomkeieff p.470)*

PUYS-ANDESITE. A term from an obsolete chemical classification, based on feldspar composition rather than SiO_2 alone, for a class of rocks in which $CaO : Na_2O : K_2O \approx 2 : 2 : 1$ and $SiO_2 \approx 63\%$. *(Lang, 1891, p.229; Chaine des Puys, Auvergne, France; Tomkeieff p.471)*

PUZZOLANA. See pozzolana.

PYCROPTOCHE. See pikroptoche.

PYNOSITE. A mnemonic name suggested for a rock consisting essentially of *py*roxene and *nos*ean. *(Belyankin, 1929, p.22)*

PYRIBOLEID. An obsolete field term for a coarse-grained igneous rock consisting almost entirely of pyroxene and/or amphibole and/or biotite. It includes the pyroxeneids and amphiboleids. *(Johannsen, 1911, p.320; Tomkeieff p.471)*

PYRITOSALITE. An obsolete term for an extremely quartz-rich rock containing pyrites. *(Brögger, 1931, p.126; Tofteholmen, Hurum, Oslo district, Norway; Tröger 3; Tomkeieff p.471)*

PYROCLASTIC. Adjective applied to fragmental rocks produced as the result of volcanic eruption. *(Teall, 1887, p.493; from the Greek pyr = fire, clastos = broken; Johannsen v.1 (2nd Edn), p.230; Tomkeieff p.472)*

PYROCLASTIC BRECCIA. Now defined in the pyroclastic classification (section 2.2.2, p.7 and Fig. 2.1, p.8) as a pyroclastic rock in which blocks > 75%. Cf. agglomerate.

PYROCLASTIC DEPOSIT. Now defined in the pyroclastic classification (section 2.2.2, p.7) as a general term to include both consolidated and unconsolidated assemblages of pyroclasts, which must make up more than 75% of the rock. *(Schmid, 1981, p.42)*

PYROCLASTIC ROCK. A collective term used in the pyroclastic classification (section 2.2.2, p.7) for pyroclastic deposits that are predominatly consolidated. Cf. tephra.

PYROCLASTS. Now defined in the pyroclastic classification (section 2.2.1, p.7) as fragments generated by disruption as a direct result of volcanic action.

PYROLITE. A term originally proposed for rocks formed at elevated temperatures within the Earth. Ringwood (1962) later used the term for an assumed Upper Mantle composition of one part basalt and four parts dunite. *(Preobrazhenskii, 1956, p.322; Tomkeieff p.473)*

PYROMERIDE. Originally applied to an orbicular diorite from Corsica described from the collection of Haüy. Later applied to spherulitic devitrified rhyolites having a nodular appearance. *(Monteiro, 1814, p.359, cites Brongniart as the original author of the name but gives no reference; from the Greek pyr = fire, meros = part, referring to the fact that feldspar is easily melted under the blowpipe, but quartz is not; Tröger 45; Tomkeieff p.473)*

PYROXENE HORNBLENDE GABBRO. Now defined modally in the gabbroic rock classification (Fig. 2.6, p.25) as a variety of gabbro in which hornblende is between 5% and 85%. *(Streckeisen, 1973, p.27)*

PYROXENE HORNBLENDE GABBRONORITE. Now defined modally in the gabbroic rock classification (Fig. 2.6, p.25) as a variety of gabbronorite in which hornblende is between 5% and 85%. *(Streckeisen, 1973, p.27)*

PYROXENE HORNBLENDE NORITE. Now defined modally in the gabbroic rock classification (Fig. 2.6, p.25) as a variety of norite in which hornblende is between 5% and 85%. *(Streckeisen, 1973, p.27)*

PYROXENE HORNBLENDE PERIDOTITE. An ultramafic plutonic rock consisting of 40%–90% olivine and various amounts of pyroxene and amphibole. Now defined modally in the ultramafic rock classification (Fig. 2.9, p.28). *(Streckeisen, 1973, p.26)*

PYROXENE HORNBLENDITE. A term for ultramafic plutonic rocks composed mainly of amphibole with up to 50% pyroxene. Now defined modally in the ultramafic rock classification (Fig. 2.9, p.28). *(Streckeisen, 1973, p.26)*

PYROXENE MELILITOLITE. Originally defined as an ultramafic plutonic rock consisting essentially of melilite and clinopyroxene with minor olivine (see 1st Edition, Fig. B.3, p.12), but not required by the new melilite-bearing rocks classification (section 2.4.1, p.11). *(Streckeisen, 1978, p.13)*

PYROXENE OLIVINE MELILITOLITE. Originally defined as an ultramafic plutonic rock consisting essentially of melilite, olivine and lesser amounts of clinopyroxene (see 1st Edition, Fig. B.3, p.12), but not required by the new melilite-bearing rocks classification (section 2.4.1, p.11). *(Streckeisen, 1978, p.13)*

PYROXENE PERIDOTITE. A term for ultramafic plutonic rocks composed mainly of olivine with up to 50% pyroxene. Now defined modally in the ultramafic rock classification (Fig. 2.9, p.28). *(Wyllie, 1967, p.2)*

PYROXENEID. An obsolete field term for a coarse-grained igneous rock consisting almost entirely of pyroxene. *(Johannsen, 1911, p.320; Tomkeieff p.474)*

PYROXENIDE. A revised spelling recommended to replace the field term pyroxeneid. Now obsolete. *(Johannsen, 1926, p.182; Johannsen v.1, p.57)*

PYROXENITE. A collective term named simultaneously in 1857 by Senft and Coquand for ultramafic plutonic rocks composed almost entirely of one or more pyroxenes and occasionally biotite, hornblende and olivine. Now defined modally in the ultramafic rock classification (Fig. 2.9, p.28). *(Senft, 1857, p.42, and Coquand, 1857, p.114; Tröger 673; Johannsen v.4, p.400; Tomkeieff p.474)*

PYROXENOLITE. A term synonymous with pyroxenite. *(Lacroix, 1895, p.752; Tröger 693; Johannsen v.4, p.400; Tomkeieff p.474)*

PYTERLITE. A local name for a variety of rapakivi granite in which the ovoids of orthoclase are not mantled by plagioclase. *(Wahl, 1925, p.60; Pyterlahti, Virolahti, Finland; Tröger 79; Tomkeieff p.474)*

QUARTZ ALKALI FELDSPAR SYENITE. A felsic plutonic rock composed mainly of alkali feldspar, quartz and mafic minerals. Now defined modally in QAPF field 6* (Fig. 2.4, p.22). *(Streckeisen, 1973, p.26)*

QUARTZ ALKALI FELDSPAR TRACHYTE. A felsic volcanic rock composed mainly of alkali feldspar, quartz and mafic minerals. Now defined modally in QAPF field 6* (Fig.

2.11, p.31). *(Streckeisen, 1978, p.4)*

QUARTZ ANORTHOSITE. A leucocratic plutonic rock consisting essentially of calcic plagioclase, quartz and small amounts of pyroxene. Now defined modally in QAPF field 10* (Fig. 2.4, p.22). *(Loughlin, 1912, p.108; Preston, Connecticut, USA; Tröger 128)*

QUARTZ DIOREID. An obsolete field term for a variety of dioreid containing quartz. *(Johannsen, 1911, p.320)*

QUARTZ DIORIDE. An obsolete field term for a variety of dioride containing quartz. *(Johannsen, 1931, p.57)*

QUARTZ DIORITE. The term was originally used for plutonic rocks consisting essentially of plagioclase, quartz and mafic minerals falling in field 5 of the QAPF, which are now called tonalite. Now defined modally in QAPF field 10* (Fig. 2.4, p.22). *(Zirkel, 1866b, p.4; Tröger 131; Johannsen v.2, p.378)*

QUARTZ DOLEREID. An obsolete field term for a variety of dolereid containing quartz. *(Johannsen, 1911, p.320)*

QUARTZ DOLERITE. A variety of dolerite composed mainly of plagioclase and pyroxenes with interstitial quartz or micropegmatite. The rock has tholeiitic affinities and its pyroxenes are usually subcalcic augite accompanied by pigeonite or orthopyroxene. *(Tyrrell, 1926, p.120; Tröger 151)*

QUARTZ GABBREID. An obsolete field term for a variety of gabbreid containing quartz. *(Johannsen, 1911, p.320)*

QUARTZ GABBRIDE. An obsolete field term for a variety of gabbride containing quartz. *(Johannsen, 1931, p.57)*

QUARTZ GABBRO. A plutonic rock composed mainly of calcic plagioclase, clinopyroxene and quartz. Now defined modally in QAPF field 10* (Fig. 2.4, p.22). *(Johannsen, 1932, p.409; Tröger 133; Tomkeieff p.476)*

QUARTZ LATITE. A term originally used for a volcanic rock composed of phenocrysts of quartz, plagioclase, biotite and hornblende in a glassy matrix potentially of quartz and alkali feldspar. Now commonly used for volcanic rocks composed of alkali feldspar and plagioclase in roughly equal amounts, quartz and mafic minerals. Now defined modally in QAPF field 8* (Fig. 2.11, p.31). *(Ransome, 1898, p.372; Tröger 100)*

QUARTZ LEUCOPHYRIDE. An obsolete field term for a variety of leucophyride containing quartz. *(Johannsen, 1931, p.58)*

QUARTZ MONZODIORITE. A plutonic rock consisting essentially of sodic plagioclase, alkali feldspar, quartz and mafic minerals. Now defined modally in QAPF field 9* (Fig. 2.4, p.22). *(Streckeisen, 1973, p.26)*

QUARTZ MONZOGABBRO. A plutonic rock consisting essentially of calcic plagioclase, alkali feldspar, mafic minerals and quartz. Now defined modally in QAPF field 9* (Fig. 2.4, p.22). *(Streckeisen, 1973, p.26)*

QUARTZ MONZONITE. A plutonic rock consisting of approximately equal amounts of alkali feldspar and plagioclase and with essential quartz (5%–20% of felsic minerals) but not enough to make the rock a granite. Now defined modally in QAPF field 8* (Fig. 2.4, p.22). The term was formerly used for granites of QAPF field 3b. *(Brögger, 1895, p.61; Tröger 86; Tomkeieff p.476)*

QUARTZ NORITE. A plutonic rock composed essentially of calcic plagioclase, quartz and orthopyroxene. Now defined modally as a variety of gabbro in QAPF field 10* (Fig. 2.4, p.22). *(Johannsen, 1932, p.409; Tröger 134)*

QUARTZPORPHYRITE. An aphanitic rock of dacite composition containing phenocrysts of quartz and plagioclase in a glassy groundmass. Originally used by European petrologists only if such rocks were pre-Tertiary. *(Brögger, 1895,* p.60; Tröger 149; Johannsen v.2, p.396)

QUARTZPORPHYRY. An aphanitic rock of rhyolite composition containing phenocrysts of quartz and orthoclase in a glassy groundmass. Originally used by European petrologists only if such rocks were pre-Tertiary. *(Durocher, 1845, p.1281; Messangé, Brittany, France; Tröger 41; Johannsen v.2, p.286; Tomkeieff p.477)*

QUARTZ SYENITE. A plutonic rock consisting essentially of alkali feldspar, quartz and mafic minerals. Now defined modally in QAPF field 7* (Fig. 2.4, p.22). *(Streckeisen, 1973, p.26; Tröger 239)*

QUARTZ TRACHYTE. A volcanic rock consisting of phenocrysts of alkali feldspar and quartz in a cryptocrystalline or glassy matrix. It is the volcanic equivalent of quartz syenite. Now defined modally in QAPF field 7* (Fig. 2.11, p.31). *(Hauer & Stache, 1863, p.70; Tröger 50; Johannsen v.2, p.266; Tomkeieff p.477)*

QUARTZ-RICH GRANITOID. A collective term for granitic rocks having a quartz content greater than 60% of the felsic minerals. Now defined modally in QAPF field 1b (Fig. 2.4, p.22). *(Streckeisen, 1973, p.26)*

QUARTZOLITE. A collective term for plutonic rocks in which the quartz content is more than 90% of the felsic minerals. Now defined modally in QAPF field 1a (Fig. 2.4, p.22). *(Streckeisen, 1973, p.26; Tomkeieff p.476)*

RAABSITE. A local name for a variety of minette (a lamprophyre) consisting of microcline, sodic amphibole, biotite and olivine. *(Hackl & Waldmann, 1935, p.272; Raabs, Lower Austria, Austria; Tröger(38) 229½; Tomkeieff p.479)*

RADIOPHYRE. An obsolete term for a porphyry with a subspherulitic texture. *(Bořický, 1882, p.74; Tomkeieff p.480)*

RADIOPHYRITE. An obsolete term for a radiophyre in which Na > K. *(Bořický, 1882,* p.119; Tomkeieff p.480)

RAFAELITE. A variety of analcime-bearing microsyenite containing amphibole and labradorite, but no nepheline. *(Johannsen, 1938, p.177; San Rafael Swell, Emery, Utah, USA; Tröger(38) 508½; Tomkeieff p.480)*

RAGLANITE. A leucocratic variety of corundum-bearing nepheline diorite with oligoclase and biotite. *(Adams & Barlow, 1908, p.62; Raglan, Renfrew County, Ontario, Canada; Tröger 549; Johannsen v.4, p.213; Tomkeieff p.480)*

RAPAKIVI. A term, used as both an adjective and noun, for a variety of granite usually containing amphibole and biotite and characterized by the presence of large oval grains (ovoids) of orthoclase which are usually mantled by plagioclase. Two generations of quartz are often present. *(Hjärne, 1694, p.V, No. 11; from the Finnish word for "rotten stone"; Johannsen v.2, p.243; Tomkeieff p.481)*

RAQQAITE. A local term for an extrusive rock composed of phenocrysts of diopside and hypersthene (= enstatite), olivine, magnetite and apatite in a light-coloured analcime-like mesostasis containing some needles of natrolite. May be related to meimechite. *(Eigenfeld, 1965, p.46; Raqqa, E. of Aleppo, Syria)*

RAUHAUGITE. A poorly defined local name for a carbonatite composed mainly of dolomite. Now called dolomite carbonatite. *(Brögger, 1921, p.252; Rauhaug, Fen Complex, Telemark, Norway; Tröger 758; Tomkeieff p.481)*

REDWITZITE. A local group name for lamprophyric rocks of variable composition characterized by the presence of large biotite phenocrysts. *(Willmann, 1919, p.5; Redwitz, Fichtelgebirge, Germany; Tröger 975; Tomkeieff p.483)*

REGADITE. An obsolete name for a variety of biotite clinopyroxenite which occurs as dykes and consists of diallage (= altered diopside) and biotite in equal amounts. *(Cotelo Neiva, 1947b, p.127; A Regada, Bragança district, Portugal; Tomkeieff p.484)*

REGENPORPHYRY. An obsolete local term for a granite porphyry containing pseudomorphs after cordierite. *(Gümbel, 1868, p.420; Regen, Bavaria, Germany; Tröger 959)*

RESINITE. An obsolete name for a variety of obsidian. More usually used for a component of coal. *(Haüy, 1822, p.580; Tenerife, Canary Islands, North Atlantic Ocean; Tomkeieff p.487)*

RETICULITE. An extremely vesiculated variety of pumice consisting of a broken network of glass threads remaining after the vesicle walls collapse. *(Wentworth & Williams, 1932, p.47; from the Latin reticulum = small net; Tomkeieff p.488)*

RETINITE. An early term for obsidian, preceding the use of the term for fossil resin. Now used for water-rich rhyolite glasses which may have expanding properties. *(Dolomieu, 1794, p.103; from the Greek retina = resin; Tomkeieff p.488)*

RHAZITE. An obsolete term for a fine-grained basalt containing a ferromagnesian mineral. *(Pinkerton, 1811b, p.45; named after Rhazes, AD 900)*

RHENOAPLITE. An erroneous spelling of rhenopalite. *(Tomkeieff et al., 1983, p.489; Tomkeieff p.489)*

RHENOPALITE. An obsolete group name for various aplitic rocks cutting gabbros, perhaps as magmatic segregation products. *(Schuster, 1907, p.43; named after Rheinpfalz, Germany; Tröger 976)*

RHOMBPORPHYRY. A term applied to rocks which vary in composition from porphyritic trachytes to trachyandesites and carry rhomb-shaped phenocrysts of ternary feldspar. *(Buch, 1810, p.106; Oslo region, Norway; Tröger 203; Johannsen v.4, p.26; Tomkeieff p.490)*

RHÖN BASALT. A term from an obsolete chemical classification, based on feldspar composition rather than SiO_2 alone, for basaltic rocks in which $CaO : Na_2O : K_2O \approx 10 : 2.3 : 1$ and $SiO_2 \approx 40\%$. *(Lang, 1891, p.237; Tomkeieff p.490)*

RHYOBASALT. A term suggested for the volcanic equivalent of a granogabbro in order to maintain similar constructions between volcanic and plutonic rock names. *(Johannsen, 1920c, p.211; Tröger 119; Johannsen v.2, p.369; Tomkeieff p.490)*

RHYODACITE. A term used for volcanic rocks intermediate between rhyolite and dacite, usually consisting of phenocrysts of quartz, plagioclase and a few ferromagnesian minerals in a microcrystalline groundmass. *(Winchell, 1913, p.214; Tröger 118; Johannsen v.2, p.356; Tomkeieff p.490)*

RHYOLITE. A collective term for silicic volcanic rocks consisting of phenocrysts of quartz and alkali feldspar, often with minor plagioclase and biotite, in a microcrystalline or glassy groundmass and having the chemical composition of granite. Now defined modally in QAPF field 3 (Fig. 2.11, p.31) and, if modes are not available, chemically in TAS field R (Fig. 2.14, p.35). *(Richthofen, 1860, p.156; from the Greek rhein = to flow; Tröger 40; Johannsen v.2, p.265; Tomkeieff p.490)*

RHYOLITOID. Originally used as a rock with the chemical composition of rhyolite but without modal quartz. Now proposed for preliminary use in the QAPF "field" classification (Fig. 2.19, p.39) for volcanic rocks tentatively identified as rhyolite. *(Lacroix, 1923, p.2; Tomkeieff p.491)*

RIACOLITE. A local name for a sanidine trachyte. *(Tomkeieff et al., 1983, p.491; Puy de Dôme, Auvergne, France)*

RICOLETTAITE. A local name for a slightly alkaline variety of gabbro with very calcic plagioclase, titanian augite, a little biotite, olivine and opaques. *(Johannsen, 1920c,*

p.224; Ricoletta, Mt Monzoni, Italy; Tröger 359; Johannsen v.3, p.131; Tomkeieff p.491)

RIEDENITE. A local name for a melanocratic variety of foidolite (noseanolite) consisting of large biotite tablets in a granular aggregate of aegirine-augite, nosean and biotite. Cf. boderite and rodderite. *(Brauns, 1922, p.76; Rieden, Eifel district, near Koblenz, Germany; Tröger 637; Johannsen v.4, p.332; Tomkeieff p.492)*

RIKOTITE. A local name for a melanocratic banded variety of diorite composed mainly of clinopyroxene and anorthoclase with minor andesine, hornblende, biotite and opaques. *(Belyankin & Petrov, 1945, p.180; Rikotsky Gorge, River Dzirula, Georgia; Tomkeieff p.492)*

RINGITE. A local name for a coarse-grained carbonatite containing aegirine and alkali feldspar, considered to be a mixture of carbonatite and syenitic fenite. *(Brögger, 1921, p.199; Ringsevja, Fen Complex, Telemark, Norway; Tröger 754; Tomkeieff p.493)*

RISCHORRITE. A variety of biotite-bearing nepheline syenite in which the nepheline crystals are poikilitically enclosed in microcline perthite. Aegirine-augite, apatite and opaques are often abundant. *(Kupletskii, 1932, p.36; Rischorr Plateau, Khibina, Kola Peninsula, Russian Federation; Tröger(38) 413½; Tomkeieff p.493)*

RIZZONITE. An obsolete name for a limburgitic dyke rock containing phenocrysts of titanian augite, olivine and opaques in a plentiful glassy base. *(Doelter, 1902, p.977; Mt Rizzoni, Mt Monzoni, Italy; Tröger 375; Tomkeieff p.494)*

ROCKALLITE. A local name for a variety of peralkaline granite containing abundant aegirine-augite and albite. *(Judd, 1897, p.57; named after the Island of Rockall, North Atlantic Ocean; Tröger 58; Johannsen v.2, p.41; Tomkeieff p.494)*

RØDBERG. An altered carbonatite composed of hematite-stained granular calcite, dolomite and sometimes ankerite. *(Vogt, 1918, p.78; from the Norwegian rødberg = red rock; Fen Complex, Telemark, Norway; Tomkeieff p.495)*

RODDERITE. A term proposed for a melanocratic variety of nosean phonolite or noseanolite, with more nosean than alkali feldspar, biotite and augite. Cf. riedenite and boderite. Originally called a nosean sanidine biotite augitite. *(Taylor et al., 1967, p.409, and Frechen, 1971, p.42 for the description; Rodderhöfen, near Rieden, Eifel district, near Koblenz, Germany)*

RODINGITE. A metasomatically altered dyke rock, probably originally a dolerite, and composed mainly of augite and hydrogrossular. Associated with serpentinized peridotites and other rocks of the ophiolite suite. *(Bell et al., 1911, p.31; Roding River, Dun Mountain, Nelson, New Zealand; Tröger 718; Tomkeieff p.495)*

ROMANITE. An obsolete name suggested as a regional term for ciminites and vulsinites from the Roman province. *(Preller, 1924, p.154; Tomkeieff p.495)*

RONGSTOCKITE. A coarse-grained variety of foid-bearing monzodiorite with abundant titanian augite and some biotite or brown hornblende. The foid is usually nepheline or cancrinite. *(Tröger, 1935, p.124; Rongstock (now Roztoky), Ústí nad Labem, N. Bohemia, Czech Republic; Tröger 282; Johannsen v.4, p.50; Tomkeieff p.495)*

ROTENBURGITE. A name suggested for a group of amphibole diorites. *(Marcet Riba, 1925, p.293; Rotenburg, Kyffhäuser, S. of Kassel, Germany; Tomkeieff p.496)*

ROUGEMONTITE. A variety of gabbro in which the plagioclase is anorthite and titanian augite is abundant. *(O'Neill, 1914, p.64; Rougemont Mt, Quebec, Canada; Tröger 360; Johannsen v.3, p.339; Tomkeieff p.496)*

ROUTIVARITE. A local name for a variety of anorthosite composed of labradorite and minor almandine. *(Sjögren, 1893, p.62; Routivare, Norrbotten, Lapland, Sweden; Tröger 297; Johannsen v.3, p.201; Tomkeieff p.497)*

ROUVILLITE. A leucocratic variety of theralite with up to 80% plagioclase zoned from labradorite to bytownite and nepheline. *(O'Neill, 1914, p.28; Rouville County, Quebec, Canada; Tröger 551; Johannsen v.4, p.219; Tomkeieff p.497)*

RUNITE. An obsolete name for graphic granite. *(Pinkerton, 1811b, p.85; named after the texture which resembles runic characters; Tröger 38; Johannsen v.2, p.84; Tomkeieff p.498)*

RUSHAYITE. A variety of olivine melilitite rich in phenocrysts of corroded forsteritic olivine in a groundmass of abundant melilite with olivine, perovskite and ore minerals with minor nepheline and augite. *(Denaeyer, 1965, p.2119; Chaîne du Rushayo, Nyiragongo Volcano, Democratic Republic of Congo)*

RUTTERITE. A medium-grained variety of nepheline-bearing syenite, with albite and perthitic microcline and subordinate hornblende. *(Quirke, 1936, p.179; Rutter, Bigwood Township, Sudbury district, Ontario, Canada; Tröger(38) 178⅔; Johannsen v.4, p.46; Tomkeieff p.498)*

S-TYPE GRANITE. A general term for a range of granitic rocks, mainly peraluminous granodiorites and granites, characterized by the presence of muscovite, alumino-silicates, garnet and/or cordierite in addition to essential quartz, alkali feldspar and plagiolcase. Hornblende is rare. The prefix S implies that the source rocks have sedimentary (pelitic) compositions. *(Chappell & White, 1974, p.173)*

SABAROVITE. A leucocratic variety of charnoenderbite of QAPF field 4 (Fig. 2.4, p.22). It is an extreme member of the bugite series with high SiO_2, low K_2O, oligoclase-andesine antiperthite, quartz and hypersthene (= enstatite). It is suggested (Streckeisen, 1974, p.358) that this term should be abandoned. *(Bezborod'ko, 1931, p.142; Sabarovo village, near Vinnitsa, Podolia, Ukraine; Tröger(38) 130½; Tomkeieff p.499)*

SAGVANDITE. A local name for a variety of orthopyroxenite composed of bronzite (= enstatite) and small amounts of magnesite. *(Pettersen, 1883, p.247; Lake Sagvandet, Balsfjord, Tromsø, Norway; Tröger 677; Johannsen v.4, p.459; Tomkeieff p.499)*

SAIBARITE (SAJBARITE). A local name for a variety of nepheline syenite with trachytoidal texture and banded structure consisting of nepheline, aegirine, alkali feldspar and albite, with minor alkali amphibole. *(Edel'shtein, 1930, p.23; Mt Saibar, Minusinsk Depression, W. Siberia, Russian Federation)*

SAKALAVITE. A local name for a quartz normative basaltic rock containing intermediate plagioclase, augite, opaques and plentiful glass. *(Lacroix, 1923, p.15; Sakalava, Madagascar; Tröger 159; Tomkeieff p.500)*

SALIC. A name used in the CIPW normative classification for one of two major groups of normative minerals which includes quartz, feldspars and feldspathoids, as well as zircon, corundum and the sodium salts. The other group is named femic. Cf. felsic. *(Cross et al., 1902, p.573; Johannsen v.1, p.191; Tomkeieff p.500)*

SALITRITE. A local name for a variety of alkali pyroxenite consisting essentially of aegirine-augite and titanite with minor microcline. *(Tröger, 1928, p.202; Salitre Mts, Minas Gerais, Brazil; Tröger 695; Johannsen v.3, p.40; Tomkeieff p.500)*

SANCYITE. A local name for a variety of trachyte that contains many large phenocrysts of sanidine and often tridymite. *(Lacroix, 1923, p.10; Aiguilles du Sancy, Mont Dore, Auvergne, France; Tröger 52; Tomkeieff p.502)*

SANDYITE. A melanocratic variety of foid-bearing alkali feldspar syenite with hastingsite, aegirine-augite, titanite, apatite, calcite, garnet and ilmenite. Later said not to be an igneous rock but a metasomatized calcareous xenolith. *(Zavaritskii & Krizhanovskii, 1937, p.9; Sandy-Elga River, Ilmen Mts, Urals, Russian Federation; Tomkeieff p.503)*

SANIDINITE. The term has been used in igneous nomenclature for a rock consisting almost entirely of sanidine. To avoid conflict with the metamorphic usage it should not be used for igneous rocks. *(Nose, 1808, p.156; Tröger 978; Johannsen v.3, p.5; Tomkeieff p.504)*

SANIDOPHYRE. A local name for a variety of liparite with large phenocrysts of sanidine. *(Dechen, 1861, p.106; Siebengebirge, near Bonn, Germany; Tröger 980; Tomkeieff p.504)*

SANNAITE. A variety of lamprophyre composed of combinations of olivine, titanian augite, kaersutite and Ti-rich biotite phenocrysts with alkali feldspar dominating over plagioclase in the groundmass which also contains nepheline. Now defined in the lamprophyre classification (Table 2.9, p.19). *(Brögger, 1921, p.180; Sannavand, Fen Complex, Telemark, Norway; Tröger 498; Johannsen v.4, p.159; Tomkeieff p.504)*

SANTORINITE. An obsolete term originally defined as an andesitic rock with calcic plagioclase and SiO_2 65%–69% but later as a variety of hypersthene (= enstatite) andesite (Becke, 1899). *(Washington, 1897b, p.368; Santorini, Greece; Tröger 154; Johannsen v.3, p.174; Tomkeieff p.504)*

SANUKITE. A name given to a rock which was originally described as a bronzite (= enstatite) andesite consisting of needles of bronzite in a groundmass of clear glass and abundant magnetite grains. Though considered sometimes as a member of the boninite group, it differs by TiO_2 being > 0.5 wt %. *(Weinschenk, 1891, p.150; Sanuki, Shikoko Island,*

Japan; Tröger 123; Johannsen v.3, p.172; Tomkeieff p.505)

SANUKITOID. A term originally given to all textural modifications of the sanukite magma type. Then, used as a synonym for orthoandesite. Now commonly used as a plutonic equivalent of sanukite or Archean high-Mg quartz monzodiorite and granodiorite (Shirey & Hanson, 1984, p.223). *(Kotô, 1916b, p.101; Johannsen v.3, p.173; Tomkeieff p.505)*

SÄRNAITE. A leucocratic variety of cancrinite nepheline syenite with tablets of perthitic orthoclase, sometimes trachytoidal, and containing prisms of aegirine-augite. *(Brögger, 1890, p.244; Särna, Dalarne, Sweden; Tröger 426; Johannsen v.4, p.108; Tomkeieff p.506)*

SATURATED. A term applied to igneous rocks which are neither oversaturated nor undersaturated with respect to silica, i.e. they have no silica minerals or foids in the mode or norm. *(Shand, 1913, p.508; Tomkeieff p.408)*

SAXONITE. A variety of harzburgite composed of olivine and enstatite. *(Wadsworth, 1884, p.85; named after Saxony, Germany; Tröger 731; Johannsen v.4, p.434; Tomkeieff p.507)*

SCANOITE. An obsolete name for a volcanic rock consisting of phenocrysts of titanian augite and olivine in a groundmass of colourless glass with minor analcime and microlites of augite and olivine. Cf. ghizite. *(Lacroix, 1924, p.531; Scano flow, Mt Ferru, Sardinia, Italy; Tröger 600; Johannsen v.4, p.218; Tomkeieff p.508)*

SCHALSTEIN. An old term originally used for diabase tuffs and altered diabases but now restricted to bedded Palaeozoic diabase tuffs. *(Cotta, 1855, p.53; from the German schal = flat, Stein = rock; Tröger 981; Tomkeieff p.508)*

SCHILLERFELS. An obsolete term for a rock described as a bastite (= altered enstatite) peridotite. *(Raumer, 1819, p.40; Tröger 709; Tomkeieff p.508)*

SCHÖNFELSITE. An obsolete name for a variety of picritic basalt containing phenocrysts of serpentinized olivine and augite in a groundmass of augite, bronzite (= enstatite), biotite, bytownite and abundant glass. *(Uhlemann, 1909, p.434; Altschönfels, Zwickau, Saxony, Germany; Tröger 411; Johannsen v.3, p.306; Tomkeieff p.510)*

SCHORENBERGITE. A local name for a variety of leucitite containing phenocrysts of nosean and sometimes leucite in a groundmass of abundant leucite, nepheline and aegirine. Feldspar is entirely absent. *(Brauns, 1922, p.46; Schorenberg, Rieden, Laacher See, near Koblenz, Germany; Tröger 638; Johannsen v.4, p.380; Tomkeieff p.510)*

SCHRIESHEIMITE. An obsolete name for a variety of pyroxene hornblende peridotite composed of large crystals of hornblende poikilitically enclosing olivine, which is altered to serpentine and talc, with minor phlogopite and diopside. *(Rosenbusch, 1896, p.348; Schriesheim, Odenwald, Germany; Tröger 709; Johannsen v.4, p.428; Tomkeieff p.510)*

SCORIA. A highly vesiculated lava or tephra which resembles clinker. Usually of basic composition. *(Cotta, 1866, p.97; Tomkeieff p.511)*

SCYELITE. An obsolete name for a variety of olivine hornblendite consisting of hornblende, olivine and phlogopite. *(Judd, 1885, p.401; Loch Scye, Scotland, UK; Tröger 706; Johannsen v.4, p.424; Tomkeieff p.512)*

SEBASTIANITE. A local name for fragments of plutonic rocks ejected from Mt Somma, Italy, composed mainly of anorthite and biotite with minor augite and apatite. Heteromorph of puglianite with biotite instead of leucite. *(Lacroix, 1917b, p.210; San Sebastiano, Mt Somma, Naples, Italy; Tröger 363; Johannsen v.4, p.75; Tomkeieff p.512)*

SELAGITE. An obsolete name used for a variety of trachyte containing biotite. *(Haüy, 1822,* p.544; from the Greek selagein = to beam brightly; Tröger 232; Johannsen v.3, p.79; Tomkeieff p.515)

SELBERGITE. A fine-grained variety of nosean leucite syenite consisting of phenocrysts of leucite, nosean (some haüyne), sanidine and aegirine-augite in a groundmass of nepheline, alkali feldspar and aegirine. (Brauns, 1922, p.47; Selberg, Laacher See, near Koblenz, Germany; Tröger 453; Johannsen v.4, p.275; Tomkeieff p.516)

SEMEITAVITE (SEMEJTAVITE, SEMEITOVITE). A variety of quartz alkali feldspar syenite containing anorthoclase, quartz, ilmenite, augite and riebeckite. (Gornostaev, 1933, p.180; Semeitau Mts, Kazakhstan; Tomkeieff p.516)

SERPENTINITE. A rock composed almost entirely of serpentine minerals. Relics of original olivine and/or pyroxene may be present and chromite or chrome spinels are commonly present. An altered peridotite. (Wells, 1948, p.95; from the mineral serpentine named after the French serpent = snake, on account of the appearence; Tomkeieff p.518)

SESSERALITE. A local name for a variety of gabbro consisting essentially of plagioclase and hornblende with some corundum. (Millosevich, 1927, p.30; Sessera Valley, Piedmont, Italy; Tröger 350; Tomkeieff p.519)

SHACKANITE. A porphyritic variety of analcime phonolite consisting of phenocrysts of rhomb-shaped anorthoclase, augite and altered olivine in a groundmass of analcime, anorthoclase, biotite and glass. (Daly, 1912, p.411; Shackan, British Columbia, Canada; Tröger 482; Johannsen v.4, p.136; Tomkeieff p.519)

SHASTAITE. A local term for a glassy dacite containing normative andesine. Cf. ungaite. (Iddings, 1913, p.106; Mt Shasta, California, USA; Tröger 156; Tomkeieff p.520)

SHASTALITE. A local name for fresh andesite glass – the altered variety has been called weiselbergite. (Wadsworth, 1884, p.97; Mt

Shasta, California, USA; Tröger 983; Johannsen v.3, p.170; Tomkeieff p.520)

SHIHLUNITE. A local name for a variety of trachyte containing augite, olivine and biotite. Some varieties contain small amounts of leucite. (Ogura et al., 1936, p.92; 1720 Shihlung lava flow, Lung-chiang Province, Manchuria, China; Tröger(38) 232½; Tomkeieff p.520)

SHISHIMSKITE. A local name for an ultramafic rock consisting essentially of magnetite with some perovskite and spinel. Cf. kiirunavaarite. (Shilin, 1940, p.350; Shishim Mts, Urals, Russian Federation; Tomkeieff p.521)

SHONKINITE. A coarse-grained rock with abundant augite, some olivine, biotite or hornblende, and essential alkali feldspar and foids, usually nepheline. Now defined modally as a melanocratic variety of foid syenite in QAPF field 11 (Fig. 2.8, p.27). (Weed & Pirsson, 1895a, p.415; from Shonkin, the Indian name for the Highwood Mts, Montana, USA; Tröger 489; Johannsen v.4, p.13; Tomkeieff p.521)

SHOSHONITE. A collective term, related to absarokite and banakite, for a trachyandesitic rock originally loosely described as an orthoclase-bearing basalt. It has since been used in several ways for potassic basaltic and intermediate volcanic rocks, e.g. Joplin, 1964. Now defined chemically as the potassic variety of basaltic trachyandesite in TAS field S2 (Fig. 2.14, p.35). (Iddings, 1895a, p.943; Shoshone River, Yellowstone National Park, Wyoming, USA; Tröger 269; Johannsen v.4, p.44; Tomkeieff p.521)

SIDEROMELANE. The extremely common transparent glass of basalts formed during submarine eruption or any magma/water interaction. (Waltershausen, 1853, p.202; from the Greek sideros = iron, melas = dark; Johannsen v.3, p.324; Tomkeieff p.522)

SIEVITE. An obsolete name given to a series of glassy rocks originally described as andesitic.

(*Marzari Pencati, 1819; Sieva, Euganean Hills, Italy; Tröger 984; Tomkeieff p.522)

SILEXITE. An old term for an igneous rock consisting almost entirely of quartz. It is recommended that it is replaced by the term quartzolite, which is etymologically more correct. (Miller, 1919, p.30; from the Latin silex = flint; Tröger 2; Johannsen v.2, p.11; Tomkeieff p.523)

SILICOCARBONATITE. An igneous rock comprising major amounts of both silicates and carbonates, but with silicates in excess of carbonates. The silicates may be sodic pyroxenes and amphiboles, biotite, phlogopite, olivine or feldspars. Now defined chemically as a carbonatitie in which $SiO_2 > 50\%$ (section 2.3, p.10). (Brögger, 1921, p.350; Fen Complex, Telemark, Norway; Tomkeieff p.525)

SILICOGRANITONE. A variety of gabbro impregnated with silica. (Mazzuoli & Issel, 1881, p.326; Tomkeieff p.525)

SILICOTELITE. An obsolete name for rocks containing over 50% of non-silicate minerals. (Rinne, 1921, p.144; Tomkeieff p.525)

SILLAR. A local Peruvian name for a variety of ignimbrite which has been indurated primarily by pneumatolytic processes. (Fenner, 1948, p.883; Arequipa, Peru; Tomkeieff p.525)

SILLITE. An obsolete name for a local rock variously called mica diorite, mica syenite, diabase porphyry and gabbro. (Gümbel, 1861, p.187; Sill-Berge, near Berchtesgaden, Germany; Tröger 985; Tomkeieff p.526)

SINAITE. A term suggested (but never adopted) for plutonic alkali feldspar rocks to replace syenite – the type rock from Syene having been shown to be a hornblende granite. (Rozière, 1826, p.306; Mt Sinai, Egypt; Tröger(38) 985½; Tomkeieff p.527)

SIZUNITE. A local term for a variety of minette characterized by high K_2O (> 10%) and P_2O_5 (2.5%) and consisting of microcline, biotite,

apatite and opaques. (Cogné & Giot, 1961, p.2569; Cap Sizun, Brittany, France)

SKEDOPHYRE. An obsolete term for a porphyritic rock with evenly spaced phenocrysts (skedophyric texture). (Iddings, 1909, p.224; Tomkeieff p.528)

SKOMERITE. A local term for a volcanic rock composed of phenocrysts of albite-oligoclase, augite and minor olivine in a groundmass of albite, chlorite and opaques. (Thomas, 1911, p.196; Skomer Island, Dyfed, Wales, UK; Tröger 215; Johannsen v.3, p.175; Tomkeieff p.529)

SMALTO. A local Italian name for a glassy rhyolite from Lipari Island. (Tomkeieff et al., 1983, p.531)

SNOBITE. A local name for a hybrid variety of hypersthene (= enstatite) dacite with inclusions of acid porphyry. (Hills, 1958, p.551; Snob's Creek, near Eildon, Victoria, Australia)

SODALITE BASALT. A term used for a volcanic rock consisting essentially of sodalite with minor amounts of olivine and other mafics. The name should not be used as the term basalt is now restricted to a rock containing essential plagioclase. As the rock is a variety of foidite it should be given the appropriate name, e.g. olivine-bearing sodalitite. (Johannsen, 1938, p.346)

SODALITE DIORITE. Now defined in QAPF field 14 (Fig. 2.4, p.22) as a variety of foid diorite in which sodalite is the most abundant foid.

SODALITE GABBRO. Now defined in QAPF field 14 (Fig. 2.4, p.22) as a variety of foid gabbro in which sodalite is the most abundant foid.

SODALITE MONZODIORITE. Now defined in QAPF field 13 (Fig. 2.4, p.22) as a variety of foid monzodiorite in which sodalite is the most abundant foid.

SODALITE MONZOGABBRO. Now defined in QAPF field 13 (Fig. 2.4, p.22) as a variety of foid monzogabbro in which sodalite is the most abundant foid.

SODALITE MONZOSYENITE. Now defined in QAPF field 12 (Fig. 2.4, p.22) as a variety of foid monzosyenite in which sodalite is the most abundant foid. The term is synonymous with sodalite plagisyenite.

SODALITE PLAGISYENITE. Now defined in QAPF field 12 (Fig. 2.4, p.22) as a variety of foid plagisyenite in which sodalite is the most abundant foid. The term is synonymous with sodalite monzosyenite.

SODALITE SYENITE. Now defined in QAPF field 11 (Fig. 2.4, p.22) as a variety of foid syenite in which sodalite is the most abundant foid. *(Steenstrup, 1881, p.34; Tröger 438; Johannsen v.4, p.113; Tomkeieff p.532)*

SODALITHOLITH (SODALITHITE). The spellings adopted by Tröger (1935) and Johannsen (1938) for the plutonic rock that Ussing (1912) had called sodalitite. *(Original reference uncertain; Tröger 632; Johannsen v.4, p.346)*

SODALITITE. A term originally proposed for a leucocratic plutonic rock almost wholly composed of sodalite crystals together with minor alkali feldspar, aegirine and eudialyte. Cf. naujaite. However, the term is now used for volcanic rocks and is defined as a variety of foidite in QAPF field 15c (Fig. 2.11, p.31). *(Ussing, 1912, p.156; Ilímaussaq, Greenland; Tomkeieff p.532)*

SODALITOLITE. Now defined in QAPF field 15 (Fig. 2.4, p.22) as a variety of foidolite in which sodalite is the most abundant foid. Cf. sodalitite. Examples are naujaite and tawite.

SODALITOPHYRE. A porphyritic rock consisting essentially of phenocrysts of sodalite, augite and hornblende in a glassy matrix. *(Hibsch, 1902, p.526; Böhmisches Mittelgebirge (now České Středohoří), N. Bohemia, Czech Republic; Tröger 986; Tomkeieff p.532)*

SOEVITE. An alternative spelling of sövite.

SOGGENDALITE. A local name for a pyroxene-rich variety of dolerite. *(Kolderup, 1896, p.159; Soggendal, Norway; Tröger 988; Johannsen v.3, p.332; Tomkeieff p.533)*

SÖLVSBERGITE. A variety of peralkaline microsyenite or peralkaline trachyte, often occurring as minor intrusions, consisting essentially of alkali feldspar with minor alkali pyroxene and/or alkali amphibole. *(Brögger, 1894, p.67; Sölvsberg, Gran, Oslo Igneous Province, Norway; Tröger 192; Johannsen v.3, p.107; Tomkeieff p.535)*

SOMMAITE. A medium- to coarse-grained variety of leucite monzosyenite consisting of phenocrysts of titanian augite and lesser olivine in a groundmass of sanidine, labradorite and leucite. It occurrs abundantly among the ejected blocks of Mt Somma. *(Lacroix, 1905, p.1189; Mt Somma, Naples, Italy; Tröger 511; Johannsen v.4, p.193; Tomkeieff p.536)*

SORDAWALITE (SORDAVALITE). An obsolete term originally used to describe a mineral and later redefined as a vitreous selvage to an olivine-rich dyke rock, i.e. a tachylyte. *(Nordenskjöld, 1820, p.86; Sordavala, Lake Ladoga, near St Petersburg, Russian Federation; Tröger 989; Johannsen v.3, p.323; Tomkeieff p.536)*

SÖRKEDALITE. A local name for a variety of olivine monzodiorite consisting of abundant antiperthitic andesine, which may be mantled by anorthoclase, and by olivine, opaques and apatite with minor clinopyroxene and biotite. *(Brögger, 1933, p.35; Kjelsås, Sörkedal, Oslo district, Norway; Tröger 286; Johannsen v.4, p.54; Tomkeieff p.536)*

SÖVITE (SOEVITE). A special term in the carbonatite classification for the coarse-grained variety of calcite carbonatite in which the most abundant carbonate is calcite (section 2.3, p.10). Biotite and apatite are also frequently present. *(Brögger, 1921, p.246; Söve, Fen Complex, Telemark, Norway; Tröger 753; Tomkeieff p.537)*

SPERONE. A local name for a yellowish, porous

leucitite from the Alban Hills, Italy. *(Gmelin, 1814; Tröger 990; Johannsen v.4, p.363; Tomkeieff p.538)*

SPESSARTITE. A variety of lamprophyre consisting of phenocrysts of hornblende with or without biotite, olivine or pyroxene in a groundmass of the same minerals plus plagioclase and subordinate alkali feldspar. Now defined in the lamprophyre classification (Table 2.9, p.19). *(Rosenbusch, 1896, p.532; Spessart, Bavaria, Germany; Tröger 318; Johannsen v.3, p.191; Tomkeieff p.538)*

SPHENITITE. A variety of pyroxenite, similar to jacupirangite, with up to 50% sphene (now called titanite). *(Allen, 1914, p.155; Ice River, British Columbia, Canada; Tröger 696; Johannsen v.4, p.464; Tomkeieff p.538)*

SPIEMONTITE. An obsolete term for a variety of fine-grained diorite. *(Steininger, 1841, p.30; Spiemont Mt, Nahe district, Saarland, Germany)*

SPILITE. A term originally used for altered phenocryst-poor or aphyric basaltic rocks but later (Flett, 1907) used for altered (often albitized) basaltic lavas. *(Brongniart, 1827, p.98; Tröger 329; Johannsen v.3, p.299; Tomkeieff p.539)*

SPINELLITE. A rock in which spinel is the predominant mineral. However, Johannsen, to whom the term is attributed, only used the term in conjunction with a mineral qualifier, e.g. magnetite spinellite. *(Tomkeieff et al., 1983, p.540)*

SPODITE. An obsolete term for volcanic ashes. *(Cordier, 1816, p.381; Tomkeieff p.541)*

SPUMULITE. A little-used name for pumice stone. The reference implies the term was used earlier by Zavaritskii. *(Lebedinsky & Chu Tszya-Syan, 1958, p.15; Tomkeieff p.542)*

SPURINE. An obsolete name for a variety of porphyry containing quartz, feldspar and talc. *(Jurine, 1806, p.375; Mont Blanc, France; Tomkeieff p.542)*

STAVRITE. A local name for a dyke rock consisting of large amounts of amphibole needles with interstitial biotite and minor quartz, opaques and apatite. The rock may be metamorphic. *(Eckermann, 1928a, p.405; Stavreviken, Alnö Island, Västernorrland, Sweden; Tröger 713; Johannsen v.4, p.445; Tomkeieff p.544)*

STIGMITE. An obsolete name for a porphyritic pitchstone or obsidian. *(Brongniart, 1813, p.44; Tomkeieff p.545)*

STUBACHITE. An obsolete term originally used for an igneous rock composed of olivine and antigorite with some chrome spinel and pyroxene, but later shown to be a partially altered olivine pyroxene rock. *(Weinschenk, 1895, p.701; Stubachtal, Tyrol, Austria; Tomkeieff p.550)*

SUBALKALI. A term used for rocks that are not alkaline in character. *(Iddings, 1895b, p.183)*

SUBALKALI BASALT. Now defined chemically as a variety of basalt in TAS field B (p.36) which do not contain normative nepheline. Cf. alkali basalt. *(Chayes, 1966, p.137)*

SUBPHONOLITE. A variety of phonolite with a texture which suggests that it crystallized at depths below which phonolite would normally crystallize but not deep enough to be a nepheline syenite. *(Adamson, 1944, p.241; Tomkeieff p.552)*

SUDBURITE. A local name for a melanocratic volcanic rock containing hypersthene (= enstatite) and augite phenocrysts in an equigranular groundmass of these minerals, bytownite and biotite. The original rock is probably a hornfels. *(Coleman, 1914, p.215; Sudbury, Ontario, Canada; Tröger 393; Johannsen v.3, p.305; Tomkeieff p.553)*

SUEVITE. A glassy material found in breccia or tuff around the Ries meteorite impact crater and containing the high-pressure polymorphs of silica, coesite and stishovite. *(Sauer, 1919, p.15; named after the Suevi who lived around Nördlinger Ries, Bavaria, Germany in Roman times; Tröger 992; Tomkeieff p.553)*

SULDENITE. An obsolete name for a variety of quartz microgabbro (monzogabbro) containing phenocrysts of labradorite, hornblende and quartz in a groundmass of these minerals with orthoclase, diopside and biotite. *(Stache & John, 1879, p.382; Suldenferner, Ortler Alps, Alto Adige, Italy; Tröger 117; Tomkeieff p.554)*

SUMACOITE. An obsolete name for a variety of porphyritic trachyandesite containing abundant phenocrysts of intermediate plagioclase and augite in a groundmass of augite, sodic plagioclase, orthoclase and minor nepheline, haüyne and altered sodalite. The rock was originally described as an andesitic tephrite. *(Johannsen, 1938, p.188; Sumaco Volcano, Ecuador; Tröger(38) 267; Tomkeieff p.554)*

SUSSEXITE. A leucocratic medium-grained variety of nepheline syenite with abundant large crystals of nepheline in a matrix of nepheline, aegirine-augite needles and alkali feldspar. Katophorite and biotite may also be present. *(Brögger, 1894, p.173; Beemerville, Sussex County, New Jersey, USA; Tröger 462; Johannsen v.4, p.269; Tomkeieff p.556)*

SUZORITE. A local name for a coarse-grained rock consisting essentially of biotite, with small amounts of red orthoclase, augite and apatite. *(Faessler, 1939, p.47; Suzor Township, Laviolette County, Quebec, Canada; Tomkeieff p.556)*

SVARTVIKITE. A variety of microsyenite consisting essentially of albite, in which abundant diopsidic pyroxene grains occur, intergrown with muscovite and calcite. Originally described as an albitophyre with a chemical similarity to holyokeite. *(Eckermann, 1938b, p.425; Svartviken Cove, N. Ulfö Island, Sweden; Tomkeieff p.556)*

SVIATONOSSITE. A local term for syenites characterized by the presence of andradite garnet.

Probably hybrid rocks. *(Eskola, 1921, p.41; Sviatoi Nos, Transbaikalia, Siberia, Russian Federation; Tröger 188; Johannsen v.3, p.111; Tomkeieff p.556)*

SYENEID. An obsolete field term for a coarse-grained igneous rock consisting of feldspar with biotite and/or amphibole and/or pyroxene. *(Johannsen, 1911, p.320; Tomkeieff p.557)*

SYENIDE. A revised spelling recommended to replace the field term syeneid. Now obsolete. *(Johannsen, 1926, p.182; Johannsen v.1, p.57)*

SYENILITE. An obsolete name for an amphibole granite. *(Cordier, 1868, p.78; Syene, now Aswân, Egypt; Tomkeieff p.557)*

SYENITE. A plutonic rock consisting mainly of alkali feldspar with subordinate sodic plagioclase, biotite, pyroxene, amphibole and occasional fayalite. Minor quartz or nepheline may also be present. Now defined modally in QAPF field 7 (Fig. 2.4, p.22). *(A term of great antiquity usually attributed to Pliny, AD 77 – see Johannsen for further discussion; Syene, now Aswân, Egypt; Tröger 240; Johannsen v.3, p.52; Tomkeieff p.557)*

SYENITELLE. An obsolete name for a variety of banded syenite. *(Rozière, 1826, p.250; Syene, now Aswân, Egypt; Tomkeieff p.557)*

SYENITITE. A term proposed (but not adopted) for both biotite-bearing syenitic aplites and plagioclase-bearing syenites. *(Polenov, 1899, p.464; Tröger 993; Johannsen v.3, p.61; Tomkeieff p.558)*

SYENITOID. Used proposed for preliminary use in the QAPF "field" classification (Fig. 2.10, p.29) for plutonic rocks tentatively identified as syenite or monzonite. *(Streckeisen, 1973, p.28)*

SYENODIORITE. May be used as a comprehensive term for plutonic rocks intermediate between syenite and diorite (p.23). *(Johannsen, 1917, p.89; Tröger 994; Johannsen v.3, p.110; Tomkeieff p.558)*

SYENOGABBRO. May be used as a comprehensive term for plutonic rocks intermediate between syenite and gabbro (p.23). *(Johannsen, 1917, p.89; Tröger 995; Johannsen v.3, p.126; Tomkeieff p.558)*

SYENOGRANITE. An optional term for a variety of granite in QAPF field 3a (Fig. 2.4, p.22) consisting of alkali feldspar with subordinate plagioclase. *(Streckeisen, 1967, p.166)*

SYENOID. An obsolete term proposed for all syenites containing feldspathoids. *(Shand, 1910, p.377; Tröger 996; Johannsen v.4, p.78; Tomkeieff p.558)*

SYNNYRITE. A kalsilite syenite, with a micropegmatitic texture, composed of K-feldspar and kalsilite with biotite, albite, minor aegirine-augite, titanite, apatite, fluorite, magnetite, garnet and alkali amphibole. *(Zhidkov, 1962, p.33; Synnyrmassive, Baikal Rift, Siberia, Russian Federation)*

SYNTECTITE. An igneous rock produced by the contamination of primary magmas with crustal rocks or melts. Cf. anatectite and prototectite. *(Loewinson-Lessing, 1934, p.7; Tomkeieff p.560)*

TABONA. A local name used on the island of Tenerife for an obsidian without phenocrysts. *(Tomkeieff et al., 1983, p.562)*

TACHYLYTE. The dust-laden basaltic glass usually containing magnetite microlites and generally occurring as the selvages of dykes and sills. *(Breithaupt, 1826, p.112; from the Greek tachys = rapid, lytos = soluble; Tröger 387; Johannsen v.3, p.290; Tomkeieff p.562)*

TAHITITE. A volcanic rock consisting of phenocrysts of haüyne in a glassy groundmass with augite microlites, titanomagnetite, haüyne, and occasional orthoclase and leucite. Although regarded as an extrusive equivalent of a nepheline monzonite (Johannsen, 1938), with which it is associated, it would be classified chemically in TAS as a haüyne tephriphonolite or phonotephrite. *(Lacroix, 1917a, p.583; Papenoo, Tahiti, Pacific Ocean; Tröger 537; Johannsen v.4, p.189; Tomkeieff p.562)*

TAIMYRITE. A local name for a variety of nosean phonolite consisting essentially of anorthoclase and nosean. *(Chrustschoff, 1894, p.427; Taimyr River, N. Siberia, Russian Federation; Tröger 434; Johannsen v.4, p.134; Tomkeieff p.563)*

TAIWANITE. A local name for an intrusive magnesium-rich basaltic glass containing phenocrysts of labradorite and olivine. *(Juan et al., 1953, p.1; named after Taiwan; Tomkeieff p.563)*

TALZASTITE. A local name for a coarse-grained variety of ijolite containing titanian augite. The nepheline is altered to hydronepheline and cancrinite. *(Termier et al., 1948, p.81; N'Talzast Volcano, Azrou, Morocco; Tomkeieff p.563)*

TAMARAITE. A local name for a mafic lamprophyric rock consisting of augite, brown hornblende, biotite, and titanite with nepheline (or cancrinite and analcime) and small amounts of plagioclase. *(Lacroix, 1918, p.544; Cape Topsail, Tamara Island, Îles de Los, Conakry, Guinea; Tröger 523; Johannsen v.4, p.297; Tomkeieff p.563)*

TANDILEOFITE. A dyke rock composed of phenocrysts of labradorite, microcline and amphibole in a groundmass of quartz, microcline, albite and labradorite. *(Pasotti, 1954, p.5; Cerro Tandileofú, Buenos Aires, Argentina)*

TANNBUSCHITE. A local name for a melanocratic variety of olivine nephelinite consisting largely of pyroxene with smaller amounts of nepheline and olivine. *(Johannsen, 1938, p.364; Mt Tannbusch (now Jedlová), near Ústí nad Labem, N. Bohemia, Czech Republic; Tröger(38) 623½; Tomkeieff p.563)*

TARANTULITE. A local name for a quartz-rich alaskite. *(Johannsen, 1920a, p.54; Missouri Mine, Tarantula Spring, Nevada, USA; Tröger 8; Johannsen v.2, p.35; Tomkeieff p.564)*

TASMANITE. A zeolitized variety of ijolite with

nepheline and titanian augite, lesser amounts of opaques, melilite and olivine, and minor amounts of apatite and perovskite. *(Johannsen, 1938, p.318; Shannon Tier, near Bothwell, Tasmania, Australia; Tröger(38) 607½; Tomkeieff p.564)*

TAURITE. An obsolete local name for a variety of spherulitic alkali feldspar rhyolite containing phenocrysts of anorthoclase in a groundmass of quartz, orthoclase, aegirine-augite and arfvedsonite. *(Lagorio, 1897, p.5; Taurida, Crimea, Ukraine; Tröger 74; Tomkeieff p.564)*

TAUTIRITE. A local name for a variety of tephritic phonolite consisting of small phenocrysts of hornblende, titanite, augite and biotite in a groundmass of alkali feldspar and andesine with some nepheline and sodalite. Cf. pollenite. *(Iddings & Morley, 1918, p.117; Tautira Beach, Taiarapu, Tahiti, Pacific Ocean; Tröger 532; Johannsen v.4, p.166; Tomkeieff p.564)*

TAVOLATITE. A variety of leucite phonolite essentially composed of leucite with subordinate alkali feldspar, nepheline, haüyne, plagioclase and clinopyroxene. *(Washington, 1906, p.50; Osteria del Tavolato, Alban Hills, near Rome, Italy; Tröger 530; Johannsen v.4, p.285; Tomkeieff p.565)*

TAWITE. A variety of foidolite composed largely of crystals of sodalite together with aegirine. Some nepheline, alkali feldspar and eudialyte are usually present. Feldspathic varieties have been called beloeilite and feldspattavite. *(Ramsay & Hackman, 1894, p.93; Tawajok Valley, Lovozero complex, Kola Peninsula, Russian Federation; Tröger 636; Johannsen v.4, p.319; Tomkeieff p.565)*

TEPHRA. A collective term used in the pyroclastic classification (section 2.2.2, p.7) for pyroclastic deposits that are predominatly unconsolidated. Cf. pyroclastic rock. *(Thorarinsson, 1944, p.6; from the Greek tephra = ashes; Tomkeieff p.567)*

TEPHRILEUCITITE. A synonym for tephritic leucitite of QAPF field 15b (Fig. 2.11, p.31). *(Rittmann, 1973, p.135)*

TEPHRINE. An obsolete term for tephrite. *(Brongniart, 1813, p.40; Tomkeieff p.567)*

TEPHRINEPHELINITE. A synonym for tephritic nephelinite of QAPF field 15b (Fig. 2.11, p.31). *(Rittmann, 1973, p.135)*

TEPHRIPHONOLITE. A synonym for tephritic phonolite of QAPF field 12 (Fig. 2.11, p.31), and also defined chemically in TAS field U3 (Fig. 2.14, p.35) *(Rittmann, 1973, p.134)*

TEPHRITE. An alkaline volcanic rock composed essentially of calcic plagioclase, clinopyroxene and feldspathoid. Now defined modally in QAPF field 14 (Fig. 2.11, p.31) and, if modes are not available, chemically in TAS field U1 (Fig. 2.14, p.35). *(A term of great antiquity usually attributed to Pliny, AD 77, – see Johannsen for further discussion; from the Greek tephra = ashes; Tröger 999; Johannsen v.4, p.230; Tomkeieff p.567)*

TEPHRITIC FOIDITE. A collective term for alkaline volcanic rocks consisting of foids with some plagioclase as defined modally in QAPF field 15b (Fig. 2.11, p.31). It is distinguished from basanitic foidite by having less than 10% modal olivine. If possible the most abundant foid should be used in the name, e.g. tephritic nephelinite, tephritic leucitite etc. (Table 2.8, p.18). *(Streckeisen, 1978, p.7)*

TEPHRITIC LEUCITITE. Now defined modally in the leucite-bearing rock classification (section 2.8, p.18) as a volcanic rock falling into QAPF field 15b and consisting of leucite, clinopyroxene, minor olivine and with plagioclase > sanidine. Other foids may be present in minor amounts. *(Streckeisen, 1978, p.7)*

TEPHRITIC PHONOLITE. A collective term for alkaline volcanic rocks consisting of alkali feldspar, sodic plagioclase, feldspathoid and various mafic minerals. Now defined modally in QAPF field 12 (Fig. 2.11, p.31). *(Streckeisen, 1978, p.6)*

TEPHRITOID. A term originally proposed for rocks intermediate between olivine-free basalt and tephrite. Later used as an adjective to describe tephrite-like rocks. Now proposed for preliminary use in the QAPF "field" classification (Fig. 2.19, p.39) for volcanic rocks thought to contain essential foids and in which plagioclase is thought to be more abundant than alkali feldspar. *(Bücking, 1881, p.157; Johannsen v.4, p.69; Tomkeieff p.568)*

TEREKTITE. A local name for an effusive equivalent of semeitavite, which is a variety of quartz alkali feldspar syenite. *(Gornostaev, 1933, p.191; Terekty Hill, Semeitau Mts, Russian Federation; Tomkeieff p.568)*

TERZONTLI. An erroneous spelling of tezontli. *(Tomkeieff et al., 1983, p.569)*

TESCHENITE (TESCHINITE). A variety of analcime gabbro consisting of olivine, titanian augite, labradorite and analcime. Originally spelt teschinite; Zirkel (1866b, p.318) changed the spelling to teschenite. May be used as a synonym for analcime gabbro of QAPF field 14 (p.24). *(Hohenegger, 1861, p.43; Teschen (now divided into Český Těšín, Czech Republic and Cieszyn, Poland); Tröger 565; Johannsen v.4, p.226; Tomkeieff p.569)*

TETIN. A local name for a volcanic ash used in the Azores for cement making. *(Lea & Desch, 1935, p.244; Tomkeieff p.570)*

TEXONTLI. A Mexican name for a cellular amygdaloidal lava or pumice. *(Humboldt, 1823, p.358; Tomkeieff p.570)*

TEZONTLI. A local term, probable Aztec in origin, originally applied to scoriaceous basalt, but now applied to any other type of rock except granite or marble. *(Ives, 1956, p.122)*

THERALITE. A variety of nepheline gabbro consisting essentially of titanian augite, labradorite and nepheline. Olivine is a vari-

able but sometimes major constituent. May be used as a synonym for nepheline gabbro of QAPF field 14 (p.24). *(Rosenbusch, 1887, p.247; from the Greek therein = to search for; Tröger 514; Johannsen v.4, p.222; Tomkeieff p.571)*

THEROLITE. According to Zirkel the correct spelling of theralite. *(Zirkel, 1894b, p.800; Kola Peninsula, Russian Federation; Tomkeieff p.571)*

THOLEIITE (THOLEITE, THOLEYITE). This term has caused considerable confusion. It was originally used for a "doleritic trapp" said to consist of albite and ilmenite. Later Rosenbusch (1887) redefined it as an olivine-poor or olivine-free plagioclase augite rock with intersertal texture. It then became used as a common variety of basalt which Yoder & Tilley (1962) later defined chemically as a hypersthene normative basalt. However, Jung (1958) had shown the type rock was not a tholeiite, as chemically defined above, but a leucocratic subvolcanic variety of monzodiorite for which he proposed the name tholeyite. The Subcommission recommend that this term should be replaced by tholeiitic basalt. *(Steininger, 1840, p.99; Tholey, Nahe district, Saarland, Germany; Tröger 344; Johannsen v.3, p.298; Tomkeieff p.572)*

THOLEIITIC BASALT. A common variety of basalt composed of labradorite, augite, hypersthene (= enstatite) or pigeonite, with olivine (often showing a reaction relationship) or quartz, and often with interstitial glass. The Subcommission recommends that this term should be used instead of tholeiite.

THOLERITE. An obsolete early form of dolerite. *(Leonhard, 1823a, p.118; from the Greek tholeros = dirty, gloomy; Tomkeieff p.572)*

THURESITE. A local term for a variety of alkali feldspar syenite composed of microcline, Na-amphibole and hornblende with augite cores. *(Hackl & Waldmann, 1935, p.260; Thures, near Raabs, Lower Austria, Austria;*

Tröger(38) 178⅓; Tomkeieff p.572)

TILAITE. A mafic variety of gabbro composed of green chrome-diopside, olivine and minor highly calcic plagioclase. *(Duparc & Pearce, 1905, p.1614; Tilai-Kamen, Koswa region, N. Urals, Russian Federation; Tröger 399; Johannsen v.3, p.245; Tomkeieff p.572)*

TIMAZITE. An obsolete local term for an altered hornblende biotite andesite. *(Breithaupt, 1861, p.51; Timok River, Serbia, Yugoslavia; Tröger 1000; Johannsen v.3, p.170; Tomkeieff p.573)*

TINGUAITE. A variety of phonolite consisting of alkali feldspar, nepheline with or without other foids, aegirine and sometimes biotite and characterized by "tinguaitic texture" in which needles of aegirine occur interstitially in a mozaic of alkali feldspar and foids. *(Rosenbusch, 1887, p.628; Serra de Tingua, Rio de Janeiro, Brazil; Tröger 445; Johannsen v.4, p.145; Tomkeieff p.573)*

TIPYNEITE. A mnemonic name suggested for a rock consisting essentially of *ti*taniferous *py*roxene and *ne*pheline. *(Belyankin, 1929, p.22)*

TIRILITE. A local term for a rapakivi-diabase hybrid of granodiorite composition consisting of andesine, microcline-perthite, hornblende, biotite and serpentinized pyroxene. *(Wahl, 1925, p.69; Tirilä, near Lappeenranta, Finland; Tröger 81; Tomkeieff p.573)*

TITANOLITE. A name for a variety of alkali pyroxenite rich in titanite with magnetite and calcite. Tröger (1938) suggests the rock may be a skarn. *(Kretschmer, 1917, p.189; Tröger 1001; Tomkeieff p.573)*

TJOSITE. A local name for a lamprophyric rock or micromelteigite similar to jacupirangite with phenocrysts of augite, abundant magnetite, ilmenite and apatite with some biotite in a groundmass of anorthoclase and nepheline. Similar to cocite but with nepheline instead of leucite. *(Brögger, 1906, p.128; Tjose, near Larvik, Oslo Igneous Province,*

Norway; Tröger 495; Johannsen v.4, p.277; Tomkeieff p.574)

TOADSTONE. An obsolete term for an amygdaloidal basalt named because it can look like the skin of a toad. *(Pinkerton, 1811a, p.93; Johannsen v.3, p.281; Tomkeieff p.574)*

TOELLITE. See töllite.

TÖIENITE. A name proposed for a rock which was later found to be identical to windsorite. The name was later withdrawn. *(Brögger, 1931, p.65; Tröger 1002)*

TOKÉITE. A local name for a melanocratic variety of basalt containing abundant phenocrysts of augite and some olivine and opaques in a fine-grained groundmass of these minerals, labradorite and biotite. *(Duparc & Molly, 1928, p.24; Toké-Grat, Gouder Valley, near Addis Ababa, Ethiopia; Tröger 407; Johannsen v.3, p.306; Tomkeieff p.574)*

TÖLLITE (TOELLITE). An obsolete local name for a garnet-bearing variety of quartz diorite porphyry. *(Pichler, 1875, p.926; Töll, near Merano, Alto Adige, Italy; Tröger 115; Johannsen v.2, p.400; Tomkeieff p.574)*

TONALITE. A plutonic rock consisting essentially of quartz and intermediate plagioclase, usually with biotite and amphibole. Now defined modally in QAPF field 5 (Fig. 2.4, p.22). *(Rath, 1864, p.249; Tonale Pass, Adamello, Alto Adige, Italy; Tröger 132; Johannsen v.2, p.378; Tomkeieff p.574)*

TÖNSBERGITE. A local term for a variety of alkali feldspar syenite in which the alkali feldspars are rhomb-shaped. Occurs as a red altered variety of larvikite. *(Brögger, 1898, p.328; Tønsberg, Oslo Igneous Province, Norway; Tröger 184; Tomkeieff p.574)*

TOPATOURBIOLILEPIQUORTHITE. An unwieldy name constructed by Johannsen to illustrate some of the possibilities of the mnemonic classification of Belyankin (1929) for a variety of granite consisting of *topaz, tourma-*

line, *bi*otite, *oli*goclase, *lepi*dolite, *qu*artz and *orth*oclase. Cf. biquahororthandite and hobiquandorthite. *(Johannsen, 1931, p.125)*

TOPAZ RHYOLITE. A variety of rhyolite rich in fluorine, silica and the lithophile elements, characterized by the presence of topaz in gas cavities. Although not present in all cavities, the topaz can usually be found in 15 to 30 minutes diligent searching with a hand lens – not the most suitable characteristic for naming a rock! Stated to be synonymous with ongonite. *(Burt et al., 1982, p.1818)*

TOPAZITE. A term suggested for rocks consisting of topaz and quartz. *(Johannsen, 1920a, p.53; Tröger 6; Johannsen v.2, p.21; Tomkeieff p.575)*

TOPSAILITE. A local name for a dyke rock containing phenocrysts of plagioclase, augite, apatite, and titanomagnetite in a groundmass of andesine, biotite, barkevikite (see p.44), augite and titanite. *(Lacroix, 1911b, p.79; Cape Topsail, Tamara Island, Îles de Los, Conakry, Guinea; Tröger 558; Tomkeieff p.575)*

TORDRILLITE. An obsolete group name proposed for quartz alkali feldspar lavas and chemically equivalent to alaskite. *(Spurr, 1900a, p.189; Tordrillo Mts, Alaska, USA; Tröger 44; Johannsen v.2, p.113; Tomkeieff p.576)*

TORRICELLITE. An obsolete term for a fine-grained basalt containing quartz. *(Pinkerton, 1811b, p.50; named after Torricelli, 1640; Tomkeieff p.576)*

TORYHILLITE. A variety of nepheline syenite composed mainly of nepheline with albite, aegirine-augite and minor calcite, but without K-feldspar. *(Johannsen, 1920b, p.163; Toryhill, Monmouth Township, Ontario, Canada; Tröger 423; Johannsen v.4, p.298; Tomkeieff p.576)*

TOSCANITE. A name for a variety of rhyodacite (a rock intermediate between rhyolite and dacite) containing calcic plagioclase,

orthoclase, hypersthene (= enstatite) and biotite in a glassy matrix of rhyolite composition. Cf. dellenite. *(Washington, 1897a, p.37; Mt Amiata, Tuscany, Italy; Tröger 101; Johannsen v.2, p.310; Tomkeieff p.576)*

TOURMALITE. A term suggested for rocks consisting of tourmaline and quartz only. *(Johannsen, 1920a, p.53; Johannsen v.2, p.22; Tomkeieff p.577)*

TRACHIVICOITE. An obsolete name for a variety of vicoite rich in sanidine. *(Washington, 1906, p.57; Tomkeieff p.577)*

TRACHORHEITE. An old field term to cover andesite, trachyte and rhyolite when they are difficult to distinguish in the field. *(Endlich, 1874, p.319; Tomkeieff p.577)*

TRACHYANDESITE. A term originally used for volcanic rocks intermediate in composition between trachyte and andesite and containing approximately equal amounts of alkali feldspar and plagioclase. Later used for volcanic rocks containing foids as well as alkali feldspar and plagioclase (Rosenbusch, 1908). Now defined chemically in TAS field S3 (Fig. 2.14, p.35). *(Michel-Lévy, 1894, p.8; Tröger 1003; Johannsen v.3, p.118; Tomkeieff p.577)*

TRACHYBASALT. Although originally used for foidal rocks, the term has mainly been used for basaltic volcanic rocks containing labradorite and alkali feldspar. Now defined chemically in TAS field S1 (Fig. 2.14, p.35). *(Bořický, 1874, p.44; Tröger 1004; Johannsen v.3, p.128; Tomkeieff p.577)*

TRACHYDACITE. A term originally used for a variety of rhyolite containing bronzite (= enstatite) and an alkali feldspar to oligoclase ratio of 2:1. Now defined chemically as rocks with more than 20% normative quartz in TAS field T (Fig. 2.14, p.35). *(Millosevich, 1908, p.418; Riu Mannu, Sassari, Sardinia, Italy; Tröger 94; Tomkeieff p.578)*

TRACHYDOLERITE. A term mainly used for volcanic rocks containing both orthoclase and labradorite. Cf. trachybasalt. *(Abich, 1841, p.100; Tröger 1005; Johannsen v.3, p.128; Tomkeieff p.578)*

TRACHYLABRADORITE. A term for a leucotrachybasalt in which the plagioclase is more calcic than An$_{50}$. *(Jung & Brousse, 1959, p.112)*

TRACHYLIPARITE. A term originally used for a variety of rhyolite with an alkali feldspar to oligoclase ratio of 3:1. *(Derwies, 1905, p.71; Medovka, Pyatigorsk, N. Caucasus, Russian Federation; Tröger 93; Tomkeieff p.578)*

TRACHYPHONOLITE. An obsolete term for a phonolite with trachytic texture. *(Bořický, 1873, p.18; Tröger 1006)*

TRACHYTE. A volcanic rock consisting essentially of alkali feldspar. Now defined modally in QAPF field 7 (Fig. 2.11, p.31) and, if modes are not available, chemically in TAS field T (Fig. 2.14, p.35). *(Brongniart, 1813, p.43; from the Greek trachys = rough; Tröger 251; Johannsen v.3, p.67; Tomkeieff p.578)*

TRACHYTE-ANDESITE. An obsolete term for the extrusive equivalents of monzonite, later superseded by the term latite. *(Brögger, 1895, p.60; Tröger 1007; Johannsen v.3, p.100)*

TRACHYTOID. A term proposed for volcanic rocks consisting mainly of sanidine, plagioclase, hornblende or mica. Later used as a textural term (Johannsen, 1931). Now proposed for preliminary use in the QAPF "field" classification (Fig. 2.19, p.39) for volcanic rocks tentatively identified as trachyte. *(Gümbel, 1888, p.86; Tomkeieff p.579)*

TRANSITIONAL BASALT. A variety of basalt transitional between typical tholeiitic basalt and alkali basalt. It consists of olivine, Ca-rich augite, plagioclase and titanomagnetite plus variable, but small, amounts of alkali feldspar. Ca-poor pyroxenes are absent. It is often associated with peralkaline rhyolite and peralkaline trachyte. In Russian the term mid-alkali basalt is used as a synonym. *(Kuno, 1960, p.121)*

TRAPP (TRAP). A term that came to be used imprecisely for volcanic and medium-grained rocks of basaltic composition. Now sometimes used for plateau basalts. *(Rinman, 1754, p.293; from the Swedish trappar = steps, i.e. jointing; Tröger 1009; Johannsen v.3, p.247; Tomkeieff p.580)*

TRAPPIDE. A name recommended to replace the field term anameseid. Now obsolete. *(Johannsen, 1926, p.181; Johannsen v.1, p.58; Tomkeieff p.580)*

TRAPPITE. An obsolete term for trapp. *(Brongniart, 1813, p.40; Tomkeieff p.580)*

TRASS. A local Italian name for a non-stratified tuff consisting of small fragments of altered trachytic pumice. *(Leonhard, 1823b, p.692; Johannsen v.3, p.20; Tomkeieff p.580)*

TRASSOITE. An obsolete term for a volcanic tuff consisting of fragments of feldspar and glass. *(Cordier, 1816, p.366; Tomkeieff p.580)*

TREMATODE. An obsolete name for a vesicular andesite. *(Haüy, 1822, p.578; Volvic, Auvergne, France; Tomkeieff p.581)*

TRIAPHYRE. An obsolete term proposed for igneous rocks intruding Triassic formations. *(Ebray, 1875, p.291; Tomkeieff p.581)*

TRISTANITE. Originally described as a member of a potassic volcanic rock series falling in Daly's compositional gap between trachyandesite and trachyte, with a differentiation index between 60 and 75. It consists essentially of phenocrysts of plagioclase, zoned from labradorite to oligoclase and often rimmed with alkali feldspar, Fe-olivine and pyroxene in a groundmass of Fe-olivine and titanian augite with andesine and interstitial alkali feldpar. *(Tilley & Muir, 1963, p.439; Tristan da Cunha, South Atlantic Ocean)*

TROCTOLITE. A variety of gabbro composed essentially of highly calcic plagioclase and olivine with little or no pyroxene. Now defined modally in the gabbroic rock classification (Fig. 2.6, p.25). *(Lasaulx, 1875, p.317; from the Greek troktes = trout; Tröger 353; Johannsen v.3, p.225; Tomkeieff p.581)*

TRONDHJEMITE. A leucocratic variety of tonalite consisting essentially of sodic plagioclase and quartz with minor biotite. Orthoclase is characteristically absent and hornblende is rare. May be used as a synonym for plagiogranite and leucocratic tonalite of QAPF field 5 (p.23). *(Goldschmidt, 1916, p.75; Trondhjem, now Trondheim, Norway; Tröger 129; Johannsen v.2, p.387; Tomkeieff p.582)*

TROWLESWORTHITE. An obsolete name for a coarse-grained pneumatolytic vein found in granite consisting of red orthoclase, tourmaline and fluorite with minor quartz. *(Worth, 1884, p.177; Trowlesworthy Tor, Cornwall, England, UK; Johannsen v.3, p.5; Tomkeieff p.582)*

TSINGTAUITE. A local name for a variety of granite porphyry containing phenocrysts of microperthite and sodic plagioclase. *(Rinne, 1904, p.142; Tsingtau, Shantung, China; Tröger 88; Johannsen v.2, p.300; Tomkeieff p.582)*

TUFAITE. An obsolete term for a tuff consisting of fragments of pyroxene and other components. *(Cordier, 1816, p.366; Tomkeieff p.582)*

TUFF. Now defined in the pyroclastic classification (section 2.2.2, p.8 and Fig. 2.1, p.8) as a pyroclastic rock in which ash > 75%. The term is synonymous with ash tuff. See also coarse (ash) tuff, fine (ash) tuff and dust tuff. *(Phillips, 1815, p.172; Tomkeieff p.583)*

TUFF BRECCIA. Now defined in the pyroclastic classification (section 2.2.2, p.8 and Fig. 2.1, p.8) as a pyroclastic rock in which bombs and/or blocks range from 25% to 75%.

TUFFISITE. An intrusive tuff occurring in pipes, dykes and sills in which the particles may consist totally of magmatic material or of fragments derived from the walls of the conduit, or a mixture of both. *(Cloos, 1941; Swabia, Germany)*

TUFFITE. A term used in the pyroclastic classification (Table 2.4, p.9) for rocks consisting of mixtures of pyroclasts and epiclasts. *(Schmid, 1981, p.43)*

TURJAITE. A name for a variety of melilitolite mainly consisting of melilite, biotite and nepheline with minor perovskite, melanite garnet and apatite. May be used as an optional term in the melilitic rocks classification if nepheline > 10% and melilite > nepheline (section 2.4, p.11). *(Ramsay, 1921, p.489; Turja, now Cape Turii, Kola Peninsula, Russian Federation; Tröger 659; Johannsen v.4, p.323; Tomkeieff p.585)*

TURJITE. A melanocratic lamprophyre rich in biotite, analcime, calcite and melanite garnet. *(Belyankin & Kupletskii, 1924, p.35; Turja, now Cape Turii, Kola Peninsula, Russian Federation; Tröger 640; Tomkeieff p.585)*

TUSCULITE. A local name for a leucocratic variety of melilite leucitite composed mainly of leucite, minor melilite, pyroxene, alkali feldspar and magnetite. *(Cordier, 1868, p.118; Tusculum, Frascati, Alban Hills, near Rome, Italy; Tröger 1011; Tomkeieff p.585)*

TUTVETITE. A local name for a variety of trachyte. *(Johannsen, 1938, p.49; Tutvet, Hedrum, Oslo district, Norway; Tröger(38) 171½; Tomkeieff p.585)*

TUVINITE. A local name for a variety of urtite in which the predominant mineral is nepheline but in which calcite is also present. *(Yashina, 1957, p.35; Tuva region, S. Siberia, Russian Federation; Tomkeieff p.585)*

TVEITÅSITE. A local name for a melanocratic variety of fenite composed mainly of aegirine-augite and sometimes perthite. Titanite, apatite and sometimes nepheline may be present. *(Brögger, 1921, p.155; Tveitåsen, Fen Complex, Telemark, Norway; Tröger 226;*

Johannsen v.4, p.33; Tomkeieff p.585)

UGANDITE. A melanocratic variety of leucitite largely composed of clinopyroxene, olivine and leucite with subordinate plagioclase in a glassy matrix. *(Holmes & Harwood, 1937, p.11; Bufumbira, Uganda; Tröger(38) 642⅔; Johannsen v.1 (2nd Edn), p.285; Tomkeieff p.587)*

UKRAINITE. A local name for a quartz monzonite. *(Bezborod'ko, 1935, p.198; Mariupol, Sea of Azov, Ukraine; Tröger(38) 86½; Tomkeieff p.587)*

ULRICHITE. A porphyritic variety of micro nepheline syenite or phonolite containing phenocrysts of alkali feldspar, nepheline and barkevikite (see p.44) in a groundmass of alkali feldspar, Na-rich amphibole and pyroxene. *(Marshall, 1906, p.397; named after Ulrich; Dunedin, New Zealand; Tröger 455; Johannsen v.4, p.153; Tomkeieff p.587)*

ULTRABASIC. A commonly used chemical term now defined in the TAS classification (Fig. 2.14, p.35) as a rock containing less than 45% SiO_2. See also basic, intermediate and acid. *(Judd, 1881, p.317; Johannsen v.1, p.194; Tomkeieff p.588)*

ULTRABASITE. A collective term that had been incorrectly used as a synonym for ultramafic rocks. Streckeisen (1967) suggested replacing the term with mafitite, which was later withdrawn in favour of ultramafitite, now replaced by ultramafic rock. *(Original reference uncertain)*

ULTRAMAFIC ROCK. Originally a term for a rock consisting essentially of mafic minerals, e.g. peridotite, dunite. Now defined as a rock with M > 90% (section 2.11.2, p.28). *(Hess, 1937, p.263; Tomkeieff p.588)*

ULTRAMAFITE. A term suggested as a less correct alternative to ultramafitolite, i.e. a collective name for ultramafic plutonic rocks. However, as a mafite is a mineral it is now recommended that it should not be used. *(Streckeisen, 1976, p.15)*

ULTRAMAFITITE. A term suggested for volcanic rocks with M > 90%, but no longer regarded as necessary. *(Streckeisen, 1978, p.7)*

ULTRAMAFITOLITE. A term suggested for a plutonic rock with M > 90%, but now replaced by the term ultramafic rock (section 2.11.2, p.28). *(Streckeisen, 1976, p.15)*

ULTRAMELILITOLITE. A name proposed for a plutonic rock in which melilite is in excess of 65%. *(Dunworth & Bell, 1998, p.899)*

UMPTEKITE. A variety of alkali feldspar syenite consisting of microperthite, arfvedsonite and aegirine. Nepheline is sometimes present. *(Ramsay & Hackman, 1894, p.81; Umptek, Khibina complex, Kola Peninsula, Russian Federation; Tröger 181; Johannsen v.4, p.8; Tomkeieff p.588)*

UNAKYTE (UNAKITE). A local term for a variety of granite containing appreciable amounts of epidote. *(Bradley, 1874, p.519; Unaka Range, Great Smoky Mts, N. Carolina – Tennessee, USA; Tröger 70; Johannsen v.2, p.59; Tomkeieff p.589)*

UNCOMPAHGRITE. A variety of pyroxene melilitolite consisting of more than 65% melilite with pyroxene. May be used as an optional term in the melilitic rocks classification if pyroxene > 10% (section 2.4, p.11). Synonymous with pyroxene melilitolite. *(Larsen & Hunter, 1914, p.473; Uncompahgre, Colorado, USA; Tröger 745; Johannsen v.4, p.320; Tomkeieff p.589)*

UNDERSATURATED. A term applied to igneous rocks which are undersaturated with respect to silica, i.e. they have foids or Mg-olivine in the mode or norm. *(Shand, 1913, p.510; Tomkeieff p.408)*

UNGAITE. A local term for a glassy dacite containing normative oligoclase. Cf. shastaite. *(Iddings, 1913, p.106; Unga Island, Kamchatka, Russian Federation; Tröger 122; Tomkeieff p.589)*

URALITITE. An obsolete term for a diabase in which all the pyroxene has been altered to uralite (= actinolite pseudomorph). *(Kloos, 1885, p.87; Tröger 1012; Johannsen v.3, p.319)*

URBAINITE. An obsolete term for a rock consisting essentially of ilmenite and rutile with less hematite and sapphirine occurring as dykes in an anorthosite. *(Warren, 1912, p.276; St Urbain, Quebec, Canada; Tröger 769; Johannsen v.4, p.470; Tomkeieff p.590)*

URTITE. A plutonic rock consisting of over 70% nepheline with some aegirine-augite, but no feldspar. Now defined modally as a leucocratic variety of foidolite in QAPF field 15 (Fig. 2.8, p.27) in which the foid is predominantly nepheline. *(Ramsay, 1896, p.463; Lujavr-Urt, Lovozero complex, Kola Peninsula, Russian Federation; Tröger 604; Johannsen v.4, p.316; Tomkeieff p.590)*

USSURITE. A local name for a volcanic rock with an unusual poikilitic texture composed of large crystals of oligoclase (36%) with numerous inclusions of microlites of augite (40%) and olivine (16%). Small amounts of analcime are present. *(Gapeeva, 1959, p.157; Ussuri River, E. Siberia, Russian Federation)*

VALAMITE. A local name for a variety of orthopyroxene quartz dolerite with abundant iron ores. Iron-rich orthopyroxene is the only mafic silicate mineral. *(Wahl, 1907, p.69; Valamo Island, Lake Ladoga, near St Petersburg, Russian Federation; Tröger(38) 342½; Tomkeieff p.591)*

VALBELLITE. An obsolete name for a variety of peridotite consisting of olivine, bronzite (= enstatite), hornblende, magnetite and some pyrrhotite. Cf. weigelite. *(Schaefer, 1898, p.501; Val Bella, Ivrea zone, Piedmont, Italy; Tröger 736; Johannsen v.4, p.430; Tomkeieff p.591)*

VALLEVARITE. An obsolete local name for a variety of anorthosite, consisting essentially of andesine and microcline antiperthite with minor diopside and biotite. *(Gavelin, 1915, p.19; Vallevara, Routivare, Norrbotten,* Lapland, Sweden; Tröger 295; Tomkeieff p.591)

VARIOLITE. An old term that came to be used for basaltic rocks with a pock-marked appearance, composed of radial aggregates of feldspar and pyroxene microlites in a microcrystalline or devitrified glassy base. The variolitic texture denotes rapid cooling and the quench growth of the crystals, such as is often found in submarine basalts, and should not be confused with spherulitic texture which indicates devitrification. *(Aldrovandi, 1648, p.882; from the Latin variola = smallpox; Tröger 1014; Johannsen v.3, p.300; Tomkeieff p.591)*

VÄRNSINGITE. A local name for a coarse-grained albite dolerite pegmatite dyke rock with some augite, minor hornblende and chlorite. *(Sobral, 1913, p.169; Västra Värnsingen Island, Nordingrå, Sweden; Tröger 198; Johannsen v.3, p.143; Tomkeieff p.592)*

VAUGNERITE. A dyke rock consisting of major amounts of biotite, hornblende and plagioclase and a little orthoclase. *(Fournet, 1861, p.606; Vaugneray, near Lyon, France; Tröger 109; Johannsen v.2, p.405; Tomkeieff p.592)*

VENANZITE. A melanocratic variety of leucite olivine melilitite largely composed of melilite, leucite, kalsilite and olivine. Rosenbusch (1899) called this rock euktolite unaware that it had already been named venanzite. A kamafugitic rock and now regarded as a kalsilite-phlogopite-olivine-leucite melilitite (Tables 2.5 & 2.6, p.12). *(Sabatini, 1899, p.60; San Venanzo, Umbria, Italy; Tröger 672; Johannsen v.4, p.361; Tomkeieff p.593)*

VENTRALLITE. A consistent misspelling of vetrallite in Johannsen on p.174 and in the index. *(Johannsen, 1938, p.174)*

VERITE. Originally described as a black pitch-like rock, associated with an extrusive mass of fortunite, and consisting of phenocrysts of phlogopite and olivine in a glassy matrix containing microlites of diopside and phlogopite. Now regarded as a hyalo-olivine-diopside-phlogopite lamproite (Table 2.7, p.17). *(Osann, 1889, p.311; Vera, Cabo de Gata, Spain; Tröger 234; Johannsen v.3, p.21; Tomkeieff p.593)*

VESBITE. A variety of foidite largely composed of leucite with subordinate clinopyroxene and melilite. *(Washington, 1920, p.46; from Mons Vesbius, the Latin name for Mt Vesuvius, Naples, Italy; Tröger 656; Johannsen v.4, p.360; Tomkeieff p.594)*

VESECITE. A local name for a variety of polzenite consisting of olivine, considerable amounts of monticellite, and melilite in a matrix of monticellite, phlogopite and nepheline. *(Scheumann, 1922, p.496; Vesec, N. Bohemia, Czech Republic; Tröger 663; Johannsen v.4, p.388; Tomkeieff p.594)*

VESUVITE. A variety of tephritic leucitite largely composed of leucite, clinopyroxene and subordinate plagioclase. *(Lacroix, 1917d, p.483; Mt Vesuvius, Naples, Italy; Tröger 582; Tomkeieff p.594)*

VETRALLITE. A variety of tephritic phonolite containing essential labradorite. *(Johannsen, 1938, p.173; Vetralla, Vico Volcano, near Viterbo, Italy; Tröger(38) 525½; Tomkeieff p.594)*

VIBETOITE (VIPETOITE). A local name for a variety of pyroxenite containing abundant titanian augite and hornblende with biotite, primary calcite, and occasional albite and nepheline. As pointed out by Sæther (1957) the name was actually misspelt by Brögger in the belief that the locality was named Vibeto instead of Vipeto. The correct spelling should have been vipetoite but vibetoite is in common usage. *(Brögger, 1921, p.76; Vibeto, Fen Complex, Telemark, Norway; Tröger 711; Tomkeieff p.594)*

VIBORGITE. See wiborgite.

VICOITE. A volcanic rock, close to the boundary between phonolitic tephrite and tephritic phonolite, largely composed of leucite, with lesser equal amounts of alkali feldspar and plagioclase and subordinate clinopyroxene. Cf. orvietite. *(Washington, 1906, p.57; Vico Volcano, near Viterbo, Italy; Tröger 539; Johannsen v.4, p.293; Tomkeieff p.594)*

VINTLITE. An obsolete local name for a variety of microdiorite or tonalite containing phenocrysts of hornblende, quartz and oligoclase. *(Pichler, 1875, p.927; Vintl, near Bressanone, Alto Adige, Italy; Tröger 144; Johannsen v.2, p.399; Tomkeieff p.595)*

VIPETOITE. See vibetoite.

VITERBITE. A variety of tephritic phonolite consisting of alkali feldspar, leucite, labradorite and augite. *(Washington, 1906, p.40; Viterbo, Italy; Tröger 528; Johannsen v.4, p.175; Tomkeieff p.595)*

VITRIC TUFF. Now defined in the pyroclastic classification (section 2.2.2, p.8) as a variety of tuff in which pumice and glass fragments predominate. *(Pirsson, 1915, p.193; Tomkeieff p.595)*

VITROPHYRE. A term for variety of porphyry in which the groundmass is glassy. Also applied to the basal portions of many welded ignimbrites. *(Vogelsang, 1872, p.534; Tröger 1015; Johannsen v.2, p.275; Tomkeieff p.597)*

VITROPHYREID. An obsolete field term for a porphyritic glass. *(Johannsen, 1911, p.322; Tomkeieff p.597)*

VITROPHYRIDE. A revised spelling recommended to replace the field term vitrophyreid. Now obsolete. *(Johannsen, 1926, p.182; Johannsen v.1, p.58)*

VITROPHYRITE. An obsolete term for non-porphyritic rocks with a glassy texture. Cf. pitchstone. *(Vogelsang, 1872, p.534; Tröger 1016; Tomkeieff p.597)*

VOGESITE. A variety of lamprophyre in which amphibole is more abundant than biotite and alkali feldspar is more abundant than plagioclase. Augite is frequently present. Now defined in the lamprophyre classification (Table 2.9, p.19). *(Rosenbusch, 1887, p.319;*

Vosges, France; Tröger 249; Johannsen v.3, p.37; Tomkeieff p.597)

VOLCANIC. A loosely defined term pertaining to those igneous processes that occur on or very close to the surface of the Earth. Volcanic rocks are usually fine-grained.

Volcanite. A name proposed for porphyritic material found in some of the bombs ejected from Vulcano in the Lipari Islands. The phenocrysts were described as anorthoclase, andesine and augite. *(Hobbs, 1893, p.602; Vulcano, Lipari Islands, Italy; Johannsen v.1 (2nd Edn), p.287; Tomkeieff p.599)*

Volhynite (Wolhynite). A local name, originally spelt wolhynite, for a variety of microgranodiorite which occurs as dykes and contains phenocrysts of labradorite, hornblende and biotite in a groundmass of andesine, orthoclase and quartz. *(Kroustchoff, 1885, p.441; Volhynian Province, Ukraine; Tröger 339; Tomkeieff p.600)*

Vredefortite. A porphyritic granogabbro containing phenocrysts of labradorite, hypersthene (= enstatite), and biotite in a matrix of quartz and K-feldspar, corresponding to the vredefortitic magma-type of Niggli (1936, p.369). *(Tröger, 1938, p.56; Vredefort, South Africa; Tröger(38) 115½; Tomkeieff p.600)*

Vulcanite. A term proposed for all extrusive rocks. *(Scheerer, 1862, p.138; Johannsen v.1 (2nd Edn), p.288; Tomkeieff p.600)*

Vulsinite. A term originally defined as the volcanic equivalent of monzonite. The rock consists mainly of Na-orthoclase with smaller amounts of labradorite and minor clinopyroxene and biotite. *(Washington, 1896a, p.547; from the Etruscan tribe of Vulsinii, Italy; Tröger 253; Johannsen v.4, p.42; Tomkeieff p.600)*

WEBSTERITE. A variety of pyroxenite consisting of equal amounts of orthopyroxene and clinopyroxene. Now defined modally in the ultramafic rock classification (Fig. 2.9,

p.28). *(Williams, 1890, p.44; Webster County, North Carolina, USA; Tröger 678; Johannsen v.4, p.460; Tomkeieff p.604)*

WEHRLITE. An ultramafic plutonic rock composed of olivine and clinopyroxene often with minor brown hornblende. Now defined modally in the ultramafic rock classification (Fig. 2.9, p.28). *(Kobell, 1838, p.313; named after Wehrle, who analysed the rock; Tröger 734; Johannsen v.4, p.419; Tomkeieff p.604)*

Weigelite. An obsolete name for a variety of peridotite which occurs as dykes and consists essentially of enstatite, later altered to actinolite, olivine and hornblende. Cf. valbellite. *(Kretschmer, 1917, p.113; Weigelsberg, near Habartice, N. Moravia, Czech Republic; Tröger 1017; Johannsen v.4, p.428; Tomkeieff p.604)*

Weilburgite. A local term proposed for a rock, previously called keratophyre spilite, consisting mainly of alkali feldspar, chlorite and considerable carbonate. *(Lehmann, 1949, p.80; Weilburg, Lahn district, Hessen, Germany; Tomkeieff p.604)*

Weiselbergite. An obsolete term originally used for a palaeovolcanic augite andesite which was usually altered. Later used for an altered glassy andesite (Wadsworth, 1884) – the fresh variety has been called shastalite. *(Rosenbusch, 1887, p.501; Weiselberg, St Wendel, Saarland, Germany; Tröger 155; Johannsen v.3, p.170; Tomkeieff p.604)*

Wennebergite. An obsolete local name for a variety of trachyandesite containing phenocrysts of sanidine, chloritized biotite and quartz in a groundmass of the same minerals with oligoclase. *(Schowalter, 1904, p.33; Wenneberg, Ries, Bavaria, Germany; Tröger 104; Tomkeieff p.605)*

Wernerite. An obsolete term for coarse-grained rock consisting of feldspar with minor ferromagnesian minerals. *(Pinkerton, 1811a, p.205; named after A.G. Werner; Tomkeieff p.605)*

WESSELITE. A local name for a melanocratic variety of nephelinite which occurs as dykes and consists of abundant phenocrysts of anomite (trioctahedral mica), syntagmatite (=titanian hastingsite), barkevikite (see p.44), and titanian augite in a matrix of nepheline, analcime and haüyne. *(Scheumann, 1922, p.505; Wesseln (now Veselí), N. Bohemia, Czech Republic; Tröger 624; Johannsen v.4, p.385; Tomkeieff p.605)*

WESTERWALDITE. An obsolete name for a volcanic rock, previously called essexite-basalt, consisting of serpentinized olivine phenocrysts in a groundmass of olivine, augite, labradorite, sanidine, biotite, ore and interstitial nepheline. *(Johannsen, 1938, p.203; Stöffel, Marienberg, Westerwald, Germany; Tröger(38) 579½; Tomkeieff p.605)*

WHINSTONE. An obsolete local name for dolerite, basalt and other dark fine-grained igneous rocks. *(Original reference uncertain; Whin Intrusive Sill, Northumberland, England, UK)*

WIBORGITE (VIBORGITE). A local name for a variety of rapakivi granite in which the ovoid crystals of orthoclase are mantled by oligoclase. *(Wahl, 1925, p.42; Wiborg, now Vyborg, near St Petersburg, Russian Federation; Tröger 80; Tomkeieff p.606)*

WICHTISITE. An obsolete name for a glassy dolerite dyke or selvage to a dyke. *(Hausmann, 1847, p.551; Wichtis, now Vihti, Finland; Tröger 1018; Johannsen v.3, p.324; Tomkeieff p.606)*

WILSONITE. A term for a strongly welded rhyolitic or dacitic tuff later renamed owharoite. *(Henderson, 1913, p.70; Tröger(38) 1018½; Tomkeieff p.606)*

WINDSORITE. A local name for an aplitic rock consisting essentially of alkali feldspar and oligoclase with smaller amounts of quartz and biotite. *(Daly, 1903, p.48; Windsor, Vermont, USA; Tröger 91; Johannsen v.2, p.303; Tomkeieff p.606)*

WOLGIDITE. A name for a rock largely composed of leucite, mica and amphibole with subordinate clinopyroxene in a serpentine-rich matrix. Now regarded as a diopside-leucite-richterite madupitic lamproite (Table 2.7, p.17). *(Wade & Prider, 1940, p.50; Wolgidee Hills, Kimberley district, West Australia, Australia; Tomkeieff p.607)*

WOLHYNITE. See volhynite.

WOODENDITE. An obsolete name for a volcanic rock composed of phenocrysts of augite, enstatite and serpentinized olivine in an abundant matrix of glass. Although the rock is rich in alkalis, it is devoid of modal feldspar. *(Skeats & Summers, 1912, p.29; Woodend, Victoria, Australia; Tröger 272; Johannsen v.3, p.120; Tomkeieff p.607)*

WYOMINGITE. Originally described as a variety of leucitite composed of clinopyroxene, mica and leucite in a glassy matrix. Madupite is a more mafic variety. Now regarded as a diopside-leucite-phlogopite lamproite (Table 2.7, p.17). *(Cross, 1897, p.120; Leucite Hills, Wyoming, USA; Tröger 503; Johannsen v.4, p.356; Tomkeieff p.607)*

XENITE. An obsolete name given to veins containing feldspar, some micas and garnets. *(Rozière, 1826, p.305; from the Greek xenos = stranger; Tomkeieff p.608)*

XENOPHYRE. An obsolete name for porphyritic rocks occurring as veins. *(Rozière, 1826, p.305; Tomkeieff p.609)*

YAKUTITE. A term given to some rare kalsilite-aegirine-orthoclase rocks, known only from one locality. Modally they vary widely with some samples containing up to 40% kalsilite. *(Vorob'ev et al., 1984, p.323; Malomurunskii (Murun) complex, Aldan Province, Russian Federation)*

YAMASKITE. A local name for a variety of pyroxenite, related to jacupirangite, composed of titanian augite, alkali amphibole and small amounts of anorthite. *(Young, 1906, p.16; Mt Yamaska, Montreal, Quebec,*

Canada; Tröger 690; Johannsen v.3, p.341; Tomkeieff p.610)

YATALITE. A local name for a coarse-grained uralitized mafic pegmatite composed of actinolite (after augite), albite, microcline and quartz. *(Benson, 1909, p.104; Hundred of Yatala, Houghton, South Australia, Australia; Tröger 230; Johannsen v.3, p.144; Tomkeieff p.610)*

YENTNITE. A name originally given to a rock consisting of sodic plagioclase, biotite and scapolite. However, the scapolite was shown to be quartz and the name was withdrawn. *(Spurr, 1900b, p.315; Yentna River, Alaska, USA; Tröger 1019; Johannsen v.3, p.157; Tomkeieff p.610)*

YOGOITE. A name originally defined as a variety of syenite containing equal amounts of orthoclase and augite. However, as the type rock also contained equal amounts of orthoclase and plagioclase the term was later withdrawn in favour of monzonite. *(Weed & Pirsson, 1895b, p.472; Yogo Peak, Little Belt Mts, Montana, USA; Tröger 279; Johannsen v.3, p.63; Tomkeieff p.610)*

YOSEMITITE. A term for light-coloured granites that correspond to the yosemititic magmatype of Niggli (1923, p.111). *(Tröger, 1935, p.47; El Capitan, Yosemite Valley, California, USA; Tröger 84; Tomkeieff p.610)*

YUKONITE. A leucocratic aplitic dyke rock of tonalitic composition consisting of oligoclase, quartz and biotite. *(Clarke, 1904, p.270; Yukon River, above Fort Hamlin, Alaska, USA; Tröger 136; Johannsen v.2, p.401; Tomkeieff p.611)*

ZIRKELYTE (ZIRKELITE). An obsolete general name for altered basaltic glass. *(Wadsworth, 1887, p.30; named after Zirkel; Tomkeieff p.612)*

ZOBTENITE. A local name for an augen gabbro-gneiss with knots of augite set in streams of uralite (= actinolite pseudomorph after pyroxene) embedded in a matrix of epidote and saussuritized plagioclase. *(Roth, 1887, p.611; Zobtenberg, Silesia, Poland; Tröger 1021; Johannsen v.3, p.230; Tomkeieff p.612)*

ZUTTERITE. An erroneous spelling of rutterite. *(Tomkeieff et al., 1983, p.613)*

4 Bibliography of terms

The references given in this section are either the source references for rock terms or are subsidiary references used in the petrological descriptions. Following each reference, in square brackets, is a list of the rock terms thought to have occurred first in that reference. If a rock term is printed in italics, then the reference is only to be found in the petrological description of the term and is not the source of a new occurrence of the term.

To avoid any possible ambiguity and confusion in the future, the Subcommission decided to give the references in full in the bibliography, including full journal titles. This was done because many of the references cited in the early literature were so abbreviated that a considerable amount of time and effort was wasted in trying to locate them in the various libraries throughout the world when they were being checked. The style of the journal titles is, wherever possible, that found in the Natural History Museum, London.

The 1st edition contained 791 references from 377 different journals and publishers. This edition now contains 809 references from 390 different journals. The Subcommission has always taken a great deal of effort to check as many as possible of the references, a task which has taken an enormous amount of time. In the first edition, only 15 of the references were not able to be located. In this edition the number is down to 4 – a truly remarkable effort by the editorial team. As before, these references have all been preceded with an "*" to indicate that they may not be entirely accurate. The references in question are: Bayan, 1866; Fedorov, E., 1896; Marzari Pencati, G., 1819; and Willems, H.W.V., 1934. If any reader can

throw light on any of these references the editor would be extremely grateful.

During the recalculation of the data for Tables 4.1 to 4.5 it was found that several errors had appeared in the equivalent tables in the 1st edition (Tables D.1 to D.5), mainly by counting references that had only been cited in the rock descriptions and did not contain new rock names.

4.1 BIBLIOGRAPHIC ANALYSIS

Table 4.1 shows which languages were used in the publication of the new rock names and terms contained in the glossary. As can be seen, although a total of ten different languages were involved, 90% of all the new rock names were published in three languages, English, German and French. Similarly, nearly 85% of the publications containing new rock names

Table 4.1 *Numbers of new rock terms and their references by publication language*

Language	Rocks	Acc%	Refs.	Acc%
English	741	47.5	321	41.6
German	462	77.1	230	71.4
French	201	90.0	101	84.5
Russian	80	95.1	59	92.1
Italian	32	97.2	29	95.9
Scandinavian	26	98.8	22	98.7
Spanish	14	99.7	6	99.5
Bulgarian	2	99.9	1	99.6
Dutch	1	99.9	2	99.9
Slovakian	1	100.0	1	100.0
Totals	1560		772	

are also in English, German and French.

Tables 4.2 and 4.3 give the more prolific authors of new rock names and of publications containing new rock names, respectively. Although Johannsen was by far the most prolific contributor of new rock names with 134, Lacroix was by far the most prolific author of publications containing new rock names with 36. Of these 24 were published in the *Compte Rendu*. This is nearly twice the number published by Washington with 20 and next on the list.

Similar data for some of the journals and publishing houses are shown in Tables 4.4 and 4.5. Worthy of mention are the *Journal of Geology*, *Compte Rendu* and the *Neues Jahrbuch*, for their contributions to igneous nomenclature.

Table 4.2 *Authors who introduced ten or more new rock terms; the number of references in which they were published is also given*

	Rocks	Acc%	Refs.
Johannsen	134	8.6	10
Streckeisen	97	14.8	11
Lacroix	70	19.3	36
Brögger	65	23.5	10
Rosenbusch	34	25.6	10
Iddings	27	27.4	10
Washington	27	29.1	20
Loewinson-Lessing	26	30.8	11
Pinkerton	24	32.3	2
Cordier	23	33.8	3
Tröger	22	35.2	4
Belyankin	21	36.5	8
Gümbel	19	37.8	5
Brongniart	17	38.8	4
Pirsson	16	39.9	11
Shand	16	40.9	6
Duparc	13	41.7	10
Bořický	13	42.6	4
Vogelsang	12	43.3	2
Rittmann	11	44.0	3
Leonhard	11	44.7	4
Niggli	10	45.4	3

Table 4.3 *Authors with 5 or more publications containing new rock terms; the total number of new rock terms which they contain is also given*

	Refs.	Acc%	Rocks
Lacroix	36	4.7	70
Washington	20	7.3	27
Streckeisen	11	8.7	97
Pirsson	11	10.1	16
Loewinson-Lessing	11	11.5	26
Johannsen	10	12.8	134
Brögger	10	14.1	65
Rosenbusch	10	15.4	34
Iddings	10	16.7	27
Duparc	10	18.0	13
Belyankin	8	19.0	21
Osann	7	19.9	8
Shand	6	20.7	16
Zirkel	6	21.5	11
Eckermann	6	22.3	9
Tyrrell	6	23.1	7
Judd	6	23.8	6
Hibsch	6	24.6	6
Gümbel	5	25.3	19
Törnebohm	5	25.9	5

Table 4.4 *Journals and publishers with 20 or more new rock terms; the total number of new rock terms which they contained is also given*

Rocks	Acc%	Refs.	
110	7.1	32	*Journal of Geology*
67	31.9	30	*Neues Jahrbuch*
62	11.0	4	*Chicago University Press*
48	14.1	1	*Geotimes*
47	17.1	35	*Compte Rendu Hebdomadaire des Séances de l'Académie des Sciences*
46	23.0	7	*Wiley*
46	20.1	6	*Skrifter udgit av Videnskabsselskabet i Kristiania. I. Math.-Nat.*
37	25.4	19	*Tschermaks Mineralogische und Petrographische Mitteilungen*
35	27.6	27	*American Journal of Science*
33	34.0	9	*Schweizerbart, Stuttgart*
24	37.1	13	*Zeitschrift der Deutschen Geologischen Gesellschaft*
24	35.6	2	*White, Cochrane & Co., London*
23	38.6	19	*Geological Magazine*
23	40.1	15	*Geologiska Föreningens i Stockholm Förhandlingar*
20	41.3	3	*Journal des Mines*

Table 4.5 *Journals and publishers with 10 or more publications containing new rock terms; the total number of new rock terms which they contained is also given*

Refs.	Acc%	Rocks	
35	4.5	47	*Compte Rendu Hebdomadaire des Séances de l'Académie des Sciences*
32	8.7	110	*Journal of Geology*
30	12.6	67	*Neues Jahrbuch*
27	16.1	35	*American Journal of Science*
19	18.5	37	*Tschermaks Mineralogische und Petrographische Mitteilungen*
19	21.0	23	*Geological Magazine*
15	22.9	23	*Geologiska Föreningens i Stockholm Förhandlingar*
13	24.6	24	*Zeitschrift der Deutschen Geologischen Gesellschaft*
10	25.9	17	*Quarterly Journal of the Geological Society of London*

4.2 REFERENCES

ABICH, H., 1839. Über Erhebungs Kratere und das Band inneren Zusammenhanges, welches in der Richtung bestimmter Linien, räumlich oft weit von einander getrennte vulkanische Erscheinungen und Gebilde zu ausgedehnten Zügen unter einander vereinigt [Abstract]. *Neues Jahrbuch für Mineralogie, Geognosie, Geologie und Petrefaktenkunde. Stuttgart.* Vol.10, p.334–337. [Haüynophyre]

ABICH, H., 1841. Geologische Beobachtungen über die vulkanischen Erscheinungen und Bildungen in Unter- und Mittelitalien. Vol. 1. Ueber die Natur und den Zusammenhang der vulkanischen Bildungen. *Vieweg, Braunschweig.* 134pp. [Acid, Basic, Trachydolerite]

ADAMS, F.D., 1913. Excursion A7. The Monteregian Hills. Guide Book No.3. *Department of Mines & Geological Survey, Canada. Issued as part of the 12th International Geological Congress, Toronto.* p.29–80. [Montrealite]

ADAMS, F.D. & BARLOW, A.E., 1908. The nepheline and associated alkali syenites of Eastern Ontario. *Transactions of the Royal Society of Canada.* Vol.2, Ser.3, p.3–76. [Dungannonite, Raglanite]

ADAMS, F.D. & BARLOW, A.E., 1910. Geology of the Haliburton and Bancroft areas, Province of Ontario. *Memoirs. Geological Survey, Canada. Ottawa.* Vol.6, p.1–419. [Craigmontite, Monmouthite]

ADAMS, F.D. & BARLOW, A.E., 1913. Excursion A2. The Haliburton–Bancroft area of Central Ontario. Guide Book No.2. *Department of Mines and Geological Survey, Canada. Issued as part of the 12th International Geological Congress, Toronto.* p.1–98. [Congressite]

ADAMSON, O.J., 1944. The petrology of the Norra Kärr District. An occurrence of alkaline rocks in Southern Sweden. *Geologiska Föreningens i Stockholm Förhandlingar.*

Stockholm. Vol.66, p.113–255. [Grennaite, Kaxtorpite, Subphonolite]

ADAN DE YARZA, R., 1893. Roca eruptiva de Fortuna (Murcia). *Boletin de la Comisión del Mapa Geológico de España. Madrid.* Vol.20, p.349–353. [Fortunite]

AINBERG, L.F., 1955. K voprosu genezisa charnokitov i porod charnokitovoi serii. *Izvestiia Akademii Nauk SSSR, Seriia Geologicheskaia.* No.4, p.102–120. [Kuzeevite (Kuseevite)]

AITCHISON, J., 1984. The statistical analysis of geochemical compositions. *Journal of the International Association for Mathematical Geology. New York.* Vol.16, p.531–564. [Boxite, Coxite, Hongite, Kongite]

ALDROVANDI, U., 1648. Bononiensis Musæi Metallici. *Ferronii [Bologna].* 992pp. [Variolite]

ALLEN, J.A., 1914. Geology of field map-area, B.C. and Alberta. *Memoirs. Geological Survey, Canada. Ottawa.* Vol.55, p.1–312. [Sphenitite]

ALMEIDA, F.F.M. DE, 1961. Geologia e Petrologia da Ilha da Trindade. *Monografias. Divisão de Geologia e Mineralogia, Ministério das Minas e Energia, Brasil. Rio de Janeiro.* Mono.18, 197pp. [Grazinite]

AMSTUTZ, A., 1925. Les roches éruptives des environs de Dorgali et Orosei en Sardaigne. *Schweizerische Mineralogische und Petrographische Mitteilungen. Zürich.* Vol.5, p.261–321. [Dorgalite]

ANDERSON, C.A., 1941. Volcanoes of the Medicine Lake Highland, California. *University of California Publications in Geological Sciences. Berkeley.* Vol.25, No.7, p.347–422. [Basaltic andesite]

ANDRADE, M. DE, 1954. Contribution à l'étude des roches alcalines d'Angola (note preliminaire). *19th International Geological Congress, Algiers. Comptes Rendus.* Vol.19, p.241-252. [Angolaite]

ANTONOV, L.B., 1934. The Apatite Deposits of

the Khibina Tundra. *The Khibina Apatite, Leningrad.* Vol.7, 196pp. [Lujavritite]

BABINGTON, W., 1799. A New System of Mineralogy, in the form of a catalogue, after the manner of Baron Born's Systematic Catalogue of the Collection of Fossils of Mlle Éléonore de Raab. *Bensley, London.* 279pp. [Pitchstone]

BACKLUND, H.G., 1915. Nefelinovyj bazalt (onkilonit) s severnago ledovitago okeana. *Izvestiia Imperatorskoi Akademii Nauk. Sankt Peterburg.* Vol.9, p.289–308. [Onkilonite]

BAILEY, E.B. & MAUFE, H.B., 1916. The geology of Ben Nevis and Glen Coe, and the surrounding country. *Memoirs. Geological Survey of Scotland. Edinburgh.* Sheet 53, p.1–247. [Aplo-, Appinite, Calc-tonalite]

BAILEY, E.B. & MAUFE, H.B., 1960. The geology of Ben Nevis and Glen Coe and the surrounding country. (Explanation of Sheet 53). (2nd revised Edn by E.B. Bailey.) *Memoirs. Geological Survey of Scotland. Edinburgh.* 2nd Edn, 307pp. [*Aplo-*]

BARTH, T.F.W., 1927. Die Pegmatitgänge der Kaledonischen Intrusivgesteine im Seiland-Gebiete. *Skrifter utgitt av det Norske Videnskaps-Akademi i Oslo. Mat.-naturv.Kl.* No.8, p.1–123. [Antifenitepegmatite]

BARTH, T.F.W., 1930. Pacificite, an anemousite basalt. *Journal of the Washington Academy of Sciences.* Vol.20, p.60–68. [*Kaulaite*, Olivine pacificite, Pacificite]

BARTH, T.F.W., 1944. Studies on the igneous rock complex of the Oslo Region. II. Systematic petrology of the plutonic rocks. *Skrifter utgitt av det Norske Videnskaps-Akademi i Oslo. Mat.-naturv.Kl.* No.9, p.1–104. [Apotroctolite, Hovlandite, Oslo-essexite]

BASCOM, F., 1893. The structures, origin, and nomenclature of the acid volcanic rocks of South Mountain. *Journal of Geology. Chicago.* Vol.1, p.813–832. [Apo-]

*BAYAN, 1866. Katalog des geognostischen

Museums kaukasischer Mineralwässer und Kurorte. [Beschtauite]

BAYLEY, W.S., 1892. Eleolite-syenite of Litchfield, Maine and Hawes' hornblende-syenite from Red Hill, New Hampshire. *Bulletin of the Geological Society of America. New York.* Vol.3, p.231–252. [Litchfieldite]

BECK, R., 1907. Untersuchungen über einige südafrikanische Diamantenlagerstätten. *Zeitschrift der Deutschen Geologischen Gesellschaft. Berlin.* Vol.59, No.11, p.275–307. [Griquaite]

BECKE, F., 1899. Der Hypersthen-Andesit der Insel Alboran. *Tschermaks Mineralogische und Petrographische Mitteilungen. Wien.* Vol.18, 2nd Ser., p.525–555. [Alboranite, Santorinite]

BECKER, G.F., 1888. Geology of the quicksilver deposits of the Pacific slope. *Monographs of the United States Geological Survey. Washington.* Vol.13, p.1–486. [Asperite]

BELL, J.M., CLARKE, E.C. & MARSHALL, P., 1911. The geology of the Dun Mountain Subdivision, Nelson. *Bulletin. Geological Survey of New Zealand. Wellington.* Vol.12, p.1–71. [Rodingite]

BELYANKIN, D.S., 1911. Ob albitovom diabaze iz Krasnoi Polyany i o kontakte ego so slantsami. *Izvestiia Sankt-Peterburgskago Polytekhnicheskago Instituta Imperatora Petra Velikago. Sankt Peterburg.* Vol.15. [Anorthobase, Orthobase]

BELYANKIN, D.S., 1923. K petrografii perevala Shtuli-Atsek v Tsentralnom Kavkaze. *Izvestiia Rossiiskoi Akademii Nauk. Petrograd. VI Ser.* Vol.17, p.95–102. [Cryptodacite, Introdacite, Phanerodacite]

BELYANKIN, D.S., 1924. K voprosu o vozraste nekotorykh kavkazskikh intruzii. *Izvestiia Geologicheskogo Komiteta. Sankt Peterburg.* Vol.43(3), p.409–424. [Caucasite (Kaukasite)]

BELYANKIN, D.S., 1929. On the term "rock" and on petrographical classification and nomenclature. *Trudy Mineralogicheskogo Muzeya (im A.E. Fersmana). Akademiya Nauk SSSR. Leningrad.* Vol.3, p.12–24. [Aegineite, Aegisodite, Amneite, Anam-aegisodite, Barneite, Bineite, Binemelite, *Biquahororthandite*, Dimelite, *Hobiquandorthite*, Pynosite, Tipyneite, *Topatour-biolilepiquorthite*]

BELYANKIN, D.S., 1931. K voprosu ob anemuzite (po povodu patsifitov Barta). *Doklady Akademii Nauk SSSR. Leningrad.* Ser.A, p.28–31. [Albasalt]

BELYANKIN, D.S. & KUPLETSKII, B.M., 1924. Gornye porody i poleznye iskopaemye Severnogo poberezhya i prilegayushchikh k nemu ostrovov Kandalakshskoi guby. *Trudy Severnoi Nauchno-promyshlennoi Ekspeditsii.* Vol.28, p.3–76. [Turjite]

BELYANKIN, D.S. & PETROV, V.P., 1945. Petrografiya Gruzii. *Petrografiia SSSR, Ser. I, Regionalnaia Petrografiia, No.11, Institut Geologicheskikh Nauk, Akademiia Nauk SSSR.* 394pp. [Rikotite]

BELYANKIN, D.S. & VLODAVETS, V.I., 1932. Shchelochnoi kompleks Turego mysa. *Trudy Petrograficheskogo Instituta. Akademiia Nauk SSSR. Leningrad.* Vol.2, p.45–71. [Aegiapite, Calcitite]

BENSON, W.N., 1909. Petrographical notes on certain Pre-Cambrian rocks of the Mount Lofty Ranges, with special reference to the geology of the Houghton district. *Transactions of the Royal Society of South Australia. Adelaide.* Vol.33, p.101–140. [Yatalite]

BERTELS, G.A., 1874. Ein neues vulkanisches Gesteine. *Verhandlungen der Physikalisch-Medizinischen Gesellschaft zu Würzburg.* Vol.8, Neue Folge, p.149–178. [Isenite]

BERTOLIO, S., 1895. Sulle comenditi, nuovo gruppo di rioliti con aegirina. *Atti della Reale Accademia (Nazionale) dei Lincei. Rendiconti. Classe di Scienze Fisiche, Matematiche e Naturali. Roma.* Vol.4(2), p.48–50. [Comendite]

BÉTHUNE, P. DE, 1956. La busorite, une roche feldspathoïdale nouvelle, du Kivu. *Bulletin de la Société Belge de Géologie, de Paléontologie et d'Hydrologie. Bruxelles.* Vol.65, p.394–399. [Busorite]

BEUDANT, F.S., 1822. Voyage minéralogique et géologique en Hongrie pendant l'année 1818. *Verdière Libraire, Paris.* Vol.3, 659pp. [Perlite]

BEUS, A.A., SEVEROV, E.A., SITNIN A.A. & SUBBOTIN, K.D., 1962. Albitizirovannye i greizenizirovannye granity (apogranity). *Izadatel'stvo Akademii Nauk SSSR, Moskva.* 196pp. [Apogranite]

BEZBOROD'KO, N.I., 1931. K petrogenezisu temnotsvetnykh porod Podolii i sosednykh raionov. *Trudy Mineralogicheskogo Instituta. Akademiia Nauk SSSR. Leningrad.* Vol.1, p.127–159. [Bugite, Epibugite, Katabugite, Mesobugite, Sabarovite]

BEZBOROD'KO, N.I., 1935. Montsonitovyi ryad i montsonity Ukrainy. *Trudy Petrograficheskogo Instituta. Akademiia Nauk SSSR. Leningrad.* Vol.5, p.169–207. [Ukrainite]

BLUM, R., 1861. Foyait, ein neues Gestein aus Süd-Portugal. *Neues Jahrbuch für Mineralogie, Geognosie, Geologie und Petrefaktenkunde. Stuttgart.* Vol.32, p.426–433. [Foyaite]

BOASE, H.S., 1834. A Treatise on Primary Geology. *Longman, London.* 399pp. [Felsparite]

BOETIUS DE BOOT, ANSELMUS, 1636. Gemmarvm et Lapidum Historia. *Joannis Maire, Lvgdvni Batavorvm [Rotterdam].* Editor Adrianus Toll. 2nd Edn, 576pp. [Bimsstein (Bynftein, Bymsftein)]

BOMBICCI, L., 1868. Sulla oligoclasite di Monte Cavaloro, presso Riola nel Bolognese e sulla composizione della pirite magnetica. *Memorie della Reale Accademia delle Scienze dell'Instituto di Bologna. Classe di Scienze Fisiche. Bologna.* Vol.8, p.77–114. [Oligoclasite]

BONNEY, T.G., 1899. The parent-rock of the diamond in South Africa. *Geological Magazine. London.* Vol.6, Decade 4, p.309–321. [Newlandite]

BOŘICKÝ, E., 1873. Petrographische Studien an den Phonolithgesteinen Böhmens. *Archiv für die Naturwissenschaftliche Landesdurchforschung von Böhmen. Prag.* Vol.3, Pt.II, No.1, p.1–96. [Haüyne phonolite, Trachyphonolite]

BOŘICKÝ, E., 1874. Petrographische Studien an den Basaltgesteinen Böhmens. *Archiv für die Naturwissenschaftliche Landesdurchforschung von Böhmen. Prag.* Vol.2, Pt.II, No.5, p.1–294. [Andesite-basalt, Leucitoidbasalt, Magmabasalt, Nephelinitoid, Noseanite, Oligoclase basalt, Peperin-basalt, Trachybasalt]

BOŘICKÝ, E., 1878. Der Glimmerpikrophyr, eine neue Gesteinsart und die Libsicer Felswand. *Mineralogische und Petrographische Mitteilungen. Wien.* Vol.1, p.493–516. [Picrophyre]

BOŘICKÝ, E., 1882. Petrologische Studien an den Porphyrgesteinen Böhmens. *Archiv für die Naturwissenschaftliche Landesdurchforschung von Böhmen. Prag.* Vol.4, No.4, p.1–177. [Radiophyre, Radiophyrite]

BORISOV, I., 1963. Petrografski izuchvaniya na magmatitite severno ot gr. Burgas i sravnitelna petrokhimichna kharakteristika na gornokrednite vulkanity v Balgariya. *Godishnik na Sofiiskiya Universitet, Geologo-Geografski Fakultet, Kniga 1, Geologiya.* Vol.58, p.197–233. [Balgarite, Burgasite]

BORN, T. VON, 1790. Catalogue méthodique et raisonné de la collection des fossiles de Mlle Eléonore de Raab. *Vienne.* Vol.1, 500pp. [Basaltine]

BOWEN, N.L., 1914. Art. XIX – The ternary system: diopside – forsterite – silica. *American Journal of Science. New Haven.* Vol.38, 4th Ser., p.207–264. [*Eutectite*]

BOWEN, N.L., 1915. Art. XVI – The crystallization of haplobasaltic, haplodioritic and related magmas. *American Journal of Science. New Haven.* Vol.40, 4th Ser., p.161–185. [Haplo-]

BRADLEY, F.H., 1874. Communication: On unakyte, an epidote rock from the Unaka range, on the borders of Tennessee and North Carolina. *American Journal of Science. New Haven.* Vol.7, 3rd Ser., p.519–520. [Unakyte (Unakite)]

BRAUNS, R., 1922. Die phonolitischen Gesteine des Laacher Seegebietes und ihre Beziehungen zu anderen Gesteinen dieses Gebietes. *Neues Jahrbuch für Mineralogie, Geologie und Paläontologie. Stuttgart. Referate Abt. A.* Bd.46, p.1–116. [Riedenite, Schorenbergite, Selbergite]

BREITHAUPT, A., 1826. Tachylyt, sehr wahrscheunlich eine neue Mineral-Species. *Archiv für die Gesammte Naturlehre. Nürnberg.* Vol.7, p.112–113. [Tachylyte]

BREITHAUPT, J.F.A., 1861. Timazit, eine neue Gesteinsart, und Gamsigradit, ein neuer Amphibol. *Berg- und Hüttenmännische Zeitung. Freiberg, Leipzig.* Vol.20, p.51–54. [Timazite]

BROCCHI, G.B., 1817. Catalogo Ragionato di una Raccolta di Rocce. *Dall'Imperiale regia stamperia, Milano.* Monograph, 346pp. [Necrolite]

BRÖGGER, W.C., 1890. Die Mineralien der Syenitpegmatitgänge der südnorwegischen Augit- und Nephelinsyenite. *Zeitschrift für Kristallographie und Mineralogie. Leipzig.* Vol.16, p.1–663. [Akerite, Grorudite, Lardalite (Laurdalite), Larvikite (Laurvikite), Lujavrite (Luijaurite, Lujauvrite), Nordmarkite, Särnaite]

BRÖGGER, W.C., 1894. Die Eruptivgesteine des Kristianiagebietes. I. Die Gesteine der Grorudit-Tinguait Serie. *Skrifter udgit av Videnskabsselskabet i Kristiania. I. Math.-Nat. Klasse.* No.4, p.1–206. [Hypabyssal,

Lindöite, Sölvsbergite, Sussexite]

BRÖGGER, W.C., 1895. Die Eruptivgesteine des Kristianiagebietes. II. Die Eruptionsfolge der triadischen Eruptivgesteine bei Predazzo in Südtyrol. *Skrifter udgit av Videnskabsselskabet i Kristiania. I. Math.-Nat. Klasse.* No.7, p.1–183. [Dellenite, Plauenite, Quartz monzonite, Quartz porphyrite, Trachyte-andesite]

BRÖGGER, W.C., 1898. Die Eruptivgesteine des Kristianiagebietes. III. Das Ganggefolge des Laurdalits. *Skrifter udgit av Videnskabsselskabet i Kristiania. I. Math.-Nat. Klasse.* No.6, p.1–377. [Alkaliplete, Calcioplete, Farrisite, Ferroplete, Hedrumite, Heumite, Kalioplete, Leucocrate, Leucocratic, Maenaite, Melanocrate, Melanocratic, Natrioplete, Osloporphyry, Oxyplete, Tönsbergite]

BRÖGGER, W.C., 1904. "Krageröit". Oversigt over Videnskabs-Selskabets Møder. *Forhandlinger i Videnskabsselskabet i Kristiania.* p.1–74. [Krageröite]

BRÖGGER, W.C., 1906. Eine Sammlung der Wichtigsten Typen der Eruptivgesteine des Kristianiagebietes nach ihren geologischen Verwandtschaftsbeziehungen geordnet. *Nyt Magazin for Naturvidenskaberne. Christiania.* Vol.44, Pt.2, p.113–144. [Ekerite, Essexite-melaphyre, Essexite-porphyrite, Kvellite, Tjosite]

BRÖGGER, W.C., 1921. Die Eruptivgesteine des Kristianiagebietes. IV. Das Fengebiet in Telemark, Norwegen. *Skrifter udgit av Videnskabsselskabet i Kristiania. I. Math.-Nat. Klasse.* No.9, p.1–408. [Calcitfels, Carbonatite, Damkjernite (Damtjernite), Fenite, Hollaite, Juvite, Kamperite, Kåsenite (Kosenite), Kristianite (Christianite), Melteigite, Rauhaugite, Ringite, Sannaite, Silicocarbonatite, Sövite (Soevite), Tveitåsite, Vibetoite (Vipetoite)]

BRÖGGER, W.C., 1931. Die Eruptivgesteine des Oslogebietes. V. Der grosse Hurumvulkan. *Skrifter utgitt av det Norske Videnskaps-Akademi i Oslo. Mat.-naturv.Kl.* No.6, p.1–146. [Hurumite, Pyritosalite, Töienite]

BRÖGGER, W.C., 1932. Die Eruptivgesteine des Oslogebietes. VI. Über verschiedene Ganggesteine des Oslogebietes. *Skrifter utgitt av det Norske Videnskaps-Akademi i Oslo. Mat.-naturv.Kl.* No.7, p.1–88. [Björnsjöite]

BRÖGGER, W.C., 1933. Die Eruptivgesteine des Oslogebietes. VII. Die chemische Zusammensetzung der Eruptivgesteine des Oslogebietes. *Skrifter utgitt av det Norske Videnskaps-Akademi i Oslo. Mat.-naturv.Kl.* No.1, p.1–147. [Essexite-akerite, Husebyite, Katnosite, Kjelsåsite, Modumite, *Osloessexite*, Sörkedalite]

BRONGNIART, A., 1807. Traité elémentaire de minéralogie. *Deterville, Paris.* Vol.1, 564pp. [Diabase]

BRONGNIART, A., 1813. Essai d'une classification minéralogique des roches mélangées. *Journal des Mines. Paris.* Vol.34, p.5–48. [Amphibolite, Euphotide, Eurite, Graphic granite, Hyalomicte, Melaphyre, Mimophyre, Mimose, Ophiolite, Pegmatite (Pegmatyte), Stigmite, Tephrine, Trachyte, Trappite]

BRONGNIART, A., 1824. Mémoire sur les terrains de sédiments supérieurs calcaréotrapéens du Vincentin. *Levrault, Paris.* 86pp. [Brecciole]

BRONGNIART, A., 1827. Classification et caractères minéralogiques des roches homogènes et hétérogènes. *Levrault, Paris.* 144pp. [Spilite]

BROUWER, H.A., 1909. Pienaarite, a melanocratic foyaite from Transvaal. *Proceedings of the Section of Sciences. Koninklijke (Nederlandse) Akademie van Wetenschappen te Amsterdam.* Vol.12, p.563–565. [Pienaarite]

BROUWER, H.A., 1917. On the geology of the alkali rocks in the Transvaal. *Journal of Geology. Chicago.* Vol.25, p.741–778.

[Leeuwfonteinite]

BRÜCKMANNS, U.F.B., 1778. Beiträge zu seiner Abhandlung von Edelsteinen. *Fürstl. Waisenhaus Buchhandlungen, Braunschweig.* 252pp. [Gestellstein, Granitello, Hornberg]

BUCH, L. VON, 1809. Geognostische Beobachtungen auf Reisen durch Deutschland und Italien mit ünum Anhange von mineralogischen Briefen aus Auvergne an den Geh. Ober-Bergrath Karsten von demsilben Verfasser. *Haude & Spener, Berlin.* Vol.1 & 2, p.1-224 & p.225-318. [Domite, Peperino]

BUCH, L. VON, 1810. Reise durch Norwegen und Lappland. *Nauck, Berlin.* Vol.1, 406pp. [Rhombporphyry]

BUCH, L. VON, 1824. VII. Ueber geognostische Erscheinungen im Fassa-Thale. *Taschenbuch für die Gesammte Mineralogie. Frankfurt am Main.* Vol.24, p.343–396. [Monzosyenite]

BUCH, L. VON, 1836. Ueber Erhebungscrater und Vulkane. *Annalen der Physik und Chemie. Leipzig.* Vol.37, p.169–190. [Andesite]

BÜCKING, H., 1881. Basaltische Gesteine aus der Gegend südwestlich vom Thüringer Walde und aus der Rhön. *Jahrbuch der Königlich Preussischen Geologischen Landesanstalt und Bergakademie. Berlin.* Vol.1, for 1880, p.149–189. [Basanitoid, Tephritoid]

BUGGE, J.A.W., 1940. Geological and petrographical investigations in the Arendal District. *Norsk Geologisk Tidsskrift. Oslo.* Vol.20, p.71–116. [Arendalite]

BUNSEN, R.W., 1851. Ueber die Processe der vulkanischen Gesteinsbildungen Islands. *Annalen der Physik und Chemie. Leipzig.* Vol.83, p.197–272. [Baulite]

BURRI, C. & PARGA-PONDAL, I., 1937. Die Eruptivgesteine der Insel Alborán (Provinz Almería, Spanien). *Schweizerische Mineralogische und Petrographische Mitteilungen. Zürich.* Vol.17, p.230–270. [Peralboranite]

BURT, D.M., SHERIDAN, M.F., BIKUN, J.V. & CHRISTIANSEN, E.H., 1982. Topaz rhyolites – distribution, origin, and significance for exploration. *Economic Geology and Bulletin of the Society of Economic Geologists. Lancaster, Pa.* Vol.77, p.1818–1836. [Topaz rhyolite]

BUSZ, K., 1904. Heptorit, ein Hauyn-Monchiquit aus dem Siebengebirge am Rhein. *Neues Jahrbuch für Mineralogie, Geologie und Paläontologie. Stuttgart. Referate Abt. A.* Bd.2, p.86–92. [Heptorite]

BUTAKOVA, E.L., 1956. K petrologii maimechakotuiskogo kompleksa ultraosnovnykh i shchelochnykh porod. *Trudy Nauchno-Issledovatel'skogo Instituta Geologii Arktiki. Leningrad.* Vol.89, p.201–249. [Kotuite]

CAESALPINO, A., 1596. De metallicis. *A. Zannetti, Roma.* Vol.2, Ch.11, 222pp. [Granite]

CAPELLINI, G., 1878. Inclusioni di apatite nella roccia di Monte Cavaloro. *Rendiconti delle Sessioni dell'Accademia delle Scienze dell'Instituto di Bologna.* p.122–125. [Cavalorite]

CARMICHAEL, I.S.E., 1964. The petrology of Thingmuli, a Tertiary volcano in Eastern Iceland. *Journal of Petrology. Oxford.* Vol.5, p.435–460. [Icelandite]

CATHREIN, A., 1890. Zur Dünnschliffsammlung der Tiroler Eruptivgesteine. *Neues Jahrbuch für Mineralogie, Geologie und Paläontologie. Stuttgart. Referate Abt. A.* Bd.1, p.71–82. [Adamellite]

CATHREIN, A., 1898. Dioritische Gang- und Stockgesteine aus dem Pusterthal. *Zeitschrift der Deutschen Geologischen Gesellschaft. Berlin.* Vol.50, p.257–278. [Klausenite]

CEDERSTRÖM, A., 1893. Om berggrunden på Norra delen af Ornön. *Geologiska Föreningens i Stockholm Förhandlingar. Stockholm.* Vol.15, p.103–118. [Ornöite]

CHAPPELL, B.W. & WHITE, A.J.R., 1974. Two contrasting granite types. *Pacific Geology. Tokyo.* Vol.8, p.173–174. [I-type granite, S-type granite]

CHAYES, F., 1966. Alkaline and subalkaline basalts. *American Journal of Science. New Haven.* Vol.264, p.128-145. [Subalkali basalt]

CHELIUS, C., 1892. Das Granitmassiv des Melibocus und seine Ganggesteine. *Notizblatt des Vereins für Erdkunde zu Darmstadt, des Mittelrheinischen Geologischen Vereins (und des Naturwissenschaftlichen Vereins zu Darmstadt).* Ser.4, Pt.13, p.1–13. [Alsbachite, Gabbrophyre, Luciite, Odinite, Orbite]

CHESTERMAN, C.W., 1956. Pumice, pumicite, and volcanic cinders in California. *Bulletin. California Division of Mines and Geology. San Francisco.* Vol.174, p.1–119. [Pumicite]

CHIRVINSKII, P.N., AFANASEV, M.S. & USHAKOVA, E.G., 1940. Massiv ultraosnovnykh porod pri stantsii Afrikanda. *Trudy Kolskoi Bazy Akademii Nauk SSSR.* Vol.5, p.31–70. [Afrikandite (Africandite)]

CHRUSTSCHOFF, K., 1888. Beiträge zur Petrographie Volhyniens und Russlands. *Mineralogische und Petrographische Mitteilungen. Wien.* Vol.9, p.470–527. [Perthitophyre]

CHRUSTSCHOFF, K., 1894. Über eine Gruppe eigenthümlicher Gesteine vom Taimyr-Lande aus der Middendorff'schen Sammlung. *Izvestiia Imperatorskoi Akademii Nauk. Sankt Peterburg.* Vol.35, p.421–431. [Taimyrite]

CHUDOBA, K., 1930. "Brandbergit", ein neues aplitisches Gestein aus dem Brandberg (SW-Afrika). *Centralblatt für Mineralogie, Geologie und Paläontologie. Stuttgart. Abt. A.* p.389–395. [Brandbergite]

CHUMAKOV, A.A., 1946. O nekotorykh kislykh differentsiatakh gabbro Mugodzhar. *Akademiku D.S. Beliankinu, Izdatel'stvo Akademii Nauk SSSR.* p.293–297. [Mugodzharite]

CLARKE, F.W., 1904. Analyses of rocks from the Laboratory of the United States Geological Survey 1880 to 1903. *Bulletin of the United States Geological Survey. Washington.* Vol.228, p.1–375. [Yukonite]

CLARKE, F.W., 1910. Analyses of rocks and minerals from the Laboratory of the United States Geological Survey 1880 to 1908. *Bulletin of the United States Geological Survey. Washington.* Vol.419, p.1–323. [Devonite]

CLOIZEAUX, A. DES, 1863. Note sur la classification des roches dites hypérites et euphotides. *Bulletin de la Société Géologique de France. Paris.* Vol.21, p.105–109. [Diallagite]

CLOOS, H., 1941. Bau und Tätigkeit von Tuffschloten. Untersuchungen an dem schwäbischen Vulkan. *Geologische Rundschau. Internationale Zeitschrift für Geologie. Stuttgart.* Vol.32, p.705-800. [Tuffisite]

COGNÉ, J. & GIOT, P., 1961. Observations à propos d'un Lamprophyre micacé à microcline et apatite, riche en enclaves granitiques, au flanc sud du Cap Sizun (Finistère). *Compte Rendu Hebdomadaire des Séances de l'Académie des Sciences. Paris.* Vol.252, p.2569-2571. [Sizunite]

COHEN, E., 1885. Berichtigung bezüglich des "Olivin-Diallag-Gesteins" von Schriesheim im Odenwald. *Neues Jahrbuch für Mineralogie, Geologie und Paläontologie. Stuttgart. Referate Abt. A.* Bd.1, p.242–243. [Hudsonite]

COLEMAN, A.P., 1899. A new analcite rock from Lake Superior. *Journal of Geology. Chicago.* Vol.7, p.431–436. [Heronite]

COLEMAN, A.P., 1914. The Pre-cambrian rocks north of Lake Huron with special reference to the Sudbury Series. *Ontario Bureau of Mines. Toronto.* Report 23, Pt.1, No.4, p.204–236. [Sudburite]

COLEMAN, R.G. & PETERMAN, Z.E., 1975. Oceanic plagiogranite. *Journal of Geophysical*

Research. Richmond, Va. Vol.80, p.1099–1108. [Oceanic plagiogranite]

COLONY, R.J. & SINCLAIR, J.H., 1928. The lavas of the Volcano Sumaco, Eastern Ecuador, South America. *American Journal of Science. New Haven.* Vol.16, 5th. Ser., p.299–312. [Andesite-tephrite]

COMUCCI, P., 1937. Di una singolare roccia del ghiacciaio Baltoro (Karakorum). *Atti della Reale Accademia (Nazionale) dei Lincei. Rendiconti. Classe di Scienze Fisiche, Matematiche e Naturali. Roma.* Vol.25, p.648–652 & p.734–738. [Baltorite]

CONYBEARE, J.J., 1817. XXII. Memoranda relative to the porphyritic veins, &c. of St. Agnes, in Cornwall. *Transactions of the Geological Society. London.* Vol.4, p.401–403. [Elvan]

COOMBS, D.S. & WILKINSON, J.F.G., 1969. Lineages and fractionation trends in undersaturated volcanic rocks from the East Otago Province (New Zealand) and related rocks. *Journal of Petrology. Oxford.* Vol.10, p.440–501. [Nepheline benmoreite, Nepheline hawaiite, Nepheline mugearite, Nepheline trachyandesite, Nepheline trachybasalt, Nepheline tristanite]

COQUAND, H., 1857. Traité des roches. *Baillière, Paris.* 421pp. [Albitophyre, Labradophyre, Oligophyre, Orthophyre, *Pyroxenite*]

CORDIER, P.L.A., 1816. Sur les substances minérales dites en masses qui entrent dans la composition des roches volcaniques de tous les ages. *Journal de Physique, de Chimie et d'Histoire Naturelle. Paris.* Vol.83, p.352–386. [Alloite, Asclerine, Cinerite, Pépérite, Phonolite, Pumite, Spodite, Trassoite, Tufaite]

CORDIER, P.L.A., 1842. Dictionnaire universel d'histoire naturelle (1842–1848) Editor Ch. d'Orbigny. *Savy, Paris.* 25 Vols. [Amphigenite, Coccolite, *Lhercoulite*, Mimosite, Nephelinite, Ophitone, Peridotite]

CORDIER, P.L.A., 1868. Description des roches composant l'écorce terrestre et des terrains cristallins. Editor Ch. d'Orbigny. *Savy, Paris.* 553pp. [Cecilite, Cristulite, Eurynite, Feldspathine, Leucostite, Lherzoline, Syenilite, Tusculite]

COTELO NEIVA, J.M., 1947A. Deux nouvelles roches éruptives de la famille des péridotites. *Estudos. Notas e Trabalhos do Serviço de Fomento Mineiro.* Vol.3, p.105–117. [Abessedite, Bragançaite]

COTELO NEIVA, J.M., 1947B. Nouvelles roches éruptives de la famille des pyroxénolites. *Estudos. Notas e Trabalhos do Serviço de Fomento Mineiro.* Vol.3, p.118–129. [Bogueirite, Conchaite, Regadite]

COTTA, B. VON, 1855. Die Gesteinslehre. *Engelhardt, Freiberg.* 255pp. [Basalt-wacke, Schalstein]

COTTA, B. VON, 1864A. Letter in "Mittheilungen an Professor Leonhard". *Neues Jahrbuch für Mineralogie, Geologie und Paläontologie. Stuttgart. Referate Abt. A.* Bd.35, p.822–827. [Acidite, Basite]

COTTA, B. VON, 1864B. Erzlagerstätten im Banat und in Serbien. *Braumüller, Wien.* 108pp. [Banatite]

COTTA, B. VON, 1866. Rocks Classified and Described. Trans. P.H. Lawrence. *Longmans, London.* 425pp. [Elvanite, Eucrite (Eucryte, Eukrite), Napoleonite, Scoria]

CRONSTEDT, A.F., 1758. Försök til Mineralogie, eller Mineral-Rikets Upställning. *Wildiska Tryckeriet, Stockholm.* 251pp. [Amygdaloid]

CROSS, W., 1897. Art. XVI – Igneous rocks of the Leucite Hills and Pilot Butte, Wyoming. *American Journal of Science. New Haven.* Vol.4, 4th Ser., p.115–141. [Madupite, Orendite, Wyomingite]

CROSS, W., IDDINGS, J.P., PIRSSON, L.V. & WASHINGTON, H.S., 1902. A quantitative chemico-mineralogical classification and nomenclature of igneous rocks. *Journal of Geology. Chicago.* Vol.10, p.555–690. [Femic, Salic]

CROSS, W., IDDINGS, J.P., PIRSSON, L.V. & WASH-

INGTON, H.S., 1906. The texture of igneous rocks. *Journal of Geology. Chicago*. Vol.14, p.692–707. [Graphiphyre (Graphophyre)]

CROSS, W., IDDINGS, J.P., PIRSSON, L.V. & WASH-INGTON, H.S., 1912. Modifications of the quantitative system of classification of igneous rocks. *Journal of Geology. Chicago*. Vol.20, p.550–561. [Felsic, Mafic]

D'AUBUISSON DE VOISINS, J.F., 1819. Traité de géognosie. *Levrault, Strasbourg*. Vol.2, 665pp. [Aphanite, Diorite, Dolerite]

DALY, R.A., 1903. Geology of Ascutney Mountain, Vermont. *Bulletin of the United States Geological Survey. Washington*. Vol.209, p.1–122. [Windsorite]

DALY, R.A., 1912. Geology of the North American Cordillera at the Forty-Ninth Parallel. *Memoirs. Geological Survey, Canada. Ottawa*. Vol.38, p.1–546. [Shackanite]

DANA, E.S., 1872. Art. X – Contributions from the Laboratory of the Sheffield Scientific School No.XXIII – On the composition of the Labradorite rocks of Waterville, New Hampshire. *American Journal of Science. New Haven*. Vol.3, 3rd Ser., p.48–50. [Ossipyte (Ossipite, Ossypite)]

DANNENBERG, A., 1898. Die Trachyte, Andesite und Phonolithe des Westwaldes. *Tschermaks Mineralogische und Petrographische Mitteilungen. Wien*. Vol.17, 2nd Ser., p.421–484. [*Isenite*]

DANNENBERG, A., 1900. Beiträge zur Petrographie der Kaukasusländer. *Tschermaks Mineralogische und Petrographische Mitteilungen. Wien*. Vol.19, 2nd Ser., p.218–272. [Dacite-andesite]

DAVID, T.E.W., GUTHRIE, F.B. & WOOLNOUGH, W.G., 1901. The occurrence of a variety of tinguaite at Kosciusko, N.S.Wales. *Journal of the Proceedings of the Royal Society of New South Wales. Sydney*. Vol.35, p.347–382. [Muniongite]

DAWSON, J.B. & GALE, N.H., 1970. Uranium and thorium in alkalic rocks from the active carbonatite volcano Oldoinyo Lerngai (Tanzania). *Chemical Geology*. Vol.6, p.221-231. [Lengaite (Lengaiite)]

DE-ANGELIS, M., 1925. Note di petrografia dancala. II. *Atti della Società Italiana di Scienze Naturali e del Museo Civico di Storia Naturale. Milano*. Vol.64, p.61–84. [Dancalite]

DECHEN, H. VON, 1845. Die Feldspath-Porphyre in den Lenne-Gegenden. *Karston's Archiv für Mineralogie, Geognosie, Bergbau und Hüttenkunde. Berlin*. Vol.19, p.367–452. [*Lenneporphyry*]

DECHEN, H. VON, 1861. Geognostischer Führer in das Siebengebirge am Rhein. *Henry & Cohen, Bonn*. 431pp. [Sanidophyre]

DEFANT, M.J. & DRUMMOND, M.S., 1990. Derivation of some modern arc magmas by melting of young subducted lithosphere. *Nature. London*. Vol.347, p.662-665. [Adakite]

DELAMÉTHERIE, J.C., 1795. Théorie de la terre. *Maradan, Paris*. Vol.3, 471pp. [Lherzolite, Pissite]

DELESSE, A., 1851. Mémoire sur la composition minéralogique et chimique des roches des Vosges. *Annales des Mines ou Recueil de Mémoires sur l'Exploration des Mines, et sur les Sciences qui s'y rapportent; rédigés par le Conseil Général des Mines. Paris*. Vol.19, p.149–183. [Kersantite]

DENAEYER, M.E., 1965. La rushayite, lave ultrabasique nouvelle du Nyiragongo (Virunga, Afrique centrale). *Compte Rendu Hebdomadaire des Séances de l'Académie des Sciences. Paris*. Vol.261, p.2119–2122. [Rushayite]

DERBY, O.A., 1891. Art. XXXVII – On the magnetite ore districts of Jacupiranga and Ipanema, São Paulo, Brazil. *American Journal of Science. New Haven*. Vol.41, 3rd Ser., p.311–321. [Jacupirangite]

DERWIES, V. DE, 1905. Recherches géologiques et pétrographiques sur les laccolithes des environs de Piatigorsk (Caucase du Nord).

Kündig, Genève. 84pp. [Trachyliparite]

DEWEY, H., 1910. The geology of the country around Padstow and Camelford. *Memoirs of the Geological Survey. England and Wales. London.* Sheet 335, p.1–119. [Minverite]

DOELTER, C., 1902. Der Monzoni und seine Gesteine. *Sitzungsberichte der Kaiserlichen Akademie der Wissenschaften. Mathematisch-Naturwissenschaftliche Classe. Wien.* Vol.111, p.929–986. [Rizzonite]

DOLOMIEU, D. DE, 1794. Distribution méthodique de toutes les matières dont l'accumulation forme les montagnes volcaniques ou tableau systématique dans lequel peuvent se placer toutes les substances qui ont des relations avec les feux souterrains. *Journal de Physique, de Chimie et d'Histoire Naturelle. Paris.* p.102–125. [Retinite]

DU BOIS, C.G.B., FURST, J., GUEST, N.J. & JENNINGS, D.J., 1963. Fresh natro carbonatite lava from Oldoinyo L'engai. *Nature. London.* Vol.197, p.445–446. [Natrocarbonatite]

DUMAS, E., 1846. Notice sur la constitution géologique de la région supérieure ou cévennique du départment du Gard. *Bulletin de la Société Géologique de France. Paris.* Vol.3, p.572–573. [Fraidronite (Fraidonite)]

DUNWORTH, E.A. & BELL, K., 1998. Melilitolites: a new scheme of classification. *The Canadian Mineralogist.* Vol.36, p.895-903. [Ultramelilitolite]

DUPARC, L., 1913. Sur l'ostraïte, une pyroxénite riche en spinelle. *Bulletin de la Société Française de Minéralogie. Paris.* Vol.36, p.18–20. [Ostraite]

DUPARC, L., 1926. Contribution à la connaissance de la pétrographie et des gîtes minéraux du Maroc. *Annales de la Sociéte Géologique de Belgique.* Vol.49, p.b114–b139. [Aiounite, Mestigmerite]

DUPARC, L. & GROSSET, A., 1916. Recherches géologiques et pétrographiques sur le district minier de Nikolai Pawda. *Kündig, Genève.* 294pp. [Kazanskite (Kasanskite), Pawdite]

DUPARC, L. & JERCHOFF, S., 1902. Sur les plagiaplites quartzifères du Kosswinsky. Intervention reportée dans le "Compte rendu des Séances du 6 Fevrier 1902". *Archives des Sciences Physiques et Naturelles. Genève.* Vol.13, p.307–310. [Plagiaplite]

DUPARC, L. & MOLLY, E., 1928. Sur la tokeïte, une nouvelle roche d'Abyssinie. *Compte Rendu des Séances de la Société de Physique et d'Histoire Naturelle de Genève.* Vol.45, p.24–25. [Tokéite]

DUPARC, L. & PAMPHIL, G., 1910. Sur l'issite, une nouvelle roche filonienne dans la dunite. *Compte Rendu Hebdomadaire des Séances de l'Académie des Sciences. Paris.* Vol.151, p.1136–1138. [Issite]

DUPARC, L. & PEARCE, F., 1900. Sur les plagioliparites du Cap Marsa (Algérie). *Compte Rendu Hebdomadaire des Séances de l'Académie des Sciences. Paris.* Vol.130, p.56–58. [Plagioliparite]

DUPARC, L. & PEARCE, F., 1901. Sur la koswite, une nouvelle pyroxénite de l'Oural. *Compte Rendu Hebdomadaire des Séances de l'Académie des Sciences. Paris.* Vol.132, p.892–894. [Koswite]

DUPARC, L. & PEARCE, F., 1904. Sur la garéwaïte, une nouvelle roche filonienne basique de l'Oural du Nord. *Compte Rendu Hebdomadaire des Séances de l'Académie des Sciences. Paris.* Vol.139, p.154–155. [Garewaite]

DUPARC, L. & PEARCE, F., 1905. Sur la gladkaïte, nouvelle roche filonienne dans la dunite. *Compte Rendu Hebdomadaire des Séances de l'Académie des Sciences. Paris.* Vol.140, p.1614–1616. [Gladkaite, Tilaite]

DUROCHER, J., 1845. Sur l'origine des roches granitiques. *Compte Rendu Hebdomadaire des Séances de l'Académie des Sciences. Paris.* Vol.20, p.1275–1284. [Quartz porphyry]

EBERT, H. VON, 1968. Suprakrustale Glieder der Charnockit-Familie in Nordwestsachsen.

Geologie. Zeitschrift für das Gesamtgebiet der Geologie und Mineralogie sowie der Angewandten Geophysik. Beiheft. Berlin. Vol.17, p.1031–1050. [Grimmaite]

EBRAY, TH., 1875. Quelques remarques sur les granulites et les minettes. Nouvelle classification des roches éruptives. *Bulletin de la Société Géologique de France. Paris.* Vol.3, p.287–291. [Anthraphyre, Carbophyre, Kohliphyre, Triaphyre]

ECKERMANN, H. VON, 1928A. Dikes belonging to the Alnö formation in the cuttings of the East Coast Railway. *Geologiska Föreningens i Stockholm Förhandlingar. Stockholm.* Vol.50, p.381–412. [Beforsite, Stavrite]

ECKERMANN, H. VON, 1928B. Hamrongite, a new Swedish alkaline mica lamprophyre. *Fennia: Bulletin de la Société de Géographie de Finlande.* Vol.50, p.1–21. [Hamrångite, Hamrongite]

ECKERMANN, H. VON, 1938A. The anorthosite and kenningite of the Nordingrå–Rödö region. *Geologiska Föreningens i Stockholm Förhandlingar. Stockholm.* Vol.60, p.243–284. [Kenningite]

ECKERMANN, H. VON, 1938B. A contribution to the knowledge of the late sodic differentiates of basic eruptives. *Journal of Geology. Chicago.* Vol.46, p.412–437. [Svartvikite]

ECKERMANN, H. VON, 1942. Ett preliminärt meddelande om nya forskningsrön inom Alnö alkalina område. *Geologiska Föreningens i Stockholm Förhandlingar. Stockholm.* Vol.64, p.399–455. [Alkorthosite, Alvikite, Hartungite]

ECKERMANN, H. VON, 1960. Borengite – a new ultra-potassic rock from Alnö Island. *Arkiv för Mineralogi och Geologi. Stockholm.* Vol.2, No.39, p.519–528. [Borengite]

EDEL'SHTEIN, YA., S., 1930. O novoi oblasti razvitiya shchelochnykh (nefelino-egirinovykh) porod v Yuzhnoi Sibiri. *Geologicheskii Vestnik.* Vol.UP, No.1-3, p.15-23. [Saibarite (Sajbarite)]

EGOROV, L.S., 1969. Melilitovie porodi Maimecha–Kotuiskoi Provintzii. *Trudy Nauchno-Issledovatel'skogo Instituta Geologii Arktiki. Leningrad.* Vol.159, 247pp. [Kugdite]

EIGENFELD, R., 1965. Raqqait, ein holo-melanokrates Lavagestein von pyroxenitischem Magmacharakter. *43. Jahrestagung der Deutschen Mineralogischen Gesellschaft, Kurzreferate der Vorträge. Hannover.* p.46. [Raqqaite]

ELIE DE BEAUMONT, M.L., 1822. Sur les mines de fer et les forges de Framont et de Rothau. *Annales des Mines ou Recueil de Mémoires sur l'Exploration des Mines, et sur les Sciences qui s'y rapportent; rédigés par le Conseil Général des Mines. Paris.* Vol.7, p.521–554. [Minette]

ELISEEV, N.A., 1937. Struktury rudnykh polei v pervichno rassloennykh plutonakh Kolskogo poluostrova. *Izvestiia Akademii Nauk SSSR, Seriia Geologicheskaia.* No.6, p.1085-1104. [Eudialytite]

ELLIS, W., 1825. A journal of a tour around Hawaii, the largest of the Sandwich islands. *Crocker & Brewster, Boston; J.P. Haven, New York.* 264pp. [Pélé's hair]

EMERSON, B.K., 1902. Holyokeite, a purely feldspathic diabase from the Trias of Massachusetts. *Journal of Geology. Chicago.* Vol.10, p.508–512. [Holyokeite]

EMERSON, B.K., 1915. Art. XIX – Northfieldite, pegmatite, and pegmatite schist. *American Journal of Science. New Haven.* Vol.40, 4th Ser., p.212–217. [Northfieldite]

ENDLICH, F.M., 1874. Report of F.M. Endlich S.N.D. *Report of the United States Geological (and Geographical) Survey of the Territories. Washington.* p.275–361. [Trachorheite]

ERDMANN, A., 1849. Letter in "Mittheilungen an den Geheimenrath v. Leonhard gerichtet". *Neues Jahrbuch für Mineralogie, Geognosie, Geologie und Petrefaktenkunde. Stuttgart.* Vol.20, p.837–838. [Eulysite]

ERDMANNSDÖRFFER, O.H., 1907. Über Vertreter der Essexit-Theralithreihe unter den diabasartigen Gesteinen der Deutschen Mittelgebirge. *Zeitschrift der Deutschen Geologischen Gesellschaft. Berlin.* Vol.59, Monatsberichte No.2, p.16–23. [Essexitediabase]

ERDMANNSDÖRFFER, O.H., 1928. Ueber ein sibirisches Nephelingestein. *Festschrift Victor Goldschmidt. Heidelberg.* p.85–88. [Bjerezite]

ESCHWEGE, W.L.C. VON, 1832. Beiträge zur Gebirgskunde Brasiliens. *Reimer, Berlin.* 488pp. [Carvoeira]

ESKOLA, P., 1921. On the igneous rocks of Sviatoy Noss in Transbaikalia. *Ofversigt av Finska Vetenskaps-Societetens Förhandlingar. Helsingfors.* Vol.63, afd.A,1, p.1–99. [Sviatonossite]

ESMARK, J., 1823. Om Norit-Formationen. *Magazin for Naturvidenskaberne. Christiania.* Vol.1, No.2, p.205–215. [Norite]

FAESSLER, C., 1939. The stock of "suzorite" in Suzor Township, Quebec. *University of Toronto Studies. Geological Series. Toronto.* Vol.42, p.47–52. [Suzorite]

FAUJAS DE SAINT-FOND, B.F., 1778. Recherches sur les volcans éteints du Vivarais et du Velay. *Cuchet, Grenoble.* 460pp. [Gallinace (Gallinazo)]

FEDERICO, M., 1976. On a kalsilitolite from the Alban Hills, Italy. *Periodico di Mineralogia. Rome.* Vol.45, p.5–12. [Kalsilitolite]

*FEDOROV, E., 1896. O novoi gruppe izverzhennykh porod. *Izvestiia Moskovskago Sel'skokhozyaistvennago Instituta. Moskva.* No.1. [Drusite]

FEDOROV, E.S., 1901. Geologicheskiya izsledovaniya letom 1900 g. *Ezhegodnik po Geologii i Mineralogii Rossii. Varshava.* Vol.4, No.6, p.135–140. [Kedabekite]

FEDOROV, E.S., 1903. Kratkoe soobshchenie o rezultatakh mineralogicheskago i petrograficheskago izsledovaniya beregov Belago morya. *Zapiski Imperatorskogo (Sankt Peterburgskogo) Mineralogicheskogo Obshchestva.* Vol.40, p.211–220. [Kovdite]

FENNER, C.N., 1948. Incandescent tuff flows in southern Peru. *Bulletin of the Geological Society of America. New York.* Vol.59, p.879–893. [Sillar]

FERSMAN, A., 1929. Geochemische Migration der Elemente und deren wissenschaftliche und wirtschaftliche Bedeutung. *Abhandlungen zur praktischen Geologie und Bergwirtschaftslehre. Halle (Salle).* Vol.18, p.1–116. [Miaskitic (Miascitic)]

FICHTEL, J.E. VON, 1791. Mineralogische Bemerkungen von den Karpathen. *Edlen von Kurzbeck, Wien.* Vol.1, 411pp. [Perlstein]

FIEDLER, A., 1936. Über Verflößungserscheinungen von Amphibolit mit diatektischen Lösungen im östlichen Erzgebirge. *Zeitschrift für Kristallographie, Mineralogie und Petrographie. Leipzig. Abt. B.* Vol.47, p.470–516. [Diatectite (Diatexite)]

FLETT, J.S., 1907. Petrography of the igneous rocks. In: W.A.E. Ussher (Editor), The Geology of the Country around Plymouth and Liskeard. *Memoirs of the Geological Survey. England and Wales. London.* Sheet 348, p.1–147. [*Spilite*]

FLETT, J.S., 1911. In: B.N. Peach, J.S.G. Wilson, J.B. Hill, E.B. Bailey & G.W. Grabham, The Geology of Knapdale, Jura and North Kintyre. *Memoirs. Geological Survey of Scotland. Edinburgh.* Sheet 28, p.1–149. [Crinanite]

FLETT, J.S., 1935. Petrography. In: Wilson, G.V., Edwards, W., Knox, J., Jones, R.C.V. et al., The Geology of the Orkneys. *Memoirs. Geological Survey of Great Britain. Scotland.* p.173–186. [*Holmite*]

FOERSTNER, E., 1881. Nota preliminare sulla geologia dell'Isola di Pantelleria secondo gli studii fatti negli anni 1874 e 1881, dal dottor Enrico Foerstner. *Bollettino del Reale Comitato Geologico d'Italia. Firenze.* Vol.12, p.523–556. [Pantellerite]

FOURNET, J., 1845. Note sur l'état actuel des connaissances touchant les roches éruptives des environs de Lyon. *Bulletin de la Société Géologique de France. Paris.* Vol.2, p.495–506. [*Ilvaite*, Miarolite]

FOURNET, J., 1861. Géologie lyonnaise. *Lyon.* Vol.1, 744pp. [Vaugnerite]

FRECHEN, J., 1971. Siebengebirge am Rhein – Laacher Vulkangebiet – Maargebiet der Westeifel. Sammlung Geologischer Führer 56. *Borntraeger, Berlin.* 2nd Edn, 195pp. [*Boderite, Rodderite*]

FRENTZEL, A., 1911. Das Passauer Granitmassiv. *Geognostische Jahreshefte. München.* Vol.24, p.105–192. [Engelburgite, Ilzite]

FUSTER, J.M., GASTESI, P., SAGREDO, J. & FERMOSO, M.L., 1967. Las rocas lamproíticas del S.E. de España. *Estudios Geológicos. Instituto de Investigaciones Geológicas "Lucas Mallada". Madrid.* Vol.23, p.35–69. [Cancalite]

GAGEL, C., 1913. Studien über den Aufbau und die Gesteine Madeiras. *Zeitschrift der Deutschen Geologischen Gesellschaft. Berlin.* Vol.64, No.7, p.344–448. [Madeirite]

GAPEEVA, G.M., 1959. Ussurit – osobaya raznovidnost' shchelochnykh bazaltovykh porod. *Doklady Akademii Nauk SSSR. Leningrad.* Vol.126, No.1, p.157–159. [Ussurite]

GAVELIN, A., 1915. Den geologiska byggnaden inom Ruoutevare-området i Kvikkjokk, Norbottens Län. *Geologiska Föreningens i Stockholm Förhandlingar. Stockholm.* Vol.37, p.17–32. [Vallevarite]

GAVELIN, A., 1916. Über Högbomit. *Bulletin of the Geological Institution of the University of Uppsala.* Vol.15, p.289–316. [Magnetite Högbomitite]

GEIKIE, A., 1903. Text-book of Geology. *Macmillan & Co., London.* 4th Edn, Vol.2, p.703-1472. [Plateau basalt]

GEMMELLARO, C., 1845. Sul basalto decomposto dell'isola de'ciclopi. *Atti dell' Accademia Gioenia di Scienze Naturali in Catania.* Vol.2, 2nd Ser., p.309–319. [Analcimite]

GERHARD, A., 1815. Beiträge zur Gesdrichte des Weissteins, des Felsit und anderer verwandber Arten. *Abhandlungen der Königlichen Akademie der Wissenschaften. Berlin.* p.12–26. [Felsite (Felsitic)]

GILL, J, 1981. Orogenic Andesites and Plate Tectonics. *Springer-Verlag, Berlin.* 390pp. [Orogenic andesite]

GIRAUD, P., 1964. Essai de classification modale des roches à caractère charnockitique. *Bulletin du Bureau de Recherches Géologiques et Minières. Paris.* Vol.4, p.36–53. [Ankaranandite]

GLOCKER, E.F., 1831. Handbuch der Mineralogie. *Schrag, Nürnberg.* Vol.2. [Fluolite]

GMELIN, L., 1814. Observationes Oryctognosticae et Chemicae de Haüyna. *Mohr & Zimmer, Heidelberg.* 58pp. [Asprone, Sperone]

GOLDSCHMIDT, V.M., 1916. Geologisch-Petrographische Studien im Hochgebirge des Südlichen Norwegiens. IV. Übersicht der Eruptivgesteine im kaledonischen Gebirge zwischen Stavanger und Trondhjem. *Skrifter udgit av Videnskabsselskabet i Kristiania. I. Math.-Nat. Klasse.* p.1–140. [Jotun-norite, Opdalite, Trondhjemite]

GORDON, C.H., 1896. Syenite-gneiss (Leopard Rock) from the apatite region of Ottawa County, Canada. *Bulletin of the Geological Society of America. New York.* Vol.7, p.95–134. [Leopard rock]

GORNOSTAEV, N.N., 1933. Differentsirovannyi ekstruzivnyi lakkolit Kyz-Emchik v gorakh Semeitau bliz g. Semipalatinska. *Sbornik po Geologii Sibiri posviashchennyi 25-letiiu nauchno-pedagogicheskoi deiatelnosti prof. M.A.Usova. Tomsk.* p.153–223. [Semeitavite (Semejtavite, Semeitovite), Terektite]

GRAHMANN, R., 1927. Blatt Riesa-Strehla.

Erläuterungen zur Geologischen Karte von Sachsen. Leipzig. 2nd Edn, No.16, p.1-82. [Gröbaite]

GRANGE, L.I., 1934. Rhyolite sheet flows of the North Island, New Zealand. *New Zealand Journal of Science and Technology. Wellington.* Vol.16, p.57–58. [Owharoite]

GREGORY, J.W., 1900. Contributions to the geology of British East Africa – Part II, The geology of Mount Kenya. *Quarterly Journal of the Geological Society of London.* Vol.56, p.205–222. [Kenyte (Kenyite)]

GREGORY, J.W., 1902. Article XVI – The geology of Mount Macedon, Victoria. *Proceedings of the Royal Society of Victoria. Melbourne.* Vol.14, 2nd Ser., p.185–217. [Geburite-dacite]

GROOM, T.T., 1889. On the occurrence of a new form of tachylyte in association with the gabbro of Carrock Fell, in the Lake District. *Geological Magazine. London.* Vol.6, Decade 3, p.43. [Carrockite]

GROTH, P., 1877. Das Gneissgebiet von Markirch im Ober-Elsass. *Abhandlungen zur Geologischen Spiezialkarte von Elsass–Lothringen. Strassburg.* Vol.1, p.395–489. [Kammgranite]

GRUBENMANN, U., 1908. Der Granatolivinfels des Gordunotales und seine Begleitgesteine. *Vierteljahrsschrift der Naturforschenden Gesellschaft in Zürich.* Vol.53, p.129–156. [Gordunite]

GRUSS, K., 1900. Beiträge zur Kenntnis der Gesteine des Kaiserstuhlgebirges. Tephritische Strom- und Ganggesteine. *Mittheilungen der Grossherzoglich Badischen Geologischen Landesanstalt. Heidelberg.* Vol.4, p.83–144. [Mondhaldeite]

GÜMBEL, C.W. VON, 1861. Geognostische Beschreibung des bayerischen Alpengebirges und sienes Vorlandes. *Geognostische Beschreibung des Königreiches Bayern. Gotha.* p.1–948. [Sillite]

GÜMBEL, C.W. VON, 1865. Geognostische Beschreibung der fränkischen Alb (Frankenjura). *Geognostische Beschreibung des Königreiches Bayern. Gotha.* Vol.4, p.1–763. [Aschaffite]

GÜMBEL, C.W. VON, 1868. Geognostische Beschreibung des ostbayerischen Grenzgebirges oder des bayerischen und oberpfälzer Waldgebirges. *Geognostische Beschreibung des Königreiches Bayern. Gotha.* p.1–968. [Nadeldiorite, Regenporphyry]

GÜMBEL, C.W. VON, 1874. Die paläolithischen Eruptivgesteine des Fichtelgebirges. *Weiss, Munich.* 50pp. [Epidiorite, Keratophyre, Lamprophyre, Leucophyre, Palaeophyre, Palaeopicrite, Proterobase]

GÜMBEL, C.W. VON, 1888. Geologie von Bayern. I. Grundzüge der Geologie. *Fischer, Kassel.* Vol.1, 1144pp. [Dioritoid, Gabbroid, Granitophyre, Hyalopsite, Kokkite (Coccite), Peridotoid, Phonolitoid, Trachytoid]

HACKL, O. & WALDMANN, L., 1935. Ganggesteine der Kalireihe aus dem neiderösterreichischen Waldviertel. *Jahrbuch der Geologischen Bundesanstalt. Wien.* Vol.85, p.259–285. [Karlsteinite, Raabsite, Thuresite]

HAIDINGER, W., 1861. Meeting/session report from 28.5.1861, without title. *Verhandlungen der Geologischen Reichsanstalt (Staatanstalt –Landesanstalt). Wien.* Vol.12, p.64. [Haüynfels]

HALL, A.L., 1922. On the marundites and allied corundum-bearing rocks in the Leydsdorp district of the Eastern Transvaal. *Transactions of the Geological Society of South Africa. Johannesburg.* Vol.25, p.43–67. [Marundite]

HALL, T.C., 1915. Summary of progress of the Geological Survey of Great Britain and the Museum of Practical Geology for 1914. *Memoirs of the Geological Survey. England and Wales. London.* p.1–84. [Lundyite]

HANSKI, E.J., 1992. Petrology of the Pechenga ferropicrites and cogenetic, Ni-bearing gabbro wehrlite intrusions, Kola peninsula, Russia. *Geological Survey of Finland, Bulletin.* Vol.367, p.1-196. [*Ferropicrite*]

HANSKI, E.J. & SMOLKIN, V.F, 1989. Pechenga ferropicrites and other early Proterozoic picrites in the eastern part of the Baltic Shield. *Precambrian Research.* Vol.45, p.63-82. [Ferropicrite]

HARKER, A., 1904. The Tertiary igneous rocks of Skye. *Memoirs of the Geological Survey of the United Kingdom.* p.1–481. [Marscoite, Mugearite]

HARKER, A., 1908. The geology of the small Isles of Inverness-shire (Rum, Canna, Eigg, Muck etc.). *Memoirs. Geological Survey of Scotland. Edinburgh.* Sheet 60, p.1–210. [Allivalite, Harrisite]

HATCH, F.H., 1909. Text-book of Petrology. *Sonnenschein, London.* 5th Edn, 404pp. [Markfieldite]

HATCH, F.H., WELLS, A.K. & WELLS, M.K., 1949. Petrology of the Igneous Rocks. *Thomas Murby, London.* 10th Edn, Vol.1, 469pp. [Orthogabbro, *Orthosyenite, Orthotrachyte*]

HATCH, F.H., WELLS, A.K. & WELLS, M.K., 1961. Petrology of the Igneous Rocks. *Thomas Murby, London.* 12th Edn, Vol.1, 515pp. [Kalmafite, Leumafite, Melmafite, Nemafite]

HAUER F. VON & STACHE, G., 1863. Geologie Siebenbürgens. Editor Graeser. *Braumüller, Wien.* 636pp. [Dacite, Quartz trachyte]

HAUSMANN, J., 1847. Handbuch der Mineralogie. *Vandenhoek & Ruprecht, Göttingen.* 2nd Edn, Pt.2, Vol.1, 896pp. [Hyalomelane, Wichtisite]

HAÜY, A., 1822. Traité de minéralogie. *Bachelier & Huzard, Paris.* 2nd Edn, Vol.4, 604pp. [Basaltoid, Pogonite, Resinite, Selagite, Trematode]

HEDDLE, M.F., 1897. On the crystalline forms of riebeckite. *Transactions of the Edinburgh Geological Society.* Vol.7, p.265–267.

[Ailsyte (Ailsite)]

HENDERSON, J.A.L., 1898. Petrographical and Geological Investigations of Certain Transvaal Norites, Gabbros and Pyroxenites. *Dulau, London.* 56pp. [Hatherlite, Pilandite]

HENDERSON, J.A.L., 1913. The geology of the Aroha Subdivision, Hauraki, Auckland. *Bulletin. Geological Survey of New Zealand. Wellington.* Vol.16, p.1–127. [Wilsonite]

HENNIG, A., 1899. Kullens kristalliniska bergarter. II. Den postsiluriska gångformationen. *Acta Universitatis Lundensis. Lund.* Vol.35, Pt.2, No.5, p.1–34. [Kullaite]

HESS, H.H., 1937. Island arcs, gravity anomalies and serpentine intrusions; a contribution to the ophiolite problem. *International Geological Congress, Report of the XVII Session, Moscow.* Vol.2, p.263–283. [Ultramafic rock]

HIBSCH, J.E., 1896. Erläuterungen zur geologischen Specialkarte des böhmischen Mittelgebirges. I. Umgebung von Tetschen. *Tschermaks Mineralogische und Petrographische Mitteilungen. Wien.* Vol.15, 2nd Ser., p.201–290. [Ash tuff]

HIBSCH, J.E., 1898. Erläuterungen zur geologischen Karte des böhmischen Mittelgebirges. III. Bensen. *Tschermaks Mineralogische und Petrographische Mitteilungen. Wien.* Vol.17, 2nd Ser., p.1–96. [Gauteite]

HIBSCH, J.E., 1899. Die Tiefengesteine des böhmischen Mittelgebirges. *Sitzungsberichte des Deutschen Naturwissenschaftlich-Medicinischen Vereines für Böhmen "Lotos" in Prag.* Vol.19, 2nd Ser., p.68–72. [Analcime syenite]

HIBSCH, J.E., 1902. Geologische Karte des böhmischen Mittelgebirges. V. Grosspriesen. *Tschermaks Mineralogische und Petrographische Mitteilungen. Wien.* Vol.21, 2nd Ser., p.465–588. [Sodalitophyre]

HIBSCH, J.E., 1910. Geologische Karte des

böhmischen Mittelgebirges. VI. Wernstadt–Zinkenstein. *Tschermaks Mineralogische und Petrographische Mitteilungen. Wien.* Vol.29, 2nd Ser., p.381–438. [Alkali basalt]

HIBSCH, J.E., 1920. Geologische Karte des böhmischen Mittelgebirges. XIV. Meronitz–Trebnitz. *Prag.* 120pp. [Analcime basanite]

HILL, J.B. & KYNASTON, H., 1900. On kentallenite and its relations to other igneous rocks in Argyllshire. *Quarterly Journal of the Geological Society of London.* Vol.56, p.531–558. [Kentallenite]

HILLS, E.S., 1958. Cauldron subsidence, granitic rocks, and crustal fracturing in S.E. Australia. *Geologische Rundschau. Internationale Zeitschrift für Geologie. Stuttgart.* Vol.47, p.543–561. [Snobite]

HJÄRNE, U., 1694. En kort Anledning till Atskillge Malm och Bergarters, Mineraliers, Wäxters och Jordeslags, samt flere sällsamme Tings effterspöriande och angiftvande. *Stockholm.* [Rapakivi]

HOBBS, W.H., 1893. Volcanite, an anorthoclase-augite rock chemically like dacites. *Bulletin of the Geological Society of America. New York.* Vol.5, p.598–602. [Volcanite]

HOCHSTETTER, F., 1859. Lecture on the geology of the Province of Nelson. *New Zealand Government Gazette.* No.39, p.269–281. [Dunite]

HØDAL, J., 1945. Rocks of the anorthosite kindred in Vossestrand (Norway). *Norsk Geologisk Tidsskrift. Oslo.* Vol.24, p.129–243. [Jotunite]

HÖGBOM, A.G., 1905. Zur Petrographie der kleinen Antillen. *Bulletin of the Geological Institution of the University of Upsala.* Vol.6, p.214–233. [Plagioclase granite]

HOHENEGGER, L., 1861. Die geognostischen Verhältnisse der Nordkarpathen in Schlesien und den angrenzenden Theilen von Mähren und Galizien. *Perthes, Gotha.* 50pp. [Teschenite (Teschinite)]

HOLLAND, T.H., 1900. The charnockite series, a group of Archaean hypersthenic rocks in Peninsular India. *Memoirs of the Geological Survey of India. Calcutta.* Vol.28, p.119–249. [Charnockite]

HOLLAND, T.H., 1907. General Report of the Geological Survey of India for the year 1906. *Records of the Geological Survey of India. Calcutta.* Vol.35, p.1–61. [Kodurite]

HOLMES, A., 1937. The petrology of katungite. *Geological Magazine. London.* Vol.74, p.200–219. [Katungite]

HOLMES, A., 1942. A suite of volcanic rocks from south-west Uganda containing kalsilite (a polymorph of $KAlSiO_4$). *Mineralogical Magazine and Journal of the Mineralogical Society. London.* Vol.26, p.197–217. [Mafurite, Protokatungite]

HOLMES, A. & HARWOOD, H.F., 1937. The petrology of the volcanic area of Bufumbira. *Memoirs. Geological Survey of Uganda. Entebbe.* Vol.3, Pt.2, p.1–300. [Lutalite, Murambite, Murambitoid, Ugandite]

HOLMQUIST, P.J., 1908. Utkast till ett bergartsschema för urbergsskiffrarna. *Geologiska Föreningens i Stockholm Förhandlingar. Stockholm.* Vol.30, p.269–293. [Peridotitoid]

HOLST, N.O. & EICHSTÄDT, F., 1884. Klotdiorit från Slättmossa, Järeda socken, Kalmar län. *Geologiska Föreningens i Stockholm Förhandlingar. Stockholm.* Vol.7, p.134–142. [Klotdiorite, Klotgranite]

HOLTEDAHL, O., 1943. Studies on the igneous rock complex of the Oslo region. I. Some structural features of the district near Oslo. *Skrifter utgitt av det Norske Videnskaps-Akademi i Oslo. Mat.-naturv.Kl.* No.2, p.1–71. [Lathus porphyry, Østern porphyry]

HORNE, J. & TEALL, J.J.H., 1892. On borolanite – an igneous rock intrusive in the Cambrian limestone of Assynt, Sutherlandshire, and the Torridon sandstones of Ross-shire. *Transactions of the Royal Society of Edinburgh.*

Vol.37, p.163–178. [Borolanite]

HUMBOLDT, A. VON, 1823. Essai géognostique sur le gisement des roches dans les deux hemisphères. *Levrault, Paris.* 379pp. [Lozero (Losero), Texontli]

HUMBOLDT, H.A. VON, 1837. Geognostische und physikalische Beobachtungen über die Vulkane des Hochlandes von Quito. *Neues Jahrbuch für Mineralogie, Geognosie, Geologie und Petrefaktenkunde. Stuttgart.* Vol.8, p.253–284. [Leucitophyre]

HUME, W.F., HARWOOD, H.F. & THEOBALD, L.S., 1935. Notes on some analyses of Egyptian igneous and metamorphic rocks. *Geological Magazine. London.* Vol.72, p.8–32. [Baramite]

HUNT, T.S., 1852. Report of T.S. Hunt, Esq., chemist and mineralogist to the provincial Geological Survey. *Annual Report. Geological Survey of Canada. Montreal and Ottawa.* p.92-121. [Parophite]

HUNT, T.S., 1862. Descriptive catalogue of a collection of the crystalline rocks of Canada: Descriptive catalogue of a collection of the economic minerals of Canada and of its crystalline rocks sent to the London International Exhibition Montreal. *Geological Survey of Canada.* p.61–83. [Anorthosite]

HUNT, T.S., 1864. On the chemical and mineralogical relations of metamorphic rocks. *Journal of the Geological Society of Dublin.* Vol.10, p.85–95. [*Anortholite*]

HUNTER, C.L., 1853. Art. XLII – Notices of the rarer minerals and new localities in western North Carolina. *American Journal of Science. New Haven.* Vol.15, 2nd Ser., p.373–378. [Leopardite]

HUNTER, M. & ROSENBUSCH, H., 1890. Ueber Monchiquit, ein camptonitisches Ganggestein aus der Gefolgschaft der Eläolithsyenite. *Tschermaks Mineralogische und Petrographische Mitteilungen. Wien.* Vol.11, 2nd Ser., p.445–466. [Bostonite, Monchiquite]

IDDINGS, J.P., 1895A. Absarokite–shoshonite–banakite series. *Journal of Geology. Chicago.* Vol.3, p.935–959. [Absarokite, Banakite, Shoshonite]

IDDINGS, J.P., 1895B. The origin of igneous rocks. *Bulletin of the Philosophical Society of Washington.* Vol.12, p.89–213. [Alkali, Subalkali]

IDDINGS, J.P., 1904. Quartz-feldspar-porphyry (Graniphyro Liparose-Alaskose) from Llano, Texas. *Journal of Geology. Chicago.* Vol.12, p.225–231. [Llanite]

IDDINGS, J.P., 1909. Igneous Rocks. *Wiley, New York.* Vol.1, 464pp. [Cumulophyre, Linophyre, Planophyre, Skedophyre]

IDDINGS, J.P., 1913. Igneous Rocks. *Wiley, New York.* Vol.2, 685pp. [Andesine basalt, Bandaite, Hawaiite, Kauaiite, Kohalaite, Laugenite, Marosite, Oligoclase andesite, Shastaite, Ungaite]

IDDINGS, J.P. & MORLEY, E.W., 1917. A contribution to the petrography of Southern Celebes. *Proceedings of the National Academy of Sciences of the United States of America. Washington.* Vol.3, p.592–597. [Batukite]

IDDINGS, J.P. & MORLEY, E.W., 1918. A contribution to the petrography of the South Sea Islands. *Proceedings of the National Academy of Sciences of the United States of America. Washington.* Vol.4, p.110–117. [Tautirite]

IMA, 1978. Nomenclature of amphiboles. *Mineralogical Magazine and Journal of the Mineralogical Society. London.* Vol.42, p.553-563.

IPPEN, J.A., 1903. Ueber den Allochetit vom Monzoni. *Verhandlungen der Geologischen Reichsanstalt (Staatanstalt–Landesanstalt). Wien.* No.7 & 8, p.133–143. [Allochetite]

IRVINE, T.N. & BARAGAR, W.R.A., 1971. A guide to the chemical classification of the common volcanic rocks. *Canadian Journal of Earth Sciences, Ottawa.* Vol.8, p.523–548. [Commendite]

ISSEL, A., 1880. Osservazioni intorno a carte rocce anfiboliche della Liguria, a proposito d'una nota del Prof. Bonney concernente alcune serpentine della Liguria e della Toscana. *Bollettino del Reale Comitato Geologico d'Italia. Firenze.* Vol.11, p.183–192. [Borzolite, Coschinolite]

ISSEL, A., 1892. Liguria Geologica e Preistorica. *Donath, Genoa.* Vol.1, 440pp. [Epidiabase]

ISSEL, A., 1916. Prime linee di un ordinamento sistematico delle pietre figurate. *Memorie della Reale Accademia (Nazionale) dei Lincei. Rendiconti. Classe di Scienze Fisiche, Matematiche e Naturali. Roma.* Vol.11, p.631–667. [Peleiti]

IVES, R.L., 1956. Tezontli rock. *Rocks and Minerals. Peekskill, New York.* Vol.31, p.122–124. [Tezontli]

JENNER, G.A., 1981. Geochemistry of the high-Mg andesites from Cape Vogel, Papua New Guinea. *Chemical Geology.* Vol.33, p.307–332. [High-Mg andesite]

JENZSCH, G., 1853. Amygdalophyr, ein Felsit-Gestein mit Weissigit, einem neuen Minerale in Blasen-Räumen. *Neues Jahrbuch für Mineralogie, Geognosie, Geologie und Petrefaktenkunde. Stuttgart.* Vol.58, p.385–398. [Amygdalophyre]

JEVONS, H.S., JENSEN, H.I., TAYLOR, T.G. & SÜSSMILCH, C.A., 1912. The geology and petrology of the Prospect intrusion. *Journal of the Proceedings of the Royal Society of New South Wales. Sydney.* Vol.45, p.445–553. [Pallioessexite, Palliogranite]

JOHANNSEN, A., 1911. Petrographic terms for field use. *Journal of Geology. Chicago.* Vol.19, p.317–322. [Amphiboleid, Anameseid, Anameseid porphyry, Aphaneid, Dioreid, Dolereid, Felseid, Felseid porphyry, Gabbreid, Graneid, Leuco-aphaneid, Leucophyreid, Melano-aphaneid, Melanophyreid, Peridoteid, Phanereid, Poikileid, Pyriboleid, Pyroxeneid, Quartz dioreid, Quartz dolereid, Quartz gabbreid,

Syeneid, Vitrophyreid]

JOHANNSEN, A., 1917. Suggestions for a quantitative mineralogical classification of igneous rocks. *Journal of Geology. Chicago.* Vol.25, p.63–97. [Granogabbro, Ortho-, Syenodiorite, Syenogabbro]

JOHANNSEN, A., 1920A. A quantitative mineralogical classification of igneous rocks – Revised. Pt.I. *Journal of Geology. Chicago.* Vol.28, p.38–60. [Bytownitite, Leuco-, Mela-, Meso-, Orthogranite, Orthoshonkinite, Orthotarantulite, Tarantulite, Topazite, Tourmalite]

JOHANNSEN, A., 1920B. A quantitative mineralogical classification of igneous rocks – Revised. Pt.II. *Journal of Geology. Chicago.* Vol.28, p.158–177. [Adam-gabbro, Adam-diorite, Adam-tonalite, Andelatite, Beloeilite, Monzodiorite, Moyite, Nepheline diorite, Nepheline monzodiorite, Orthofoyaite, Orthorhyolite, Orthosyenite, Toryhillite]

JOHANNSEN, A., 1920C. A quantitative mineralogical classification of igneous rocks – Revised. Pt.III. *Journal of Geology. Chicago.* Vol.28, p.210–232. [Basalatite, Chromitite, Magnetitite, Monzogabbro, Monzonorite, Nepheline monzogabbro, Rhyobasalt, Ricolettaite]

JOHANNSEN, A., 1926. A revised field classification of igneous rocks. *Journal of Geology. Chicago.* Vol.34, p.181–182. [Amphibolide, Aphanide, Dioride, Felside, Gabbride, Granide, Leucophyride, Melanophyride, Peridotide, Perknide, Phaneride, Pyroxenide, Syenide, Trappide, Vitrophyride]

JOHANNSEN, A., 1931. A Descriptive Petrography of the Igneous Rocks. *Chicago University Press.* Vol.1, 267pp. [Biquahororthandite, Diaschistite, Hobiquandorthite, Quartz diorite, Quartz gabbride, Quartz leucophyride, Topatourbiolilepiquorthite]

JOHANNSEN, A., 1932. A Descriptive Petrography of the Igneous Rocks. *Chicago University Press.* Vol.2, 428pp. [Haplite, Hraftinna,

Pearlstone, Quartz gabbro, Quartz norite]

JOHANNSEN, A., 1937. A Descriptive Petrography of the Igneous Rocks. *Chicago University Press*. Vol.3, 360pp. [Calciclasite, Clinopyroxene norite, Elkhornite, Felsoandesite, Labradoritite, Masafuerite, Orthopyroxene gabbro]

JOHANNSEN, A., 1938. A Descriptive Petrography of the Igneous Rocks. *Chicago University Press*. Vol.4, 523pp. [Analcimolith, Baldite, Barshawite, Bogusite, Caltonite, Columbretite, Cuyamite, Deldoradite (Deldoradoite), Feldspathoidite, Fiasconite, Glenmuirite, Gooderite, Haüynitite, Haüynolith, Highwoodite, Hilairite, Holmite, Leucitite basanite, Leucitite tephrite, Leucitolith, Linosaite, Marienbergite, Martinite, Melanephelinite, Melilitholith, Nepheline andesite, Nephelinite basanite, Nephelinite tephrite, Nordsjöite, Noseanolith, Noselitite, Parchettite, Penikkavaarite, Pikeite, Rafaelite, Sodalite basalt, Sumacoite, Tannbuschite, Tasmanite, Tutvetite, Ventrallite, Vetrallite, Westerwaldite]

JOHANNSEN, A., 1939. A Descriptive Petrography of the Igneous Rocks. *Chicago University Press*. 2nd Edn, Vol.1, 318pp.

JOPLIN, G.A., 1964. A Petrography of Australian Igneous Rocks. *Angus & Robertson, Sydney*. 253pp. [*Shoshonite*]

JUAN, V.C., TAI, H. & CHANG, F.H., 1953. Taiwanite, a new basaltic glassy rock of East Coastal Range, Taiwan, and its bearing on parental magma-type. *Acta Geologica Taiwanica. Taipei*. Vol.5, p.1–25. [Taiwanite]

JUDD, J.W., 1876. On the ancient volcano of the district of Schemnitz, Hungary. *Quarterly Journal of the Geological Society of London*. Vol.32, p.292–325. [Matraite]

JUDD, J.W., 1881. Volcanoes. *Kegan Paul & Trench, London*. 2nd Edn, 381pp. [Ultrabasic]

JUDD, J.W., 1885. On the Tertiary and older peridotites of Scotland. *Quarterly Journal of the Geological Society of London*. Vol.41, p.354–422. [Scyelite]

JUDD, J.W., 1886A. On the gabbros, dolerites and basalts, of Tertiary Age, in Scotland and Ireland. *Quarterly Journal of the Geological Society of London*. Vol.42, p.49–97. [Intermediate]

JUDD, J.W., 1886B. On marekanite and its allies. *Geological Magazine. London*. Vol.3, Decade 3, p.241–248. [Marekanite]

JUDD, J.W., 1897. On the petrology of Rockall. *Transactions of the Royal Irish Academy. Dublin*. Vol.31, p.48–58. [Rockallite]

JUNG, D., 1958. Untersuchungen am Tholeyit von Tholey (Saar). *Beiträge zur Mineralogie und Petrographie. Berlin etc*. Vol.6, p.147–181. [*Tholeiite*]

JUNG, J. & BROUSSE, R., 1959. Classification modale des roches éruptives. *Masson & Cie, Paris*. 122pp. [Leucitonephelinite, Trachylabradorite]

JURINE, L., 1806. Lettre de M. le Professeur Jurine, de Genève, à M. Gillet-Laumont, Membre du Conseil des Mines, Correspondant de l'Institut. *Journal des Mines. Paris*. Vol.19, p.367–378. [Arkesine, Dolerine, Notite, Protogine, Spurine]

KAISER-GIESSEN, E., 1913. Neue nephelingesteine aus Deutsch-Südwestafrika. *Verhandlungen der Gesellschaft Deutscher Naturforscher und Ärtze*. Vol.86, Pt.2, p.595–598. [Klinghardtite]

KALB, G., 1936. Beiträge zur Kenntnis der Auswürflinge des Laacher Seegebietes. II. Zwei Arten von Umbildungen kristalliner Schiefer zu sanidiniten. *Zeitschrift für Kristallographie, Mineralogie und Petrographie. Leipzig. Abt. B*. Vol.47, p.185–210. [Laachite]

KALB, G. & BENDIG, M., 1938. Beiträge zur Kenntnis der Auswürflinge des Laacher Seegebietes. *Decheniana. Verhandlungen des Naturhistorischen Vereins der Rheinlande und Westfalens. Bonn*. Vol.98, p.1–12.

[Gleesites]

KARPINSKII, A., 1869. Avgitovye porody derevni Muldakaevoi i gory Kachkanar. *Gornyi zhurnal. Sankt Peterburg.* Vol.2, No.5, p.225–255. [Muldakaite]

KARPINSKII, A., 1903. O zamechatelnoi tak nazyvaemoi groruditovoi gornoi porode iz Zabaikalskoi oblasti. *Izvestiia Imperatorskoi Akademii Nauk. Sankt Peterburg.* Vol.19, No.2, p.1–32. [Karite]

KATO, T., 1920. A contribution to the knowledge of the cassisterite veins of pneumato-hydatogenetic or hydrothermal origin. A study of the copper–tin veins of the Akénobé district in the Province of Tajima, Japan. *Journal of the College of Science, Imperial University of Tokyo.* Vol.43, Art.5, p.17–18. [Akenobeite]

KEILLER, A., 1936. Two axes of Presely stone from Ireland. *Antiquity. A Quarterly Review of Archaeology. Gloucester.* Vol.10, p.220–221. [Preselite]

KEMP, J.F., 1890. The basic dikes occurring outside of the syenite areas of Arkansas. *Annual Report of the Geological Survey of Arkansas. Little Rock.* Vol.2, p.392–406. [Ouachitite]

KEYES, C.R., 1895. The origin and relations of Central Maryland granites. *Report of the United States Geological Survey. Washington.* p.651–755. [Binary granite]

KHAZOV, R.A., 1983. Ladogality – novye apatitonosnye shchelochnye ultraosnovnye porody. *Doklady Akademii Nauk SSSR. Leningrad.* Vol.268, p.1199–1203. [Ladogalite, Ladogite, Nevoite]

KILPATRICK, J.A. & ELLIS, D.J., 1992. C-type magmas: igneous charnockites and their extrusive equivalents. *Transactions of the Royal Society of Edinburgh: Earth Science.* Vol.83, p.155–164. [C-type Granite]

KINAHAN, G.H., 1875. On the nomenclature of rocks. *Geological Magazine. London.* Vol.2, Decade 2, p.425–426. [Bottleite]

KIRWAN, R., 1784. Elements of Mineralogy. *Elmsly, London.* 412pp. [Granitell, Granitone]

KIRWAN, R., 1794. Elements of Mineralogy. *Elmsly, London.* 2nd Edn, Vol.1, 510pp. [Ferrilite, Granatine, Granilite, Igneous rock, Plutonic]

KLIPSTEIN, A. VON, 1843. Beiträge zur geologischen Kenntnis der östlichen Alpen. *Heyer, Giessen.* Vol.1, 311pp. [Mulatoporphyry (Mulattophyre)]

KLOOS, J.H., 1885. Ein Uralitgestein von Ebersteinburg im nördlichen Schwarzwald. *Neues Jahrbuch für Mineralogie, Geologie und Paläontologie. Stuttgart. Referate Abt. A.* Bd.2, p.82–88. [Uralitite]

KNIGHT, C.W., 1905. Analcite-trachyte tuffs and breccias from southeast Alberta, Canada. *The Canadian Record of Science. Montreal.* Vol.9, p.265–278. [Blairmorite]

KNIGHT, W.C., 1898. Bentonite. *Engineering and Mining Journal.* Vol.66, p.491. [Bentonite]

KOBELL, F. VON, 1838. Grundzüge der Mineralogie. *Schrag, Nürnberg.* 348pp. [Wehrlite]

KOCH, A., 1858. Paläozoische Schichten und Grünsteine in den herzogisch nassauischen Aemtern Dillenburg und Herborn, unter Berücksichtigung allgemeiner Lagerungs-verhältnisse in angränzenden Ländertheilen. *Jahrbuch des Vereins für Naturkunde im Herzogthum Nassau. Wiesbaden.* Vol.13, p.85–326. [Lahnporphyry]

KOLDERUP, C.F., 1896. Die Labradorfelse des westlichen Norwegens. I. Das Labrador-felsgebiet bei Ekersund und Soggendal. *Bergens Museums Aarbog (Årbok) Afhandlinger og Årsberetning.* Vol.5, p.1–222. [Ilmenitite, Soggendalite]

KOLDERUP, C.F., 1898. Lofotens og Vesteraalens gabbrobergarter. *Bergens Museums Aarbog (Årbok) Afhandlinger og Årsberetning.* Vol.7, p.1–54. [*Oligoclasite*]

KOLDERUP, C.F., 1903. Die Labradorfelse des westlischen Norwegens. II. Die Labradorfelse und die mit denselben verwandten Gesteine in dem Bergensgebiete. *Bergens Museums Aarbog (Årbok) Afhandlinger og Årsberetning.* Vol.12, p.1–129. [Birkremite (Bjerkreimite), Farsundite, Mangerite]

KOLENEC, F., 1904. Über einige leukokrate Gang-Gesteine vom Monzoni und Predazzo. *Mitteilungen des Naturwissenschaftlichen Vereins für Steiermark. Graz.* Vol.40, p.161–212. [Feldspathite]

KOTÔ, B., 1909. Journeys through Korea (First contribution). *Journal of the College of Science, Imperial University of Tokyo.* Vol.26, Art.2, p.1–207. [Eutectofelsite (Eutektofelsite, Eutectophyre), Granomasanite, Masanite, Masanophyre]

KOTÔ, B., 1916A. The great eruption of Sakurajima in 1914. *Journal of the College of Science, Imperial University of Tokyo.* Vol.38, Art.3, p.1–237. [*Ceramicite*, Keramikite]

KOTÔ, B., 1916B. On the volcanoes of Japan. V. *Journal of the Geological Society of Toyko.* Vol.23, p.95–127. [Orthoandesite, Orthobasalt, Sanukitoid]

KOVALENKO, V.I., KUZMIN, M.I., ANTIPIN, V.S. & PETROV, L.L., 1971. Topassoderzhashchii kvartsevyi keratofyr (ongonit) – novaya raznovidnost subvulkanicheskikh zhilnykh magmaticheskikh porod. *Doklady Akademii Nauk SSSR. Leningrad.* Vol.199, No.2, p.430–433. [Ongonite]

KRANCK, E.H., 1939. The rock-ground of the coast of Labrador and the connection between the Pre-cambrian of Greenland and North America. *Bulletin de la Commission Géologique de la Finlande. Helsingfors.* Vol.22, No.125, p.65–86. [Aillikite]

KRENNER, M.J., 1910. Über Tephrite in Ungarn. *Compte Rendu de la XIe Session du Congrès Géologique International (Stockholm).* Vol.1, p.740. [Danubite]

KRETSCHMER, F., 1917. Der metamorphe Dioritgabbrogang nebst seinen Peridotiten und Pyroxeniten im Spieglitzer Schnee- und Bielengebirge. *Jahrbuch der Kaiserlich-Königlichen Geologischen Reichsanstalt. Wien.* Vol.67, p.1–201. [Aegirinolith (Aigirinolith, Egirinolith), Bielenite, Marchite, Minettefels, Niklesite, Titanolite, Weigelite]

KROUSTCHOFF, K. DE, 1885. Note préliminaire sur la Wolhynite de M. d Ossowski. *Bulletin de la Société Française de Minéralogie. Paris.* Vol.8, p.441–451. [Volhynite (Wolhynite)]

KTÉNAS, C.A., 1928. Sur la présence de laves alcalines dans la mer Egée. *Compte Rendu Hebdomadaire des Séances de l'Académie des Sciences. Paris.* Vol.186, p.1631–1633. [Eustratite]

KUNO, H., 1960. High-alumina basalt. *Journal of Petrology. Oxford.* Vol.1, p.121–145. [High-alumina basalt, Transitional basalt]

KUNZ, G.F., 1908. Precious stones. *Mineral Industry, its Statistics, Technology and Trade (in the United States and other Countries). New York.* Vol.16, p.790–812. [Jadeolite]

KUPLETSKII, B.M., 1932. Kukisvumchorr i prilegayushchie k nemu massivy tsentralnoi chasti Khibinskikh tundr. *Trudy Soveta po Izucheniiu Proizvoditel'nykh Sil, Seriia Kolskaia. Materialy po Petrografii i Geokhimii Kolskogo Poluostrova.* Vol.2, p.5–72. [Rischorrite]

LACROIX, A., 1893. Les enclaves des roches volcaniques. *Macon, Paris.* 710pp. [Melilitite]

LACROIX, A., 1894. Etude minéralogique de la lherzolite des Pyrénées et de ses phénomènes de contact. *Nouvelles Archives du Muséum d'Histoire Naturelle. Paris.* Vol.6, 3rd Ser., p.209–308. [Amphibololite]

LACROIX, A., 1895. Sur les roches basiques constituant des filons minces dans la lherzolite des Pyrénées. *Compte Rendu Hebdomadaire des Séances de l'Académie des Sciences. Paris.* Vol.120, p.752–755.

[Diopsidite, Pyroxenolite]

LACROIX, A., 1900. Sur un nouveau groupe d'enclaves homoeogènes des roches volcaniques, les microtinites des andésites et des téphrites. *Compte Rendu Hebdomadaire des Séances de l'Académie des Sciences. Paris.* Vol.130, p.348–351. [*Microtinite*]

LACROIX, A., 1901A. Sur un nouveau groupe de roches très basiques. *Compte Rendu Hebdomadaire des Séances de l'Académie des Sciences. Paris.* Vol.132, p.358–360. [Ariegite]

LACROIX, A., 1901B. Les roches basiques accompagnant les lherzolites et les ophites des Pyrénées. *8th International Geological Congress, Paris.* Vol.8, p.826–829. [Avezacite]

LACROIX, A., 1902. Matériaux pour la minéralogie de Madagascar – Les roches alcalines caractérisant la province pétrographique d'Ampafindava. *Nouvelles Archives du Muséum d'Histoire Naturelle. Paris.* Vol.4, 4th Ser., p.1–214. [Nepheline gabbro, Nepheline monzonite]

LACROIX, A., 1905. Sur un nouveau type pétrographique représentant la forme de profondeur de certaines leucotéphrites de la Somma. *Compte Rendu Hebdomadaire des Séances de l'Académie des Sciences. Paris.* Vol.141, p.1188–1193. [Sommaite]

LACROIX, A., 1907. Etude minéralogique des produits silicatés de l'éruption du Vésuve (Avril 1906). *Nouvelles Archives du Muséum d'Histoire Naturelle. Paris.* Vol.9, 4th Ser., p.1–172. [Pollenite]

LACROIX, A., 1909. Sur l'existence de roches grenues intrusives pliocènes dans le massif volcanique du Cantal. *Compte Rendu Hebdomadaire des Séances de l'Académie des Sciences. Paris.* Vol.149, p.540–546. [Essexite-gabbro]

LACROIX, A., 1910. Sur l'existence à la Côte d'Ivoire d'une série pétrographique comparable à celle de la charnockite. *Compte Rendu Hebdomadaire des Séances de l'Académie des Sciences. Paris.* Vol.150, p.18–22. [Ivoirite]

LACROIX, A., 1911A. Le cortège filonien des péridotites de Nouvelle-Calédonie. *Compte Rendu Hebdomadaire des Séances de l'Académie des Sciences. Paris.* Vol.152, p.816–822. [Ouenite]

LACROIX, A., 1911B. Les syénites néphéliniques de l'Archipel de Los et leurs minéraux. *Nouvelles Archives du Muséum d'Histoire Naturelle. Paris.* Vol.3, 5th Ser., p.1–131. [Topsailite]

LACROIX, A., 1914. Les roches basiques non volcaniques de Madagascar. *Compte Rendu Hebdomadaire des Séances de l'Académie des Sciences. Paris.* Vol.159, p.417–422. [Anabohitsite]

LACROIX, A., 1915. Sur un type nouveau de roche granitique alcaline renfermant une eucolite. *Compte Rendu Hebdomadaire des Séances de l'Académie des Sciences. Paris.* Vol.161, p.253–258. [Fasibitikite]

LACROIX, A., 1916A. Sur quelques roches volcaniques mélanocrates des Possessions françaises de l'Océan Indien et du Pacifique. *Compte Rendu Hebdomadaire des Séances de l'Académie des Sciences. Paris.* Vol.163, p.177–183. [Ankaramite]

LACROIX, A., 1916B. La constitution des roches volcaniques de l'extrême nord de Madagascar et de Nosy bé; les ankaratrites de Madagascar en général. *Compte Rendu Hebdomadaire des Séances de l'Académie des Sciences. Paris.* Vol.163, p.253–258. [Ankaratrite, Fasinite]

LACROIX, A., 1916C. Les syénites à riébéckite d'Alter Pedroso (Portugal), leurs formes mésocrates (lusitanites) et leur transformation en leptynites et en gneiss. *Compte Rendu Hebdomadaire des Séances de l'Académie des Sciences. Paris.* Vol.163, p.279–283. [Lusitanite]

LACROIX, A., 1917A. Les laves à haüyne

d'Auvergne et leurs enclaves homoeogènes: importance théorique de ces dernières. *Compte Rendu Hebdomadaire des Séances de l'Académie des Sciences. Paris.* Vol.164, p.581–588. [Mareugite, Ordanchite, Tahitite]

LACROIX, A., 1917B. Les roches grenues d'un magma leucitique étudiées à l'aide des blocs holocristallins de la Somma. *Compte Rendu Hebdomadaire des Séances de l'Académie des Sciences. Paris.* Vol.165, p.207–211. [Campanite, Ottajanite, Puglianite, Sebastianite]

LACROIX, A., 1917C. Les péridotites des Pyrénées et les autres roches intrusives non feldspathiques qui les accompagnent. *Compte Rendu Hebdomadaire des Séances de l'Académie des Sciences. Paris.* Vol.165, p.381–387. [Lherzite]

LACROIX, A., 1917D. Les laves leucitiques de la Somma. *Compte Rendu Hebdomadaire des Séances de l'Académie des Sciences. Paris.* Vol.165, p.481–487. [Vesuvite]

LACROIX, A., 1917E. Les formes grenues du magma leucitique du volcan Laziale. *Compte Rendu Hebdomadaire des Séances de l'Académie des Sciences. Paris.* Vol.165, p.1029–1035. [Braccianite]

LACROIX, A., 1918. Sur quelques roches filoniennes sodiques de l'Archipel de Los, Guinée française. *Compte Rendu Hebdomadaire des Séances de l'Académie des Sciences. Paris.* Vol.166, p.539–545. [Kassaite, Tamaraite]

LACROIX, A., 1919. Dacites et dacitoïdes à propos des laves de la Martinique. *Compte Rendu Hebdomadaire des Séances de l'Académie des Sciences. Paris.* Vol.168, p.297–302. [Dacitoid]

LACROIX, A., 1920. La systématique des roches grenues à plagioclases et feldspathoïdes. *Compte Rendu Hebdomadaire des Séances de l'Académie des Sciences. Paris.* Vol.170, p.20–25. [Berondrite, Luscladite, Mafraite]

LACROIX, A., 1922. Minéralogie de Madagascar. *Challamel, Paris.* Vol.2, 694pp. [Algarvite, Ampasimenite, Antsohite, Dissogenite, Finandranite, Itsindrite, Lindinosite]

LACROIX, A., 1923. Minéralogie de Madagascar. *Challamel, Paris.* Vol.3, 450pp. [Dellenitoid, Doreite, Etindite, Kivite, Oceanite, *Phonolitoid*, Rhyolitoid, Sakalavite, Sancyite]

LACROIX, A., 1924. Les laves analcimiques de l'Afrique du Nord et, d'une façon générale, la classification des laves renfermant de l'analcime. *Compte Rendu Hebdomadaire des Séances de l'Académie des Sciences. Paris.* Vol.178, p.529–535. [Scanoite]

LACROIX, A., 1925. Sur un nouveau type de roche éruptive alcaline récente. *Compte Rendu Hebdomadaire des Séances de l'Académie des Sciences. Paris.* Vol.180, p.481–484. [Ordosite]

LACROIX, A., 1926. La systématique des roches leucitiques; les types de la famille syénitique. *Compte Rendu Hebdomadaire des Séances de l'Académie des Sciences. Paris.* Vol.182, p.597–601. [Gaussbergite, Kajanite]

LACROIX, A., 1927A. Les rhyolites et les trachytes hyperalcalins quartzifères, à propos de ceux de la Corée. *Compte Rendu Hebdomadaire des Séances de l'Académie des Sciences. Paris.* Vol.185, p.1410–1415. [Hakutoite]

LACROIX, A., 1927B. La constitution lithologique des îles volcaniques de la Polynésie Australe. *Mémoires de l'Académie des Sciences (de l'Institut de France). Paris.* Vol.59, p.107. [Murite]

LACROIX, A., 1928A. La composition minéralogique et chimique des roches éruptives et particulièrement des laves mésozoïques et plus récentes de la Chine orientale. *Bulletin of the Geological Society of China. Peking.* Vol.7, p.13–59. [Mandchurite (Manchurite, Mandchourite, Mandschurite)]

LACROIX, A., 1928B. Les pegmatitoïdes des roches volcaniques à faciès basaltiques. *Compte Rendu Hebdomadaire des Séances de l'Académie des Sciences. Paris.* Vol.187, p.321–326. [Pegmatitoid]

LACROIX, A., 1933. Contribution à la connaissance de la composition chimique et minéralogique des roches éruptives de l'Indochine. *Bulletin du Service Géologique de l'Indochine. Saïgon.* Vol.20, No.3, 208pp. [Argeinite, Cocite, Florinite, Jerseyite, Kemahlite, Melfite, Melilitolite, Mikenite, *Monzogranite*, Nemite, Niligongite, Plagioclasolite]

LAGORIO, A., 1897. Itinéraire géologique d'Alouchta à Sébastopol par Yalta, Bakhtchissaraï et Mangoup-Kalé. *7th International Geological Congress, St Petersburg. Excursions Guide.* Vol.33, p.1–28. [Taurite]

LAITAKARI, A., 1918. Einige Albitepidotgesteine von Südfinnland. *Bulletin de la Commission Géologique de la Finlande. Helsingfors.* Vol.9, No.51, p.1–13. [Helsinkite]

LAMEYRE, J., 1966. Leucogranites et muscovitisation dans le Massif Central français. *Annales de la Faculté des Sciences de l'Université de Clermont, Clermont-Ferrand.* 264pp. [Episyenite, Leucogranite]

LANG, H.O., 1877. Grundriss der Gesteinskunde. *Haessel, Leipzig.* 289pp. [Hungarite, Plädorite]

LANG, H.O., 1891. Versuch einer Ordnung der Eruptivgesteine nach ihren chemischen Bestande. *Tschermaks Mineralogische und Petrographische Mitteilungen. Wien.* Vol.12, 2nd Ser., p.199–252. [Aetna-basalt, Amiatite, Bolsenite, Christianite, Puys-andesite, Rhön basalt]

LANGIUS, C.N., 1708. Historia Lapidum Figuratorum Helvetiae ejusque Viciniae. *Jacobi Tomasini, Venice.* 165pp. [Grammite]

LAPPARENT, A.A. DE, 1864. Mémoire sur la constitution géologique du Tyrol méridional. *Annales des Mines ou Recueil de Mémoires sur l'Exploration des Mines, et sur les Sciences qui s'y rapportent; rédigés par le Conseil Général des Mines. Paris.* Vol.6, p.245–314. [Monzonite]

LAPPARENT, A.A. DE, 1883. Traité de Géologie. *Savy, Paris.* 1st Edn, 1280pp. [Ortholithe (Ortholite)]

LAPPARENT, A.A. DE, 1885. Traité de Géologie. *Savy, Paris.* 2nd Edn, 1504pp. [Granulophyre]

LAPPARENT, A.A. DE, 1893. Traité de Géologie. *Masson & Cie, Paris.* 3rd Edn, 1645pp. [Granoliparite]

LAPWORTH, C., 1898. Long excursion to the Birmingham district. *Proceedings of the Geologists' Association. London.* Vol.15, p.417–428. [Anchorite]

LARSEN, E.S. & HUNTER, J.F., 1914. Melilite and other minerals from Gunnison County, Colorado. *Journal of the Washington Academy of Sciences.* Vol.4, p.473–479. [Uncompahgrite]

LARSEN, E.S. & PARDEE, J.T., 1929. The stock of alkaline rocks near Libby, Montana. *Journal of Geology. Chicago.* Vol.37, p.97–112. [Glimmerite]

LASAULX, A. VON, 1875. Elemente der Petrographie. *Strauss, Bonn.* 486pp. [Olivine gabbro, Orthophonite, Troctolite]

LASPEYRES, H., 1869. Über das Zusammenvorkommen von Magneteisen und Titaneisen in Eruptivgesteinen und über die sogenannten petrographischen Gesetze. *Neues Jahrbuch für Mineralogie, Geologie und Paläontologie. Stuttgart. Referate Abt. A.* Bd.40, p.513–531. [Palatinite]

LAWSON, A.C., 1893. The geology of Carmelo Bay. *University of California Publications. Bulletin of the Department of Geology. Berkeley.* Vol.1, p.1–59. [Carmeloite]

LAWSON, A.C., 1896. On malignite, a family of basic plutonic orthoclase rocks rich in alkalis and lime intrusive in the Coutchiching schists of Poohbah Lake. *University of California*

Publications. Bulletin of the Department of Geology. Berkeley. Vol.1, No.12, p.337–362. [Malignite]

LAWSON, A.C., 1903. Plumasite, an oligoclase-corundum rock near Spanish Peak, California. *University of California Publications. Bulletin of the Department of Geology. Berkeley.* Vol.3, No.8, p.219–229. [Plumasite]

LE BAS, M.J., 1977. Carbonatite–Nephelinite Volcanism. *Wiley, London.* 347pp. [Ferrocarbonatite]

LE BAS, M.J., LE MAITRE, R.W., STRECKEISEN, A. & ZANETTIN, B., 1986. A chemical classification of volcanic rocks based on the total alkali–silica diagram. *Journal of Petrology. Oxford.* Vol.27, p.745–750. [Basaltic trachyandesite]

LE MAITRE, R.W., 1984. A proposal by the IUGS Subcommission on the Systematics of Igneous Rocks for a chemical classification of volcanic rocks based on the total alkali silica (TAS) diagram. *Australian Journal of Earth Science. Melbourne.* Vol.31, p.243–255. [Picrobasalt, Potassic trachybasalt]

LE MAITRE, R.W. (EDITOR), BATEMAN, P., DUDEK, A., KELLER, J., LAMEYRE, M., LE BAS, M.J., SABINE, P.A., SCHMID, R., SØRENSEN, H., STRECKEISEN, A., WOOLLEY, A.R. & ZANETTIN, B., 1989. A Classification of Igneous Rocks and a Glossary of Terms. Recommendations of the International Union of Geological Sciences Subcommission on the Systematics of Igneous Rocks. *Blackwell Scientific Publications, Oxford.* p.193. [Hyalo-]

LE PUILLON DE BOBLAYE, E. & VIRLET, TH., 1833. Expédition scientifique de Morée: géologie et minéralogie. *Levrault, Paris.* Vol.2, Pt.2, p.1–375. [Prasophyre]

LEA, F.M. & DESCH, C.H., 1935. The Chemistry of Cement and Concrete. *Arnold, London.* 429pp. [Tetin]

LEAKE, B.E., WOOLLEY, A.R., ARPS, C.E.S., BIRCH, W.D., GILBERT, M.C., GRICE, J.D., HAWTHORNE, F.C., KATO, A., KISCH, H.J., KRIVOVICHEV, V.G., LINTHOUT, K., LAIRD, J., MANDRINO, J.A., MARESCH, W.V., NICKEL, E.H., ROCK, N.M.S., SCHUMACHER, J.C., SMITH, D.C., STEPHENSON, N.C.N., UNGARETTI, L., WHITTAKER, E.J.W. & YOUZHI, G., 1997. Nomenclature of amphiboles: Report of the Subcommittee on Amphiboles of the International Mineralogical Association Commission on New Minerals and Mineral Names. *Mineralogical Magazine and Journal of the Mineralogical Society. London.* Vol.61, p.295-321; *The Canadian Mineralogist.* Vol.35, p.216-246; *American Mineralogist.* Vol.82, p.1019-1031.

LEBEDEV, A.P. & VAKHRUSHEV, V.A., 1953. Yavleniya kontaminatsii v zhilnykh giperbazitakh yuzhnoi Fergany. *Izvestiia Akademii Nauk SSSR, Seriia Geologicheskaia.* No.1, p.114–131. [Ferganite]

LEBEDINSKY, V.I. & CHU TSZYA-SYAN, 1958. Ob anortoklaze v shchelochnykh bazaltakh yuzhnoi okrainy mongolskogo plato (KNR). *Zapiski Vsesoiuznogo Mineralogicheskogo Obshchestva. Moskva.* Vol.87, p.14–22. [Spumulite]

LEHMANN, E., 1924. Das Vulkangebiet am Nordende des Nyassa als magmatische Provinz. *Zeitschrift für Vulkanologie. Reimer, Berlin.* Ergänzungsband 4, 209pp. [Atlantite, Essexite-basalt]

LEHMANN, E., 1949. Das Keratophyr-Weilburgit-Problem. *Heidelberger Beiträge zur Mineralogie und Petrographie. Berlin.* Vol.2, p.1–166. [Weilburgite]

LEONHARD, K.C. VON, 1821. Handbuch der Oryktognosie. *Mohr & Winter, Heidelberg.* 720pp. [Cantalite]

LEONHARD, K.C. VON, 1823A. Charakteristik der Felsarten. *Engelmann, Heidelberg.* Vol.1, p.1–230. [Aplite, Blatterstein, Felstone, Granitite, Greisen, Kugeldiorite, Tholerite]

LEONHARD, K.C. VON, 1823B. Charakteristik

der Felsarten. *Engelmann, Heidelberg.* Vol.3, p.599–772. [Agglomerate, Trass]

LEONHARD, K.C. VON, 1832. Die Basalt-Gebilde in ihren Beziehungen zu normalen und abnormen Felsmassen. *Schweizerbart, Stuttgart.* Vol.1, 498pp. [Anamesite]

LEPSIUS, R., 1878. Das Westliche Süd-Tyrol. *Hertz, Berlin.* 372pp. [Nonesite]

LEWIS, H. C., 1888. The matrix of the diamond. *Geological Magazine. London.* Vol.5, Decade 3, p.129–131. [Kimberlite]

LINDGREN, W., 1886. Eruptive rocks. In: W.M. Davis, Relation of the coal of Montana to the older rocks. In: R. Pumpelly, Report on the Mining Industries of the United States (exclusive of the precious metals), with special investigation into the iron resources of the Republic and into the Cretaceous coal of the Northwest. *10th Decennial Census of the United States: 1880.* Vol.15, p.727. [Analcime basalt]

LINDGREN, W., 1893. Art. XXX – The auriferous veins of Meadow Lake, California. *American Journal of Science. New Haven.* Vol.46, 3rd Ser., p.201–206. [Granodiorite]

LOEWINSON-LESSING, F. YU.,[†] 1896. K voprosu o khimicheskoi klassifikatsii izverzhennykh gornykh porod. *Trudy Imperatorskago Sankt-Peterburgskago Obshchestva Estestvoispytatelei. Sankt Peterburg.* Vol.27, No.1, p.175–176. [Hyperacidite]

LOEWINSON-LESSING, F. YU., 1898. Issledovaniya po teoreticheskoi petrografii v svyazi s izucheniem izverzhennykh porod Centralnogo Kavkaza. *Trudy Imperatorskago Sankt-Peterburgskago Obshchestva Estestvoispytatelei. Sankt Peterburg.* Vol.26, No.5, p.1–104. [Hypobasite, Mesite]

LOEWINSON-LESSING, F. YU., 1900A. Kritische Beiträge zur Systematik der Eruptivgesteine. II. *Tschermaks Mineralogische und Petrographische Mitteilungen. Wien.* Vol.19, 2nd Ser., p.169–181. [Anorthophyre, Anorthosyenite]

LOEWINSON-LESSING, F. YU., 1900B. Geologicheskii ocherk Yuzhno-Zaozerskoi dachi i Denezhkina kamnya na severnom Urale. *Trudy Imperatorskago Sankt-Peterburgskago Obshchestva Estestvoispytatelei. Sankt Peterburg.* Vol.30, No.5, p.1–257. [Alkaliptoche]

LOEWINSON-LESSING, F. YU., 1901. Kritische Beiträge zur Systematik der Eruptivgesteine. IV. *Tschermaks Mineralogische und Petrographische Mitteilungen. Wien.* Vol.20, 2nd Ser., p.110–128. [Anorthoclasite, Calciptoche, Feldspathidolite, Feldspatholite, Leucolite, Leucoptoche, Melanolite, Melanoptoche, Melilithite, Microclinite, Nephelinolith, Pikroptoche (Pycroptoche)]

LOEWINSON-LESSING, F. YU., 1905A. Petrograficheskaya ekskursiya po r. Tagilu. *Izvestiia Sankt-Peterburgskago Polytekhnicheskago Instituta Imperatora Petra Velikago. Sankt Peterburg.* Vol.3, p.1–40. [Camptovogesite]

LOEWINSON-LESSING, F. YU., 1905B. Petrographische Untersuchungen im Centralen Kaukasus. *Zapiski Imperatorskogo (Sankt Peterburgskogo) Mineralogicheskogo Obshchestva.* Vol.42, p.237–280. [Dumalite]

LOEWINSON-LESSING, F. YU., 1928. Some queries on rock classification and nomenclature. *Doklady Akademii Nauk SSSR. Leningrad.* Ser.A., p.139–142. [Keratophyrite]

LOEWINSON-LESSING, F. YU., 1934. Problema genezisa magmaticheskikh porod i puti k ee razresheniyu. *Izadatel'stvo Akademii Nauk SSSR, Moskva.* 58pp. [Anatectite (Anatexite), Prototectite, Syntectite]

LOEWINSON-LESSING, F. YU., 1936. O nesilikatnykh magmakh. *Akademiku*

[†] The current transliteration of this name would be Levinson-Lessing. However, to maintain uniformity with previous publications, the older transliteration of Loewinson-Lessing has been used.

V.I.Vernadskomu k piatidesiatiletiiu nauchnoi i pedagogicheskoi deiatel'nosti. Akademiia Nauk SSSR. Vol.2, p.989–997. [Apatitolite]

LOEWINSON-LESSING, F. YU., GINZBERG, A.S. & DILAKTORSKII, N.P., 1932. Trappy Tuluno-Udinskogo i Bratskogo raionov v Vostochnoi Sibiri. *Trudy Soveta po Izucheniiu Proizvoditel'nykh Sil. Seriia Sibirskaia.* Vol.1, p.1–82. [Angarite, Anorthobase, Orthobase]

LOISELLE, M.C. & WONES, D.R., 1979. Characteristics and origin of anorogenic granites. *Abstracts of papers to be presented at the Annual Meetings of the Geological Society of America and Associated Societies, San Diego, California, November 5-8, 1979.* Vol.11, p.468. [A-type granite]

LORENZO, G. DE, 1904. The history of volcanic action in the Phlegræan Fields. *Quarterly Journal of the Geological Society of London.* Vol.60, p.296–314. [Mappamonte, Piperno]

LOSSEN, K.A., 1886. Über die verschiedene Bedeutung, welche dem Worte Palatinit von den Petrographen beigelegt worden ist, und knüpfte daran einige Mittheilungen über seine Stellung zur Mealphyr-Frage. *Zeitschrift der Deutschen Geologischen Gesellschaft. Berlin.* Vol.38, p.921–926. [Hysterobase]

LOSSEN, K.A., 1892. Vergleichende Studien über die Gesteine des Spiemonts und des Bosenbergs bei St. Wedel und verwandte benachbarte Eruptivtypen aus der Zeit des Rothliegenden. *Jahrbuch der Königlich Preussischen Geologischen Landesanstalt und Bergakademie. Berlin.* Vol.10, for 1889, p.258–321. [Pegmatophyre]

LOUGHLIN, G.F., 1912. The gabbros and associated rocks at Preston, Connecticut. *Bulletin of the United States Geological Survey. Washington.* Vol.492, p.1–158. [Quartz anorthosite]

LUCHITSKII, I.V., 1960. Vulkanizm i tektonika devonskikh vpadin Minusinskogo mezhgornogo progiba. *Izadatel'stvo Akademii Nauk SSSR, Moskva.* 176pp. [Goryachite]

LUNDQVIST, T., 1975. The "kenningite" found near Sundsvall, central Sweden. *Geologiska Föreningens i Stockholm Förhandlingar. Stockholm.* Vol.97, p.189-192. [Kenningite]

LYELL, C., 1835. Principles of Geology. *Murray, London.* 4th Edn, Vol.1, 406pp. [Lapilli]

MACDONALD, G.A., 1949. Hawaiian Petrographic Province. *Bulletin of the Geological Society of America. New York.* Vol.60, p.1541–1596. [Mimosite]

MACHATSCHKI, F., 1927. Enstatit-Hornblendit von Grönland. *Centralblatt für Mineralogie, Geologie und Paläontologie. Stuttgart. Abt.A.* p.172–175. [Grönlandite (Greenlandite)]

MARCET RIBA, J., 1925. El método notural en Petrografía. Rocas eruptivas intrusivas de la serie calco-alcalina. *Memorias de la Real Academia de Ciencias y Artes de Barcelona.* Vol.19, No.10, p.245–419. [Alentegite, Andersonite, Eichstädtite, Finlandite, Ivreite, Kolderupite, Langite, Lossenite, Rotenburgite]

MARSHALL, P., 1906. The geology of Dunedin (New Zealand). *Quarterly Journal of the Geological Society of London.* Vol.62, p.381–424. [Kaiwekite, Ulrichite]

MARSHALL, P., 1932. Notes on some volcanic rocks of the North Island of New Zealand. *New Zealand Journal of Science and Technology. Wellington.* Vol.13, p.198–200. [Ignimbrite]

MARSHALL, P., 1935. Acid rocks of the Taupo-Rotorua volcanic district. *Transactions of the Royal Society of New Zealand. Dunedin.* Vol.64, p.323–366. [Lapidite, Lenticulite, Pulverulite]

MARTIN, R. & DE SITTER-KOOMANS, C., 1955. Pseudotectites from Colombia and Peru. *Leidse Geologische Mededelingen. Leiden.* Vol.20, p.151–164. [Macusanite]

*MARZARI PENCATI, G., 1819. Cenni geologici e mineralogici sulle province venete e sul Tirolo. 1-54, Vicenza, 1820. *Supplemento al Nuovo Osservatorio Veneziano*. N. 118–127, Venezia. [Sievite]

MAWSON, D., 1906. The minerals and genesis of the veins and schlieren traversing the aegirine-syenite in the Bowral quarries. *Proceedings of the Linnean Society of New South Wales. Sydney*. Vol.31, p.579–607. [Bowralite]

MAZZUOLI, L. & ISSEL, A., 1881. Relazione degli studi fatti per un rilievo delle masse ofiolitiche nella riviera di Levante (Liguria). *Bollettino del Reale Comitato Geologico d'Italia. Firenze*. Vol.12, p.313–349. [Calcogranitone, Ophigranitone, Silico-granitone]

MCHENRY, A. & WATT, W.W., 1895. Guide to the Collection of Rocks and Fossils belonging to the Geological Survey of Ireland. *Ireland Geological Survey. Museum of Science and Art, Dublin*. 155pp. [Ivernite]

MENNELL, F.P., 1929. Some Mesozoic and Tertiary igneous rocks from Portuguese East Africa. *Geological Magazine. London*. Vol.66, p.529–540. [Lupatite]

MICHEL-LÉVY, A., 1874. Note sur une classe de roches éruptives intermédiaires entre les granites porphyroïdes et les porphyres granitoïdes groupe des granulites. *Bulletin de la Société Géologique de France. Paris*. Vol.2, p.177–189. [Granulite]

MICHEL-LÉVY, A., 1894. Etude sur la détermination des feldspaths. *Baudry, Paris*. 107pp. [Trachyandesite]

MICHEL-LÉVY, A., 1897. Mémoire sur le porphyre bleu de l'Esterel. *Bulletin des Services de la Carte Géologique de la France et des Topographies Souterraines. Paris*. Vol.9, No.57, p.1–40. [Esterellite]

MILCH, L., 1927. A. Johannsen: A revised field Classification of Igneous Rocks. (Journ. of Geol. 34. 1926. 181-182.). *Neues Jahrbuch für Mineralogie, Geologie und Paläontologie. Stuttgart. Referate Abt. A*. Bd.II, p.62–63. [Melanide]

MILLER, W.J., 1919. Pegmatite, silexite, and aplite of Northern New York. *Journal of Geology. Chicago*. Vol.27, p.28–54. [Silexite]

MILLOSEVICH, F., 1908. Studi sulle rocce vulcaniche di Sardegna. I. Le rocce di Sassari e di Porto Torres. *Atti della Reale Accademia (Nazionale) dei Lincei. Rendiconti. Classe di Scienze Fisiche, Matematiche e Naturali. Roma*. Vol.6, p.403–438. [Trachydacite]

MILLOSEVICH, F., 1927. Le rocce a corindone della Val Sessera (Prealpi Biellesi). *Atti della Reale Accademia (Nazionale) dei Lincei. Rendiconti. Classe di Scienze Fisiche, Matematiche e Naturali. Roma*. Vol.5, p.22–31. [Sesseralite]

MITCHELL, R.H., 1995. Kimberlites, Orangeites, and Related Rocks. *Plenum Press, New York and London*. 410pp. [Orangeite]

MIYASHIRO, A. & SHIDO, F., 1975. Tholeiitic and calc-alkali series in relation to the behaviours of titanium, vanadium, chromium, and nickel. *American Journal of Science. New Haven*. Vol.275, p.165-277. [Abyssal tholeiite]

MONTEIRO, M, 1814. Du pyroméride globulaire ou de la roche connue sous le nom de porphyre globuleux de Corse. *Journal des Mines. Paris*. Vol.35, p.347–360. [Pyromeride]

MOOR, G.G. & SHEINMAN, YU. M., 1946. Porody iz severnoi okrainy Sibirskoi platformy. *Doklady Akademii Nauk SSSR. Leningrad*. Vol.51, No.2, p.141–144. [Meimechite (Meymechite)]

MORIMOTO, N., FABRIES, J., FERGUSON, A.K., GINZBURG, I.V., ROSS, M., SEIFERT, F.A. & ZUSSMAN, J., 1988. Nomenclature of pyroxenes. *Mineralogical Magazine and Journal of the Mineralogical Society. London*. Vol.52, p.535-550.

MOROZEWICZ, J., 1899. Experimentelle

Untersuchungen über die Bildung der Minerale im Magma. *Tschermaks Mineralogische und Petrographische Mitteilungen. Wien.* Vol.18, 2nd Ser., p.105–240. [Kyschtymite]

MOROZEWICZ, J., 1901. Gora Magnitnaya i eya blizhaishiya okrestnosti. *Trudy Geologicheskago Komiteta. Sankt Peterburg.* Vol.18, No.1, p.1–104. [Atatschite]

MOROZEWICZ, J., 1902. Ueber Mariupolit, ein extremes Glied der Elaeolithsyenite. *Tschermaks Mineralogische und Petrographische Mitteilungen. Wien.* Vol.21, 2nd Ser., p.238–246. [Mariupolite]

MOROZOV, A.I., 1938. Zametka o kamchatskoi novoi gornoi porode. *Biulleten Vulkanologicheskoi Stantsii na Kamchatke.* No.3, p.17–20. [Kamchatite]

MOSEBACH, R., 1956. Khewraite vom Khewra Gorge, Pakistan, ein neuer Typus kalireicher Effusivgesteine. *Neues Jahrbuch für Mineralogie. Stuttgart. Abhandlungen.* Vol.89, p.182–209. [Khewraite]

MUELLER, G., 1964. Lepaigite, a new porphyritic glass from Northern Chile. *International Geological Congress. Report of the 22nd Session India. New Delhi.* Vol.16, p.374–385. [Lepaigite]

MÜGGE, O., 1893. Untersuchungen über die "Lenneporphyre" in Westfalen und den angrenzenden Gebieten. *Neues Jahrbuch für Mineralogie, Geologie und Paläontologie. Stuttgart. Referate Abt. A.* Bd.8, p.535-721. [Lenneporphyry]

NARICI, E., 1932. Contributo alla petrografia chimica della provincia magmatica compana e del Monte Vulture. *Zeitschrift für Vulkanologie. Berlin.* Vol.14, p.210–239. [Guardiaite]

NAUMANN, C.F., 1850. Lehrbuch der Geognosie. *Engelmann, Leipzig.* Vol.1, 1000pp. [Hyperite, Hypersthenite, Leucilite, Nepheline basalt]

NAUMANN, C.F., 1854. Lehrbuch der Geognosie. *Engelmann, Leipzig.* Vol.2, 1222pp. [Porphyrite]

NEMOTO, T., 1934. Preliminary note on alkaline rhyolites from Tokati, Hokkaido. *Journal of the Faculty of Science, Hokkaido University. Sapporo.* Vol.2, Ser. 4, p.299–321. [Okawaite]

NIGGLI, P., 1923. Gesteins- und Mineralprovinzen. *Borntraeger, Berlin.* Vol.1, 602pp. [Coloradoite, *Engadinite*, *Evisite*, Greenhalghite, Lamproite, Lamprosyenite, Nosykombite, Orvietite, *Peléeite*, *Yosemitite*]

NIGGLI, P., 1931. Die quantitive mineralogische Klassifikation der Eruptivgesteine. *Schweizerische Mineralogische und Petrographische Mitteilungen. Zürich.* Vol.11, p.296–364. [Alkali diorite, Alkali gabbro, Pheno-]

NIGGLI, P., 1936. Die Magmatypen. *Schweizerische Mineralogische und Petrographische Mitteilungen. Zürich.* Vol.16, p.335–399. [Achnahaite, *Kaulaite*, *Olivine pacificite*, *Vredefortite*]

NIKITIN, V. & KLEMEN, R., 1937. Dioritpirokseniti iz okolice Cizlaka na Pohorju. *Geoloshki Anali Balkanskoga Poluostrva.* Vol.14, p.149–198. [Cizlakite]

NOCKOLDS, S.R., 1954. Average chemical composition of some igneous rocks. *Bulletin of the Geological Society of America. New York.* Vol.65, p.1007–1032. [Nepheline latite]

NORDENSKJÖLD, N., 1820. Bidrag till Närmare Kännedom af Finlands Mineralier och Geognosie. *Första Häftet, Stockholm.* 103pp. [Sordawalite (Sordavalite)]

NORDENSKJÖLD, O., 1893. Ueber archaeische Ergussgesteine aus Småland. *Bulletin of the Geological Institution of the University of Upsala.* Vol.1, p.133–255. [Eorhyolite]

NOSE, C.W., 1808. Mineralogische Studien über die Gebirge am Niederrhein. Editor J. Nöggerath. *Hermann, Frankfurt am Main.* [Sanidinite]

O'NEILL, J.J., 1914. St. Hilaire (Beloeil) and Rougemont Mountains, Quebec. *Memoirs. Geological Survey, Canada. Ottawa.* Vol.43, p.1–108. [Rougemontite, Rouvillite]

OEBBEKE, K., 1881. Beiträge zur Petrographie der Philippinen und Palau-Inseln. *Neues Jahrbuch für Mineralogie, Geologie und Paläontologie. Stuttgart. Referate Abt. A.* Bd.1, p.451–501. [*Orthoandesite*]

OGURA, T., MATSUDA, K., NAKAGAWA, T., MATSUMOTO, M. & MURATA, K., 1936. Volcanoes of the Wu Ta Lien Chih district, Lung Chiang Province, Manchuria (In Japanese with English summary). *Survey Reports of Volcanoes in Manchuria. Ryojun.* No.1, p.1–96. [Shihlunite]

OSANN, A., 1889. Beiträge zur Kenntniss der Eruptivgesteine des Cabo de Gata (Prov. Almeria). *Zeitschrift der Deutschen Geologischen Gesellschaft. Berlin.* Vol.41, p.297–311. [Verite]

OSANN, A., 1892. Über dioritische Ganggesteine im Odenwald. *Mittheilungen der Grossherzoglich Badischen Geologischen Landesanstalt. Heidelberg.* Vol.2, p.380–388. [Malchite]

OSANN, A., 1893. Report on the rocks of Trans-Pecos Texas. *Report of the Geological Survey of Texas. Austin.* Vol.4, p.123–138. [Paisanite]

OSANN, A., 1896. Beiträge zur Geologie und Petrographie der Apache (Davis) Mts, Westtexas. *Tschermaks Mineralogische und Petrographische Mitteilungen. Wien.* Vol.15, 2nd Ser., p.394–456. [Apachite]

OSANN, A., 1922. H. Rosenbusch: Elemente der Gesteinslehre. *Schweizerbart, Stuttgart.* 4th Edn, 779pp. [Kasanskite, Pedrosite]

OSANN, A.H., 1902. Versuch einer chemischen Classification der Eruptivgesteine. III. Die Ganggesteine. *Tschermaks Mineralogische und Petrographische Mitteilungen. Wien.* Vol.21, 2nd Ser., p.365–448. [Katzenbuckelite]

OSANN, A.H., 1906. Über einige Alkaligesteine aus Spanien. In: E.A. Wülfing (Editor) Festschrift Harry Rosenbusch. *Schweizerbart, Stuttgart.* p.263–310. [Jumillite]

OSBORNE, F.F. & WILSON, N.L., 1934. Some dike rocks from Mount Johnson, Quebec. *Journal of Geology. Chicago.* Vol.42, p.180–187. [Monnoirite]

OSTADAL, C., 1935. Über ein calcitführendes Tiefengestein aus dem nordwestlichen Waldviertel. *Verhandlungen der Geologischen Reichsanstalt (Staatanstalt–Landesanstalt). Wien.* No.8/9, p.117–126. [Hörmannsite]

OYAWOYE, M.O., 1965. Bauchite: a new variety in the quartz monzonite series. *Nature. London.* Vol.205, p.689. [Bauchite]

PALMIERI, L. & SCACCHI, A., 1852. Della regione vulcanica del Monte Vulture. *Nobile, Naploli.* Monograph, 160pp. [Augitophyre]

PAPASTAMATIOU, J., 1939. Sur quelques minéraux types de roches à corindon de l'Ile de Naxos (Archipel Grec). *Compte Rendu Hebdomadaire des Séances de l'Académie des Sciences. Paris.* Vol.208, p.2088–2090. [Naxite]

PARGA-PONDAL, I., 1935. Quimismo de las manifestaciones magmáticas cenozoicas de la Península Iberica. *Trabajos Museo Nacional de Ciencias Naturales. Madrid. Serie Geologica.* Vol.39, p.1–174. [Cancarixite]

PASOTTI, P., 1954. Sobre una roca filoneana adamellítica del Cerro Tandileofú, Prov. de Buenos Aires, Argentina. *Publicaciones del Instituto de Fisiografia y Geologia. Universidad Nacional del Litoral. Rosario.* Vol.9, No.41, p.1–25. [Tandileofite]

PELIKAN, A., 1902. Dahamit, ein neues Ganggestein aus der Gefolgschaft des Alkaligranit (a review of another paper). *American Journal of Science. New Haven.* Vol.14, 4th Ser., p.397. [Dahamite]

PELIKAN, A., 1906. Über zwei Gesteine mit

primärem Analcim nebst Bemerkungen über die Entstehung der Zeolithe. *Tschermaks Mineralogische und Petrographische Mitteilungen. Wien.* Vol.25, 2nd Ser., p.113–126. [Analcime phonolite]

PERRET, F.A., 1913. Art. LII – Some Kilauean ejectamenta. *American Journal of Science. New Haven.* Vol.35, 4th Ser., p.611–618. [Pélé's tears]

PETERSEN, J., 1890A. Beiträge zur Petrographie von Sulphur Island, Peel Island, Hachijo und Mijakeshima. *Jahrbuch der Hamburgischen Wissenschaftlichen Anstalten. Hamburg.* Vol.8, p.1–58. [Boninite, Mijakite]

PETERSEN, J., 1890B. Der Boninit von Peel Island. *Jahrbuch der Hamburgischen Wissenschaftlichen Anstalten. Hamburg.* Vol.8, p.341–349. [*Boninite*]

PETERSEN, TH., 1869. Über den Basalt und Hydrotachylyt von Rossdorf bei Darmstadt. *Neues Jahrbuch für Mineralogie, Geologie und Paläontologie. Stuttgart. Referate Abt. A.* Bd.40, p.32–41. [Hydrotachylyte]

PETTERSEN, K., 1883. Sagvandit, eine neue enstatitführende Gebirgsart. *Neues Jahrbuch für Mineralogie, Geologie und Paläontologie. Stuttgart. Referate Abt. A.* Bd.2, p.247. [Sagvandite]

PHEMISTER, J., 1926. The geology of Strath Oykell and Lower Loch Shin (South Sutherlandshire and North Ross-shire). *Memoirs. Geological Survey of Scotland. Edinburgh.* Sheet 102, p.1–220. [Perthosite]

PHILLIPS, J., 1848. The Malvern Hills compared with the Palaeozoic districts of Abberley, &c. *Memoirs of the Geological Survey of Great Britain, and of the Museum of Practical Geology in London.* Vol.2, Pt.1, p.1–330. [Hornblendite]

PHILLIPS, W., 1815. An Outline of Mineralogy and Geology, intended for the use of those who may desire to become acquainted with the elements of those sciences; especially of young persons. *Phillips, London.* 193pp.

[Tuff]

PICHLER, A., 1875. Beiträge zur Geognosie Tirols. *Neues Jahrbuch für Mineralogie, Geologie und Paläontologie. Stuttgart. Referate Abt. A.* Bd.46, p.926–936. [Ehrwaldite, Töllite (Toellite), Vintlite]

PINKERTON, J., 1811A. Petrology. A Treatise on Rocks. *White, Cochrane & Co., London.* Vol.1, 599pp. [Amygdalite, Basaltin, Basalton, Granitel, Granitin, Granitoid, Graniton, Intrite, Lehmanite, Patrinite, Porphyrin, Porphyron, Toadstone, Wernerite]

PINKERTON, J., 1811B. Petrology. A Treatise on Rocks. *White, Cochrane & Co., London.* Vol.2, 654pp. [Bergmanite, Blacolite, Bornite, Corsilite, Firmicite, Miagite, Niolite, Rhazite, Runite, Torricellite]

PIRSSON, L.V., 1895. Art. XIII – Complementary rocks and radial dikes. *American Journal of Science. New Haven.* Vol.50, 3rd Ser., p.116–121. [Oxyphyre]

PIRSSON, L.V., 1896. On the monchiquites or analcite group of igneous rocks. *Journal of Geology. Chicago.* Vol.4, p.679–690. [Analcitite]

PIRSSON, L.V., 1905. Petrology and geology of the igneous rocks of the Highwood Mountains, Montana. *Bulletin of the United States Geological Survey. Washington.* Vol.237, p.1–208. [Cascadite, Fergusite]

PIRSSON, L.V., 1914. Art. XXVII – Geology of Bermuda Island; Petrology of the lavas. *American Journal of Science. New Haven.* Vol.38, 4th Ser., p.331–344. [Bermudite]

PIRSSON, L.V, 1915. The microscopical characters of volcanic tuffs — a study for students. *American Journal of Science. New Haven.* Vol.40, 4th Ser., p.191–211. [Crystal tuff, Lithic tuff, Vitric tuff]

PISANI, F., 1864. Sur quelques nouveaux minéraux de Cornouailles. *Compte Rendu Hebdomadaire des Séances de l'Académie des Sciences. Paris.* Vol.59, p.912–913. [Luxulianite (Luxullianite, Luxulyanite)]

PLINY, C., AD 77. Naturalis historiae. *[Various printed versions and translations are available.]* [Basalt, Ophite, Syenite, Tephrite]

POLENOV, B.K., 1899. Massivnye gornye porody severnoi chasti Vitimskogo ploskogorya. *Trudy Imperatorskago Sankt-Peterburgskago Obshchestva Estestvoispytatelei. Sankt Peterburg.* Vol.27, p.89–483. [Diabasite, Dioritite, Dioritophyrite, Gabbrite, Gabbrophyrite, Syenitite]

POLKANOV, A.A., 1940. Egirinity plutona Gremyakha-Vyrmes na Kolskom poluostrove. *Zapiski Vserossiiskogo Mineralogicheskogo Obshchestva. Moskva.* Vol.69, p.303–307. [Aegirinite]

POLKANOV, A.A. & ELISEEV, N.A., 1941. Petrologiya plutona Gremyakha-Vyrmes. *Leningradskii Gosudarstvennyi Universitet.* 244pp. [Hortonolitite]

PRATT, J.H. & LEWIS, J.V., 1905. Corundum and the peridotites of Western North Carolina. *Report North Carolina Geological and Economic Survey. Chapel Hill, N.C.* Vol.1, p.1–464. [Enstatolite]

PRELLER, C.S. DU RICHE, 1924. Italian Mountain Geology. Parts I & II. Northern Italy and Tuscany. *Wheldon & Wesley, London.* 3rd Edn, 195pp. [Romanite]

PREOBRAZHENSKII, I.A., 1956. O nekotorykh nazvaniyakh v oblasti petrografii osadochnykh porod. *Geologicheskii Sbornik. L'vovskoe Geologicheskoe Obshchestvo pri L'vovskom Gosudarstvennom Universitete im. Iv. Franko. L'vov.* No.2–3, p.320–322. [Pyrolite]

PREYER, E. & ZIRKEL, F., 1862. Reise nach Island im Sommer 1860. *Brockhaus, Leipzig.* 497pp. [Krablite]

QUENSEL, P., 1913. The alkaline rocks of Almunge. *Bulletin of the Geological Institution of the University of Upsala.* Vol.12, p.9–200. [Canadite]

QUENSEL, P.D., 1912. Die Geologie der Juan Fernandezinseln. *Bulletin of the Geological Institution of the University of Upsala.* Vol.11, p.252–290. [Picrite basalt]

QUIRKE, T.T., 1936. New nepheline syenites from Bigwood Township, Ontario. *Transactions of the Illinois State Academy of Science. Springfield.* Vol.29, p.179–185. [Bigwoodite, Rutterite]

RACHKOVSKY, J., 1911. Über Alkaligesteine aus dem Südwesten des Gouvernements Yenissej. I. Der Teschenit und seine Beziehung zu den Ergussgesteinen. *Trudy Geologicheskago (i Mineralogicheskago) Muzeya imeni Petra Velikago Imperatorskoi Akademii Nauk. Sankt Peterburg.* Vol.5, p.217–283. [Ijussite]

RAMSAY, W., 1896. Urtit, ein basisches Endglied der Augitsyenit-Nephelinsyenit-Serie. *Geologiska Föreningens i Stockholm Förhandlingar. Stockholm.* Vol.18, p.459–468. [Urtite]

RAMSAY, W., 1921. En melilitförande djupbergart från Turja på sydsidan av Kolahalvön. *Geologiska Föreningens i Stockholm Förhandlingar. Stockholm.* Vol.43, p.488–489. [Turjaite]

RAMSAY, W. & BERGHELL, H., 1891. Das Gestein vom Iiwaara in Finnland. *Geologiska Föreningens i Stockholm Förhandlingar. Stockholm.* Vol.13, p.300–312. [Ijolite]

RAMSAY, W. & HACKMAN, V., 1894. Das Nephelinsyenitgebiet auf der Halbinsel Kola. I. *Fennia: Bulletin de la Société de Géographie de Finlande.* Vol.11, No.2, p.1–225. [Imandrite, Khibinite (Chibinite), Tawite, Umptekite]

RANSOME, F.L., 1898. Art. XLV – Some lava flows of the western slope of the Sierra Nevada, California. *American Journal of Science. New Haven.* Vol.5, 4th Ser., p.355–375. [Latite, Quartz latite]

RATH, G. VOM, 1855. Chemische Untersuchung einiger Grünsteine aus Schlesien. *Annalen der Physik und Chemie. Leipzig.* Vol.95, p.533–561. [Forellenstein]

RATH, G. VOM, 1864. Beiträge zur Kenntnis der Eruptiven Gesteine der Alpen. *Zeitschrift der Deutschen Geologischen Gesellschaft. Berlin.* Vol.16, p.249–266. [Tonalite]

RATH, G. VOM, 1868. Geognostisch mineralogische Fragmente aus Italien. *Zeitschrift der Deutschen Geologischen Gesellschaft. Berlin.* Vol.20, p.265–364. [Leucite trachyte, Petrisco]

RAUMER, K. VON, 1819. Das Gebirge Niederschlesiens, der Grafschaft Glatz und eines Theiles von Böhmen und der Ober-Lausitz. *Reimer, Berlin.* 184pp. [Basaltite, Schillerfels]

REINISCH, R., 1912. Petrographisches Praktikum. *Borntraeger, Berlin.* Vol.2, 217pp. [Arsoite, Drakontite (Drakonite), Ponzaite]

REINISCH, R., 1917. Blatt Wiesenthal–Weipert. *Erläuterungen zur Geologischen Spezialkarte des Königreichs Sachsen. Leipzig.* No.147, p.1–84. [Haüynite]

REYNOLDS, D.L., 1937. Augite-biotite-diorite of the Newry complex. *Geological Magazine. London.* Vol.74, p.476-477. [Garronite]

REYNOLDS, D.L., 1958. Granite: some tectonic, petrological, and physico-chemical aspects. *Geological Magazine. London.* Vol.95, p.378–396. [Haplo-pitchstone, Minimite]

RICHTHOFEN, F. BARON, 1868. The natural system of volcanic rocks. *Memoirs of the California Academy of Sciences. San Francisco.* Vol.1, Pt.2, p.1–94. [Nevadite, Propylite]

RICHTHOFEN, F. VON, 1860. Studien aus den ungarisch-siebenbürgischen Trachytgebirgen. *Jahrbuch der Kaiserlich-Königlichen Geologischen Reichsanstalt. Wien.* Vol.11, p.153–278. [Lithoidite, Rhyolite]

RINGWOOD, A.E., 1962. A model for the Upper Mantle. *Journal of Geophysical Research. Richmond, Va.* Vol.67, No.2, p.857–867. [*Pyrolite*]

RINMAN, A., 1754. Anmärkningar angående Järnhaltiga Jord-och Stenarter. *Handlingar Kongliga Svenska Vetenskaps-Akademiens, Stockholm.* Vol.15, p.282–297. [Trapp (Trap)]

RINNE, F., 1904. Beitrag zur Gesteinskunde des Kiautschou-Schutz-Gebietes. *Zeitschrift der Deutschen Geologischen Gesellschaft. Berlin.* Vol.56, p.122–167. [Tsingtauite]

RINNE, F., 1921. Gesteinskunde. *Jänecke, Leipzig.* 6th & 7th Edns, 365pp. [Kiirunavaarite (Kirunavaarite), Peracidite, Silicotelite]

RIO, N. DA, 1822. Traité sur la structure extérieure du globe. *Giegler, Milan.* Vol.3, 557pp. [Masegna]

RITTER, E.A., 1908. The Evergreen copper-deposit, Colorado. *Transactions of the American Institute of Mining Engineers. New York.* Vol.38, p.751–765. [Evergreenite]

RITTMANN, A., 1960. Vulkane und ihre Tätigkeit. *Enke, Stuttgart.* 2nd Edn, 336pp. [Etnaite]

RITTMANN, A., 1962. Volcanoes and their Activity. Translated from the 2nd German Edn by E.A. Vincent. *Wiley, New York.* 305pp. [Hyaloclastite]

RITTMANN, A., 1973. Stable Mineral Assemblages of Igneous Rocks. *Springer-Verlag, Berlin and Heidelberg.* 262pp. [Latiandesite, Latibasalt, Phonoleucitite, Phononephelinite, Phonotephrite, Plagidacite, Tephrileucitite, Tephrinephelinite, Tephriphonolite]

RIVIÈRE, A., 1844. Mémoire minéralogique et géologique sur les roches dioritiques de la France occidentale. *Bulletin de la Société Géologique de France. Paris.* Vol.1, p.528–569. [Kersanton]

ROEVER, W.P. DE, 1940. Geological investigations in the southwestern Moetis region (Netherlands Timor). In: Brouwer, H.A. (Editor) Geological Expedition of the University of Amsterdam to the Lesser Sunda Islands in the Southeastern part of the Netherlands East

Indies 1937. *N.V. Noord-Hollandsche Uitgevers Maatschappij, Amsterdam*. Vol.2, p.97–344. [Poeneite]

ROGERS, G., SAUNDERS, A.D., TERRELL, D.J., VERMA, S.P. & MARRINER, G.F., 1985. Geochemistry of the Holocene volcanic rocks associated with ridge subduction in Baja California, Mexico. *Nature. London*. Vol.315, p.389–392. [Bajaite]

ROSE, G., 1837. Mineralogisch-geognostische Reise nach dem Ural, dem Altai und dem Kaspischen Meere. Vol.1. Reise nach dem nördlichen Ural und dem Altai. *Sandersche Buchhandlung, Berlin*. 641pp. [Beresite]

ROSE, G., 1839. Ueber die mineralogische und geognostische Beschaffenheit des Ilmengebirges. *Annalen der Physik und Chemie. Leipzig*. Vol.47, p.373–384. [Miaskite (Miascite)]

ROSE, G., 1863. Beschreibung und Eintheilung der Meteoriten auf Grund der Sammlung im mineralogischen Museum zu Berlin. *Abhandlungen der Königlichen Akademie der Wissenschaften. Berlin*. p.23–161. [*Eucrite*]

ROSENBUSCH, H., 1872. Petrographische Studien an den Gesteinen des Kaiserstuhls. *Neues Jahrbuch für Mineralogie, Geologie und Paläontologie. Stuttgart. Referate Abt. A*. Bd.43, p.35–65. [Limburgite]

ROSENBUSCH, H., 1877. Mikroskopische Physiographie der Mineralien und Gesteine. Vol.II. Massige Gesteine. *Schweizerbart, Stuttgart*. 596pp. [Leucite phonolite, Leucite tephrite, Nepheline syenite, Nepheline tephrite]

ROSENBUSCH, H., 1883. In review of C. Dölter: Die Vulkane der Capeverden und ihre Produkte. *Neues Jahrbuch für Mineralogie, Geologie und Paläontologie. Stuttgart. Referate Abt. A*. Bd.1, p.396–405. [Augitite]

ROSENBUSCH, H., 1887. Mikroskopische Physiographie der Mineralien und Gesteine. Vol.II. Massige Gesteine. *Schweizerbart, Stuttgart*. 2nd Edn, 877pp. [Alnöite, Camptonite, Cuselite (Kuselite), Harzburgite, Leucite basanite, Navite, Nepheline basanite, Olivine tholeiite, Theralite, *Tholeiite*, Tinguaite, Vogesite, Weiselbergite]

ROSENBUSCH, H., 1896. Mikroskopische Physiographie der Mineralien und Gesteine. Vol.II. Massige Gesteine. *Schweizerbart, Stuttgart*. 3rd Edn, 1360pp. [Alkali granite, Allalinite, Lestiwarite, Olivine basalt, Schriesheimite, Spessartite]

ROSENBUSCH, H., 1898. Elemente der Gesteinslehre. *Schweizerbart, Stuttgart*. 546pp. [Gabbroporphyrite]

ROSENBUSCH, H., 1899. Über Euktolith, ein neues Glied der Theralithschen Effusivmagmen. *Sitzungsberichte der Preussischen Akademie der Wissenschaften zu Berlin*. No.7, p.110–115. [Euktolite]

ROSENBUSCH, H., 1907. Mikroskopische Physiographie der Mineralien und Gesteine. Vol.II. Massige Gesteine. *Schweizerbart, Stuttgart*. 4th Edn, Pt.1, p.1–716. [Alkali syenite, Bekinkinite]

ROSENBUSCH, H., 1908. Mikroskopische Physiographie der Mineralien und Gesteine. Vol.II. Massige Gesteine. *Schweizerbart, Stuttgart*. 4th Edn, Pt.2, p.717–1592. [Alkali trachyte, Orthotrachyte, Phonolitoid tephrite, Prowersite, *Trachyandesite*]

ROTH, J., 1861. Die Gesteins-Analysen in tabellarischer Übersicht und mit kritischen Erläuterungen. *Hertz, Berlin*. 68pp. [Liparite]

ROTH, J., 1887. Über den Zobtenit. *Sitzungsberichte der Preussischen Akademie der Wissenschaften zu Berlin*. No.32, p.611–630. [Zobtenite]

ROZIÈRE, F.M. DE, 1826. Description de l'Egypte. Histoire naturelle, minéralogie, zoologie. *Paris*. 2nd Edn, Vol.21, 482pp. [Chlorophyre, Iophyre, Sinaite, Syenitelle, Xenite, Xenophyre]

RUSSELL, H.D., HEIMSTRA, S.A. & GROENEVELD,

D., 1955. The mineralogy and petrology of the carbonatite at Loolekop, Eastern Transvaal. *Transactions and Proceedings of the Geological Society of South Africa. Johannesburg.* Vol.57, p.197–208. [Phoscorite]

SABATINI, V., 1899. Relazione sul lavoro eseguito nel triennio 1896–97–98 sui vulcani dell'Italia centrale e i loro prodotti. *Bollettino del Reale Comitato Geologico d'Italia. Firenze.* Vol.30, p.30–60. [Venanzite]

SABATINI, V., 1903. La pirossenite melilitica di Coppaeli (Cittaducale). *Bollettino del Reale Comitato Geologico d'Italia. Firenze.* Vol.34, p.376–378. [Coppaelite]

SÆTHER, E., 1957. The alkaline rock province of the Fen area in southern Norway. *Det Kongelige Norske Videnskabers Selskabs Skrifter, Trondheim.* No.1, p.1–148. [*Damkjernite (Damtjernite), Vipetoite*]

SAGGERSON, E.P. & WILLIAMS, L.A.J., 1963. Ngurumanite, a new hypabyssal alkaline rock from Kenya. *Nature. London.* Vol.199, p.479. [Ngurumanite]

SAHAMA, T.G., 1974. Potassium-rich alkaline rocks. In: Sørensen, H. (Editor) The Alkaline Rocks. *Wiley, New York.* 622pp. [Kamafugite]

SAKSELA, M., 1948. Das pyroklastische Gestein von Lappajärvi und seine Verbreitung als Geschiebe. *Bulletin de la Commission Géologique de la Finlande. Helsingfors.* Vol.25, No.144, p.19–30. [Kärnäite]

SANDBERGER, F., 1872. Vorläufige Bemerkungen über den Buchonit, ein Felsart aus der Gruppe der Nephelingesteine. *Sitzungsberichte der Mathematischen-Physikalischen Classe der Königlich Bayerischen Akademie der Wissenschaften zu München.* Vol.2, p.203–208. [Buchonite]

SAUER, A., 1892. Der Granitit von Durbach im nördlichen Schwarzwalde und seine Grenzfacies von Glimmersyenit (Durbachite). *Mittheilungen der Grossherzoglich Badischen Geologischen Landesanstalt.*

Heidelberg. Vol.2, p.231–276. [Durbachite]

SAUER, A., 1919. Atlasblatt Bopfingen. *Begleitworte zu der geognostischen Spezialkarte von Württemberg. Stuttgart.* 2nd Edn, No.20, p.1–31. [Suevite]

SCHAEFER, R.W., 1898. Der basische Gesteinszug von Ivrea im Gebiet des Mastallone-Thales. *Tschermaks Mineralogische und Petrographische Mitteilungen. Wien.* Vol.17, 2nd Ser., p.495–517. [Valbellite]

SCHEERER, TH., 1862. Die Gneuse des Sächsischen Erzgebirges und verwandte Gesteine, nach ihrer chemischen Constitution und geologischen Bedeutung. *Zeitschrift der Deutschen Geologischen Gesellschaft. Berlin.* Vol.14, p.23–150. [Plutonite, Vulcanite]

SCHEERER, TH., 1864. Vorläufiger Bericht über krystallinische Silikatgesteine des Fassathales und benachbarter Gegenden Südtyrols. *Neues Jahrbuch für Mineralogie, Geologie und Paläontologie. Stuttgart. Referate Abt. A.* Bd.35, p.385–411. [Plutovolcanite]

SCHEIBE, R., 1933. Compilación de los estudios geólogicos officiales en Colombia 1917 a 1933. *Biblioteca de Departmento de Minas y Petroleo. Ministerio de Industria. Bogota.* Vol.1, p.1–470. [Corcovadite]

SCHEUMANN, K.H., 1913. Petrographische Untersuchungen an Gesteinen des Polzengebietes in Nord-Böhmen. *Abhandlungen der Königlich-Sächsischen Gesellschaft der Wissenschaften. Mathematisch-Physische Classe. Leipzig.* Vol.32, p.605–776. [Polzenite]

SCHEUMANN, K.H., 1922. Zur Genese alkalisch-lamprophyrischer Ganggesteine. *Centralblatt für Mineralogie, Geologie und Paläontologie. Stuttgart. Abt.A.* p.495–520, 521–545. [Luhite, Modlibovite, Vesecite, Wesselite]

SCHEUMANN, K.H., 1925. Ausländische Systematik, Klassifikation und Nomenklatur der Magmengesteine. *Fortschritte der*

Mineralogie, Kristallographie und Petrographie. Jena. Vol.10, p.187–310. [Mienite]

SCHMID, E.E., 1880. Die quartzfreien Porphyre des Centralen Thüringer Waldgebirges und ihre Begleiter. *Fischer, Jena.* 106pp. [Paramelaphyre]

SCHMID, R., 1981. Descriptive nomenclature and classification of pyroclstic deposits and fragments: Recommendations of the IUGS Subcommission on the Systematics of Igneous Rocks. *Geology. The Geological Society of America. Boulder, Co.* Vol.9, p.41–43. [Ash, Ash grain, Dust grain, Dust tuff, Pyroclastic deposit, Tuffite]

SCHOWALTER, E., 1904. Chemisch-geologische Studien im vulkanischen Ries bei Nördlingen. *Inaugural Dissertation. Jacob, Erlangen.* 65pp. [Wennebergite]

SCHÜLLER, A., 1949. Ein Plagioklas-Charnockit vom Typus Akoafim und seine Stellung innerhalb der Charnockit-Serie. *Heidelberger Beiträge zur Mineralogie und Petrographie. Berlin.* Vol.1, p.573–592. [Akoafimite]

SCHULZ, C.F. & POETSCH, C.G., 1759. Neue gesellschaftliche Erzählungen für die Liebhaber der Naturlehre der Haushaltungswissenschaft, der Arztneykunst und der Sitten. *Leipzig.* Vol.2. [Pechstein]

SCHUSTER, M.E., 1907. Beiträge zur mikroskopischen Kenntnis der basischen Eruptivgesteine aus der bayerischen Rheinpfalz. *Geognostische Jahreshefte. München.* Vol.18, p.1–70. [Rhenopalite]

SEARS, J.H., 1891. Elaeolite-zircon-syenites and associated granitic rocks in the vicinity of Salem, Essex County, Massachusetts. *Bulletin of the Essex Institute. Salem, Mass.* Vol.23, p.145–155. [Essexite]

SEDERHOLM, J.J., 1891. Ueber die finnländischen Rapakiwigesteine. *Tschermaks Mineralogische und Petrographische Mitteilungen. Wien.* Vol.12, 2nd Ser., p.1–31. [Anoterite]

SEDERHOLM, J.J., 1926. On migmatites and associated Pre-Cambrian rocks of southwestern Finland. II. The region around the Barösundsfjärd W. of Helsingfors and neighbouring areas. *Bulletin de la Commission Géologique de la Finlande. Helsingfors.* Vol.12, No.77, p.1–139. [Epibasite]

SEDERHOLM, J.J., 1928. On orbicular granites. Spotted and nodular granites etc. and on the rapakivi texture. *Bulletin de la Commission Géologique de la Finlande. Helsingfors.* Vol.13, No.83, p.1–105. [Esboite, Orbiculite]

SEDERHOLM, J.J., 1934. On migmatites and associated Pre-Cambrian rocks of southwestern Finland. III. The Åland Islands. *Bulletin de la Commission Géologique de la Finlande. Helsingfors.* Vol.17, No.107, p.1–68. [Kleptolith]

SENFT, F., 1857. Classification und Beschreibung der Felsarten. *Korn, Breslau.* 442pp. [Alabradorite, Labradorite, Leucitite, Melaporphyre, Orthoclasite, Pyroxenite]

SHAND, S.J., 1910. On borolanite and its associates in Assynt. *Transactions of the Edinburgh Geological Society.* Vol.9, p.376–416. [Aploid, Assyntite, Cromaltite, Dioroid, Doleroid, Ledmorite, Pegmatoid, Syenoid]

SHAND, S.J., 1913. On saturated and unsaturated igneous rocks. *Geological Magazine. London.* Vol.10, Decade 5, p.508–514. [Oversaturated, Saturated, Undersaturated]

SHAND, S.J., 1916. A recording micrometer for geometrical rock analysis. *Journal of Geology. Chicago.* Vol.24, p.394–404. [Colour index (Colour ratio)]

SHAND, S.J., 1917. A system of petrography. *Geological Magazine. London.* Vol.4, Decade 6, p.463–469. [Granodolerite]

SHAND, S.J., 1927. Eruptive Rocks. *Murby, London.* 360pp. [*Gabbroid*, Metaluminous, Peraluminous]

SHAND, S.J., 1945. Coronas and coronites. *Bulletin of the Geological Society of America. New York.* Vol.56, p.247–266. [Coronite]

SHARASKIN, A.Y., DOBRETSOV, N.L. & SOBOLEV, N.V., 1980. Marianites: the clinoenstatite-bearing pillow lavas associated with the ophiolite assemblage of the Mariana Trench. In: Panayiotou, A. (Editor) Ophiolites. *Proceedings of the International Ophiolite Conference, Cyprus, 1979. Ministry of Agriculture and Natural Resources. Geological Survey Department. Republic of Cyprus.* p.473–479. [Marianite]

SHILIN, L.L., 1940. Perovskit-shpinelevyi magnetitit iz Praskovye-Evgenevskoi kopi Shishimskikh gor na Yuzhnom Urale. *Doklady Akademii Nauk SSSR. Leningrad.* Vol.28, p.347–350. [Shishimskite]

SHIREY, S.B. & HANSON, G.N., 1984. Mantle derived Archean monzodiorites and trachyandesites. *Nature. London.* Vol.310, p.222-224. [*Sanukitoid*]

SIGMUND, A., 1902. Die Eruptivgesteine bei Gleichenberg. *Tschermaks Mineralogische und Petrographische Mitteilungen. Wien.* Vol.21, 2nd Ser., p.261–306. [Andesitoid]

SILVESTRI, O., 1888. Sopra alcune lave antiche e moderne del vulcano Kilauea nelle isole Sandwich. III. Le lave preistoriche stratificate le quali costituiscono le pareti all'intorno del grande bacino del Kilauea. *Bollettino del Reale Comitato Geologico d'Italia. Firenze.* Vol.19, p.168–196. [Kilaueite]

SJÖGREN, A., 1876. Om förekomsten af Tabergs jernmalmsfyndighet i Småland. *Geologiska Föreningens i Stockholm Förhandlingar. Stockholm.* Vol.3, p.42–62. [Olivinite]

SJÖGREN, H., 1893. En ny jernmalmstyp representerad af Routivare malmberg. *Geologiska Föreningens i Stockholm Förhandlingar. Stockholm.* Vol.15, p.55–63. [Magnetite Spinellite, Routivarite]

SKEATS, E.W., 1910. The volcanic rocks of Victoria (Presidential Address, Section C). *Report of the 12th Meeting of the Australian Association for the Advancement of Science, Brisbane 1909.* p.173–235. [Macedonite]

SKEATS, E.W. & SUMMERS, H.S., 1912. The geology and petrology of the Macedon district. *Bulletin of the Geological Survey of Victoria. Melbourne.* Vol.24, p.1–58. [Anorthoclase basalt, Woodendite]

SOBOLEV, N.D., 1959. Neivit–novaya gornaya poroda iz gruppy zhilnykh. *Izvestiia Akademii Nauk SSSR, Seriia Geologicheskaia.* No.10, p.115–120. [Neivite]

SOBRAL, J.M., 1913. Contributions to the Geology of Nordingrå Region. *Almquist & Wiksell's Boktrycherii, Uppsala.* 179pp. [Värnsingite]

SÖLLNER, J., 1913. Über Bergalith ein neues melilithreiches Ganggestein aus dem Kaiserstuhl. *Mittheilungen der Grossherzoglich Badischen Geologischen Landesanstalt. Heidelberg.* Vol.7, p.413–466. [Bergalite]

SØRENSEN, H., 1960. On the agpaitic rocks. *International Geological Congress. Report of the 21st Session Norden.* Vol.13, p.319-327. [*Agpaite*]

SØRENSEN, H. (EDITOR), 1974. The Alkaline Rocks. *Wiley, London.* 622pp. [*Fitzroydite, Fitzroyite*]

SOROTCHINSKY, S., 1934. Types pétrographiques nouveaux (kanzibite, kahusite, et quartz porphyre siliceux) provenant de l'édifice volcanique du Kahusi et du Biega (Kivu). *Bulletin de l'Académie Royale de Belgique. Classe des Sciences. Bruxelles.* Vol.20, p.183–195. [Kahusite, Kanzibite]

SOUZA-BRANDÃO, V., 1907. Les espichellites. Une nouvelle famille de roches de filons, au Cap Espichel. *Annals da Academia Polytechnica do Porto.* Vol.2, p.1–70. [Espichellite]

SPURR, J.E., 1900A. A reconnaissance in southwestern Alaska in 1898. *Report of the United States Geological Survey. Washington.* Pt.VII, p.31–264. [Alaskite, Aleutite, Belugite, Tordrillite]

SPURR, J.E., 1900B. Art. XXXI – Scapolite

rocks from Alaska. *American Journal of Science. New Haven.* Vol.10, 4th Ser., p.310–315. [Kuskite, Yentnite]

SPURR, J.E., 1906. The southern Klondike district, Esmeralda County, Nevada – a study in metalliferous quartz veins of magmatic origin. *Economic Geology and Bulletin of the Society of Economic Geologists. Lancaster, Pa.* Vol.1, p.369–382. [Esmeraldite]

SPURR, J.E. & WASHINGTON, H.S., 1917. In: Washington, H.S., 1917, Chemical analyses of igneous rocks. *Professional Paper. United States Geological Survey. Washington.* No.99, p.1–1201. [Arizonite]

STACHE, G. & JOHN, C. VON, 1877. Geologische und petrographische Beiträge zur Kenntniss der älteren Eruptiv- und Massengesteine der Mittel- und Ostalpen. I. Die Gesteine der Zwölferspitzgruppe in Westtirol. *Jahrbuch der Kaiserlich-Königlichen Geologischen Reichsanstalt. Wien.* Vol.27, p.143–242. [Haplophyre]

STACHE, G. & JOHN, C. VON, 1879. Geologische und petrographische Beiträge zur Kenntniss der älteren Eruptiv- und Massengesteine der Mittel- und Ostalpen. II. Das Cevedale-Gebiet als Hauptdistrict älterer dioritischer Porphyrite (Palaeophyrite). *Jahrbuch der Kaiserlich-Königlichen Geologischen Reichsanstalt. Wien.* Vol.29, p.317–404. [Ortlerite, Palaeophyrite, Protopylite, Suldenite]

STANSFIELD, J., 1923A. Extensions of the Monteregian petrographical province to the west and north-west. *Geological Magazine. London.* Vol.60, p.433–453. [Okaite]

STANSFIELD, J., 1923B. Nomenclature and relations of the lamprophyres. *Geological Magazine. London.* Vol.60, p.550–554. [Bizardite]

STARZYNSKI, Z., 1912. Ein Beitrag zur Kenntnis der pazifischen Andesite und der dieselben bildenden Mineralien. *Bulletin International de l'Académie des Sciences et des Lettres de Cracovie. Cracovie.* Ser.A, p.657–681. [Beringite]

STEENSTRUP, K.J.V., 1881. Bemaerkninger til et geognostisk Oversigtskaart over en del af Julianehaabs distrikt. *Meddelelser om Grønland. Kjøbehavn.* Vol.2, p.27–41. [Sodalite syenite]

STEININGER, J., 1840. Geognostische Beschreibung des Landes zwischen der untern Saar und dem Rheine. *Lintz, Trier.* 149pp. [Tholeiite (Tholeite, Tholeyite)]

STEININGER, J., 1841. Geognostische Beschreibung des Landes zwischen der untern Saar und dem Rheine. *Lintz, Trier.* Nachträge, 48pp. [Spiemontite]

STELZNER, A., 1882. Vorläufige Mittheilungen über Melilithbasalte. *Neues Jahrbuch für Mineralogie, Geologie und Paläontologie. Stuttgart. Referate Abt. A.* Bd.1, p.229–231. [Melilite basalt]

STELZNER, A.W., 1885. Beiträge zur Geologie und Palaeontologie der Argentinischen Republik. *Fischer, Cassel and Berlin.* Vol.1, 329pp. [Andendiorite, Andengranite]

STEWART, D.C. & THORNTON, C.P., 1975. Andesite in oceanic regions. *Geology. The Geological Society of America. Boulder, Co.* Vol.3, p.565-568. [Duncanite, Goughite, Jervisite]

STRECKEISEN, A., 1938. Das Nephelinsyenite-Massiv von Ditro in Rumänien als Beispiel einer kombinierten Differentiation und Assimilation. *Verhandlungen der Schweizerischen Naturforschenden Gesellschaft.* p.159–161. [Orotvite]

STRECKEISEN, A., 1952. Das Nephelinsyenit-Massiv von Ditro (Siebenbürgen). *Schweizerische Mineralogische und Petrographische Mitteilungen. Zürich.* Vol.32, p.251–308. [Ditro-essexite]

STRECKEISEN, A., 1954. Das Nephelinsyenit-Massiv von Ditro (Siebenbürgen) – Part II. *Schweizerische Mineralogische und Petrographische Mitteilungen. Zürich.*

Vol.34, p.336–409. [*Ditro-essexite*]

STRECKEISEN, A., 1965. Die Klassifikation der Eruptivegesteine. *Geologische Rundschau. Internationale Zeitschrift für Geologie. Stuttgart.* Vol.55, p.478–491. [Foidite]

STRECKEISEN, A., 1967. Classification and nomenclature of igneous rocks (Final Report of an Inquiry). *Neues Jahrbuch für Mineralogie. Stuttgart. Abhandlungen.* Vol.107, p.144–214. [Hololeucocratic, Holomelanocratic, Latite andesite, Latite basalt, Mafitite, Monzogranite, Syenogranite, *Ultrabasite*]

STRECKEISEN, A., 1973. Plutonic rocks. Classification and nomenclature recommended by the IUGS Subcommission on the Systematics of Igneous Rocks. *Geotimes.* Vol.18, No.10, p.26–30. [Alkali feldspar granite, Alkali feldspar syenite, Alpha granite, Beta granite, Foid diorite, Foid dioritoid, Foid gabbro, Foid gabbroid, Foid monzodiorite, Foid monzogabbro, Foid monzosyenite, Foid plagisyenite, Foid syenite, Foid syenitoid, Foid-bearing alkali feldspar syenite, Foid-bearing anorthosite, Foid-bearing diorite, Foid-bearing gabbro, Foid-bearing monzodiorite, Foid-bearing monzogabbro, Foid-bearing monzonite, Foid-bearing syenite, Foidolite, Gabbronorite, Olivine clinopyroxenite, Olivine gabbronorite, Olivine hornblende pyroxenite, Olivine hornblendite, Olivine orthopyroxenite, Olivine pyroxene hornblendite, Olivine pyroxenite, Olivine websterite, Plagioclase-bearing hornblende pyroxenite, Plagioclase-bearing hornblendite, Plagioclase-bearing pyroxene hornblendite, Plagioclase-bearing pyroxenite, Pyroxene hornblende gabbro, Pyroxene hornblende gabbronorite, Pyroxene hornblende norite, Pyroxene hornblende peridotite, Pyroxene hornblendite, Quartz alkali feldspar syenite, Quartz monzodiorite, Quartz monzogabbro, Quartz syenite, Quartz-rich granitoid, Quartzolite, Syenitoid]

STRECKEISEN, A., 1974. How should charnockitic rocks be named? *Centenaire de la Société Géologique de Belgique Géologie des Domaines Cristallins, Liège.* p.349–360. [*Akoafimite*, Alkali feldspar charnockite, *Amherstite*, *Ankaranandite*, *Arendalite*, *Bauchite*, *Birkremite*, *Bugite*, *Epibugite*, *Farsundite*, *Grimmaite*, *Ivoirite*, *Katabugite*, *Mesobugite*, *Sabarovite*]

STRECKEISEN, A., 1976. To each plutonic rock its proper name. *Earth Science Reviews. International Magazine for Geo-Scientists. Amsterdam.* Vol.12, p.1–33. [Cancrinite diorite, Cancrinite gabbro, Cancrinite monzodiorite, Cancrinite monzogabbro, Cancrinite monzosyenite, Cancrinite plagisyenite, Nephelinolite, Ultramafite, Ultramafitolite]

STRECKEISEN, A., 1978. IUGS Subcommission on the Systematics of Igneous Rocks. Classification and nomenclature of volcanic rocks, lamprophyres, carbonatites and melilite rocks. Recommendations and suggestions. *Neues Jahrbuch für Mineralogie. Stuttgart. Abhandlungen.* Vol.143, p.1–14. [Alkali feldspar foidite, Alkali feldspar rhyolite, Alkali feldspar trachyte, Basanitic foidite, Calcite-carbonatite, Foid-bearing alkali feldspar trachyte, Foid-bearing latite, Foid-bearing trachyte, Foiditoid, Olivine melilitite, Olivine melilitolite, Olivine pyroxene melilitolite, Olivine uncompahgrite, Phonobasanite, Phonofoidite, Phonolitic basanite, Phonolitic foidite, Phonolitic leucitite, Phonolitic nephelinite, Phonolitic tephrite, Pyroxene melilitolite, Pyroxene olivine melilitolite, Quartz alkali feldspar trachyte, Tephritic foidite, Tephritic leucitite, Tephritic phonolite, Ultramafitite]

STRENG, A., 1864. Bemerkungen über den Serpentinfels und den Gabbro von Neurode in Schlesien. *Neues Jahrbuch für Mineralogie, Geologie und Paläontologie. Stuttgart. Referate Abt. A.* Bd.35, p.257–278. [Enstatitite]

STRENG, A. & KLOOS, J.H., 1877. Ueber die krystallinischen Gesteine von Minnesota in Nord-Amerika. *Neues Jahrbuch für Mineralogie, Geologie und Paläontologie. Stuttgart. Referate Abt. A.* p.113–138. [Hornblende gabbro]

SUN, S-S., NESBIT, R.W. & SHARASKIN, A.YA., 1979. Geochemical characteristics of mid-ocean ridge basalts. *Earth and Planetary Science Letters.* Vol.44, p.119-138. [Mid-ocean ridge basalt (MORB)]

SUTHERLAND, D.S., 1965. Nomenclature of the potassic–feldspathic rocks associated with carbonatites. *Bulletin of the Geological Society of America. New York.* Vol.76, p.1409-1412. [*Orthoclasite*]

SZÁDECZKY, J., 1899. Ein neues Ganggestein aus Assuan. *Földtani Közlöny. A Magyar Földtani Tarsulat Folyoirata. Budapest.* Vol.29, p.210–216. [Josefite]

SZENTPÉTERY, S.V., 1935. Petrologische Verhältnisse des Fehérkö-Berges und die detaillierte Physiographie seiner Eruptivgesteine. *Acta Litterarum ac Scientiarum Regiae Universitatis Hungaricae Francisco-Josephinae. Sectio Chemica, Mineralogica et Physica. Szeged.* Vol.4, p.18–123. [Plagiophyrite]

TARASENKO, V.E., 1895. O khimicheskom sostave porod semeistva gabbro iz Zhitomirskogo uezda Volynskoi gubernii. *Zapiski Kievskago Obshchestva Estestvoispytatelei. Kiev.* Vol.14, p.15. [Gabbrosyenite]

TARGIONI-TOZZETTI, G., 1768. Relazioni d'Alcuni Viaggi Fatti in Diversi Parti della Toscana. *Firenze.* 2nd Edn, Vol.2, p.1–540. [Gabbro]

TAYLOR, H.P. (JR), FRECHEN, J. & DEGENS, E.T., 1967. Oxygen and carbon isotope studies of carbonatites from the Laacher See district, West Germany and the Alnö district, Sweden. *Geochimica et Cosmochimica Acta. London.* Vol.31, p.407–430. [Boderite, Rodderite]

TAYLOR, S.R., 1969. Trace element chemistry of andesites and associated calc-alkaline rocks. In: McBirney, A.R. (Editor) Proceedings of the Andesite Conference. *Department of Geology and Mineral Industries. State of Oregon. Portland.* Bulletin 65, p.43–63. [High-K, Low-K]

TEALL, J.J.H., 1887. On the origin of certain banded gneisses. *Geological Magazine. London.* Vol.4, Decade 3, p.484–493. [Pyroclastic]

TEALL, J.J.H., 1888. British Petrography: With Special Reference to the Igneous Rocks. *Dulau, London.* 469pp. [Orthofelsite]

TERMIER, H., TERMIER, G. & JOURAVSKY, G., 1948. Une roche volcanique á gros grain de la famille des Ijolites: la Talzastite. *Notes et Mémoires. Service des Mines et de la Carte Géologique du Maroc.* Vol.71, p.81–120. [Talzastite]

THEOPHRASTUS, C., 320 BC. De lapidus (On Stones). *[Various printed versions and translations are available.]* [Basanite, Obsidian, Pumice]

THOMAS, H.H., 1911. The Skomer volcanic series (Pembrokeshire). *Quarterly Journal of the Geological Society of London.* Vol.67, p.175–214. [Marloesite, Skomerite]

THOMAS, H.H. & BAILEY, E.B., 1915. Notes on the rock-species leidleite and inninmorite. In: Anderson, E.M. & Radley, E.G., The pitchstones of Mull and their genesis. *Quarterly Journal of the Geological Society of London.* Vol.71, p.205–217. [Inninmorite, Leidleite]

THOMAS, H.H. & BAILEY, E.B., 1924. Tertiary and post-Tertiary geology of Mull, Loch Aline, and Oban. *Memoirs. Geological Survey of Scotland. Edinburgh.* p.1–445. [Craignurite]

THORARINSSON, S., 1944. Tefrokronologiska studier på Island. *Akademisk Avhandling, Stockholms Högskola.* p.1-217. [Tephra]

TILLEY, C.E., 1936. Enderbite, a new member of the charnockite series. *Geological Magazine. London.* Vol.73, p.312–316. [Enderbite]

TILLEY, C.E. & MUIR, I.D., 1963. Intermediate members of the oceanic basalt–trachyte association. *Geologiska Föreningens i Stockholm Förhandlingar. Stockholm.* Vol.85, p.436–444. [Benmoreite, Tristanite]

TOBI, A.C., 1971. The nomenclature of the charnockitic rock suite. *Neues Jahrbuch für Mineralogie. Stuttgart. Monatshefte.* p.193–205. [m-Charnockite, m-Enderbite etc.]

TOBI, A.C., 1972. The nomenclature of the charnockitic rock suite: Reply to a discussion. *IUGS Subcommission on Nomenclature in Igneous Petrology, 10th Circular, Contribution No.25, May 1972.* [Charnoenberbite]

TOMKEIEFF, S.I., WALTON, E.K., RANDALL, B.A.O., BATTEY, M.H. & TOMKEIEFF, O., 1983. Dictionary of Petrology. *Wiley, New York.* 680pp. [Amphiboldite, Anortholite, Bizeul, Busonite, Ceramicite, Egerinolith, Eutectite, Florianite, Gamaicu, Ghiandone, Granitelle, Ilmen-granite, Ilvaite, Isophyre, Kammstein, Leckstone, Lhercoulite, Rhenoaplite, Riacolite, Smalto, Spinellite, Tabona, Terzontli, Zutterite]

TÖRNEBOHM, A.E., 1877A. Om Sveriges vigtigere diabas-och gabbro-arter. *Handlingar Kongliga Svenska Vetenskaps-Akademiens, Stockholm.* Vol.14, No.13, p.1–55. [Hyperitite]

TÖRNEBOHM, A.E., 1877B. Ueber die wichtigeren Diabas und Gabbro Gesteine Schwedens. *Neues Jahrbuch für Mineralogie, Geologie und Paläontologie. Stuttgart. Referate Abt. A.* p.379–393. [Gabbrodiorite]

TÖRNEBOHM, A.E., 1881. Beskrifning till blad no. 7 af Geologisk öfversigtskarta öfver mellersta Sveriges Bergslag. *Beckman, Stockholm.* Vol.7, p.1-36. [Gabbrogranite]

TÖRNEBOHM, A.E., 1883. Om den s.k. Fonoliten från Elfdalen, dess klyftort och förekomstsätt.

Geologiska Föreningens i Stockholm Förhandlingar. Stockholm. Vol.6, p.383–405. [Cancrinite syenite]

TÖRNEBOHM, A.E., 1906. Katapleiit-syenit – en nyupptäckt varietet af nefelinsyenit i Sverige. *Sveriges Geologiska Undersökning. Afhandlingar och Uppsatser. Stockholm.* Ser.C, No.199, p.1–54. [Lakarpite]

TRAVIS, C.B. (EDITOR), 1915. The igneous and pyroclastic rocks of the Berwyn Hills (North Wales). Cope Memorial Volume. *Proceedings of the Liverpool Geological Society. Liverpool.* p.1–115. [Hirnantite]

TRIMMER, J., 1841. Practical Geology and Mineralogy. *Parker, London.* 519pp. [Haüyne basalt]

TRÖGER, W.E., 1928. Alkaligesteine aus der Serra do Salitre im westlichen Minas Geraes, Brasilien. *Centralblatt für Mineralogie, Geologie und Paläontologie. Stuttgart. Abt. A.* p.202–204. [Bebedourite, Salitrite]

TRÖGER, W.E., 1931. Zur "Typenvermischung" bei Lamprophyren. *Fortschritte der Mineralogie, Kristallographie und Petrographie. Jena.* Vol.16, p.139–140. [Camptospessartite]

TRÖGER, W.E., 1935. Spezielle Petrographie der Eruptivgesteine. *Verlag der Deutschen Mineralogischen Gesellschaft, Berlin.* p.360. [Andesilabradorite, Anorthitissite, Aplosyenite, Dolomite-carbonatite, Engadinite, Essexite-diorite, Essexite-foyaite, Evisite, Feldspattavite, Granatite, Haüyne basanite, Mimesite, Nosean basanite, Peléeite, Picotitite, Rongstockite, Yosemitite]

TRÖGER, W.E., 1938. Spezielle Petrographie der Eruptivgesteine. Eruptivgesteinsnamen. *Fortschritte der Mineralogie, Kristallographie und Petrographie. Jena.* Vol.23, p.41–90. [Kaulaite, Vredefortite]

TRÖGER, W.E., 1969. Spezielle Petrographie der Eruptivgesteine. Ein Nomenklatur-Kompendium. mit 1. Nachtrag. Eruptivgesteinsnamen. *Verlag der Deutschen*

Mineralogischen Gesellschaft, Stuttgart. 360pp. + 80pp.

TSCHERMAK, G., 1866. Felsarten von ungewähnlicher Zusammensetzung in den Umgebungen von Teschen und Neutitschein. *Sitzungsberichte der Kaiserlichen Akademie der Wissenschaften. Mathematisch-Naturwissenschaftliche Classe. Wien.* Vol.53, Abt.1, p.260–286. [Picrite]

TSUBOI, S., 1918. Notes on miharite. *Journal of the Geological Society of Toyko.* Vol.25, No.302, p.47–58. [Miharaite]

TURNER, H.W., 1896. Further contributions to the geology of the Sierra Nevada. *Report of the United States Geological Survey. Washington.* Pt.1, p.521–740. [Albitite]

TURNER, H.W., 1899. The granitic rocks of the Sierra Nevada. *Journal of Geology. Chicago.* Vol.7, p.141–162. [Granolite]

TURNER, H.W., 1900. The nomenclature of feldspathic granolites. *Journal of Geology. Chicago.* Vol.8, p.105–111. [Andesinite, Anorthitite, Labradite, Oligosite, Orthosite]

TURNER, H.W., 1901. Perknite (lime – magnesia rocks). *Journal of Geology. Chicago.* Vol.9, p.507–511. [Perknite]

TYRRELL, G.W., 1911. A petrological sketch of the Carrick Hills, Ayrshire. *Transactions of the Geological Society of Glasgow.* Vol.15, p.64–83. [Plagiophyre]

TYRRELL, G.W., 1912. The Late Palaeozoic alkaline igneous rocks of the West of Scotland. *Geological Magazine. London.* Vol.9, Decade 5, p.120–131. [Kylite, Lugarite]

TYRRELL, G.W., 1917. Some Tertiary dykes of the Clyde Area. *Geological Magazine. London.* Vol.4, Decade 6, p.305–315. [Cumbraite]

TYRRELL, G.W., 1926. The Principles of Petrology. *Methuen, London.* 349pp. [Quartz dolerite]

TYRRELL, G.W., 1931. Volcanoes. *Butterworth, London.* 252pp. [Agglutinate]

TYRRELL, G.W., 1937. Flood basalts and fissure eruptions. *Bulletin Volcanologique.*

Organe de l'Association de Volcanologie de l'Union Géodésique et Géophysique Internationale. Bruxelles. (Napoli.) Vol.1, 2nd Ser., p.89–111. [Flood basalt]

UHLEMANN, A., 1909. Die Pikrite des sächsischen Vogtlandes. *Tschermaks Mineralogische und Petrographische Mitteilungen. Wien.* Vol.28, 2nd Ser., p.415–472. [Schönfelsite]

USSING, N.V., 1912. Geology of the country around Julianehaab, Greenland. *Meddelelser om Grønland. Kjøbehavn.* Vol.38, p.1–426. [Agpaite (Agpaitic), Kakortokite, Naujaite, *Sodalitholith (Sodalithite)*, Sodalitite]

VAKAR, V., 1931. Geologicheskie issledovaniya v basseine r. Berezovki Kolymskogo okruga. *Izvestiia Glavnogo Geologo-Razvedochnogo Upravleniia. Moskva.* Vol.65, p.1015–1032. [Antirapakivi]

VÄYRYNEN, H., 1938. Notes on the geology of Karelia and the Onega region in the summer of 1937. *Bulletin de la Commission Géologique de la Finlande. Helsingfors.* Vol.21, No.123, p.65–80. [Karjalite]

VERBEEK, R.D.M., 1905. Geologische Beschrijving van Ambon. *Jaarboek van het Mijnwezen in Nederlandsch Oost-Indië. Amsterdam.* Vol.34, p.1–308. [Ambonite]

VILJOEN, M.J. & VILJOEN, R.P., 1969. The geology and geochemistry of the Lower Ultramafic Unit of the Onverwacht Group and a proposed new class of igneous rocks. *Special Publication of the Geological Society of South Africa.* Vol.2, p.55–85. [Basaltic komatiite (Komatiitic basalt), Komatiite, Peridotitic komatiite]

VIOLA, C., 1892. Nota preliminare sulla regione dei gabbri e delle serpentine nell'alta valle del Sinni in Basilicata. *Bollettino del Reale Comitato Geologico d'Italia. Firenze.* Vol.23, p.105–125. [Plagioclasite]

VIOLA, C., 1901. L'augitite anfibolica di Giumarra presso Rammacca (Sicilia). *Bollettino del Reale Comitato Geologico d'Italia. Firenze.* Vol.32, p.289–312. [Giumarrite]

VIOLA, C. & STEFANO, G. DI, 1893. La Punta delle Pietre Nere presso il lago di Lesina in provincia di Foggia. *Bollettino del Reale Comitato Geologico d'Italia. Firenze.* Vol.24, p.129–143. [Garganite]

VLODAVETS, V.I., 1930. Nefelin-apatitovye mestorozhdeniya v Khibinskikh tundrakh. *Trudy Instituta po Izucheniiu Severa. Moskva.* Vol.46, p.14–60. [Apaneite, Neapite]

VLODAVETS, V.I., 1952. O svoeobraznom kaltsitovom diabaze (kabitavite) Mongolii. *Doklady Akademii Nauk SSSR. Leningrad.* Vol.87, p.657–659. [Cabytauite]

VOGELSANG, H., 1872. Ueber die Systematik der Gesteinslehre und die Eintheilung der gemengten Silikatgesteine. *Zeitschrift der Deutschen Geologischen Gesellschaft. Berlin.* Vol.24, p.507–544. [Basalt-trachyte, Basite-porphyry, Felsophyre, Felsophyrite, Granophyre, Granophyrite, Leucite-basite, Nepheline basite, Vitrophyre, Vitrophyrite]

VOGELSANG, H.P.J., 1875. Die Krystalliten. *Cohen, Bonn.* 175pp. [Felsovitrophyre, Granofelsophyre]

VOGT, J.H.L., 1918. Jernmalm og jernverk. *Norges Geologiske Undersögelse. Kristiania.* No.85, p.1–181. [Rødberg]

VOGT, TH., 1915. Petrographisch-chemische Studien an einigen Assimilations-Gesteinen der nordnorwegischen Gebirgskette. *Skrifter udgit av Videnskabsselskabet i Kristiania. I. Math.-Nat. Klasse.* No.8, p.1–34. [Hortite]

VOROB'EV, E.I., MALYSHONOK, YU. V. & KONEV, A.A., 1984. New types of useful minerals in ultrapotassic syenite massifs [translated from Russian]. *27th International Geological Congress, Moscow. Abstracts.* Vol.7, p.323-324. [Yakutite]

WADE, A. & PRIDER, R.T., 1940. The leucite-bearing rocks of the west Kimberley area, Western Australia. *Quarterly Journal of the Geological Society of London.* Vol.96, p.39–98. [Cedricite, Fitzroyite, Mamilite, Wolgidite]

WADSWORTH, M.E., 1884. Lithological studies: a description and classification of the rocks of the Cordilleras. *Memoirs of the Museum of Comparative Zoology, at Harvard College, Cambridge, Mass.* Vol.11, Pt.1, p.1–208. [Buchnerite, Cumberlandite, Saxonite, Shastalite]

WADSWORTH, M.E., 1887. Preliminary description of the peridotytes, gabbros, diabases and andesytes of Minnesota. *Bulletin of the Geological Survey of Minnesota.* Vol.2, p.1–159. [Zirkelyte (Zirkelite)]

WADSWORTH, M.E., 1893. Sketch of the geology of the iron, gold and copper districts of Michigan. *Report of the State Board of Geological Survey of Michigan for the years 1891 and 1892. Lansing.* p.75-186. [Ferrolite, Lassenite, Metabolite]

WAGER, L.R., BROWN, G.M. & WADSWORTH, W.J., 1960. Types of igneous cumulates. *Journal of Petrology. Oxford.* Vol.1, p.73–85. [Cumulate]

WAGER, L.R. & DEER, W.A., 1939. Geological investigations in East Greenland. III. The petrology of the Skaergaard Intrusion, Kangerdlugssuaq, East Greenland. *Meddelelser om Grønland. Kjøbehavn.* Vol.105, No.4, p.1–352. [Ferrogabbro]

WAGER, L.R. & VINCENT, E.A., 1962. Ferrodiorite from the Isle of Skye. *Mineralogical Magazine and Journal of the Mineralogical Society. London.* Vol.33, p.26–36. [Ferrodiorite]

WAGNER, P.A., 1928. The evidence of the kimberlite pipes on the constitution of the outer part of the Earth. *South African Journal of Science.* Vol.25, p.127-148. [Orangite]

WAHL, W., 1907. Beiträge zur Geologie der präkambrischen Bildungen im Gouvernement Olonez. II.3. Die Gesteine der Westküste des Onega-sees. *Fennia: Bulletin de la Société de*

Géographie de Finlande. Vol.24, No.3, p.1–94. [Valamite]

WAHL, W., 1925. Die Gesteine des Wiborger Rapakiwigebietes. *Fennia: Bulletin de la Société de Géographie de Finlande*. Vol.45, No.20, p.1–127. [Pyterlite, Tirilite, Wiborgite (Viborgite)]

WALKER, G.P.L. & CROASDALE, R., 1971. Characteristics of some basaltic pyroclastics. *Bulletin Volcanologique. Organe de l'Association de Volcanologie de l'Union Géodésique et Géophysique Internationale. Bruxelles. (Napoli.)* Vol.35, 2nd Ser., p.303–317. [Achnelith]

WALKER, T.L., 1931. Alexoite, a pyrrhotite-peridotite from Ontario. *University of Toronto Studies. Geological Series. Toronto.* No.30, p.5–8. [Alexoite]

WALTERSHAUSEN, W.S. VON, 1846. Ueber die submarinen vulkanischen Ausbrüche im Tertiär-Formation des Val di Noto. *Göttinger Studien*. p.371–431. [Palagonite tuff]

WALTERSHAUSEN, W.S. VON, 1853. Über die vulkanischen Gesteine in Sicilien und Island und ihre submarine Umbildung. *Dieterichschen Buchhandlung, Göttingen.* 532pp. [Sideromelane]

WARREN, C.H., 1912. Art. XXV – The ilmenite rocks near St. Urbain, Quebec; a new occurrence of rutile and sapphirine. *American Journal of Science. New Haven.* Vol.33, 4th Ser., p.263–277. [Urbainite]

WASHINGTON, H.S., 1894. Art. XV – On the basalts of Kula. *American Journal of Science. New Haven.* Vol.47, 3rd Ser., p.114–123. [Kulaite]

WASHINGTON, H.S., 1896A. Italian petrological sketches. I. The Bolsena region. *Journal of Geology. Chicago.* Vol.4, p.541–566. [Vulsinite]

WASHINGTON, H.S., 1896B. Italian petrological sketches. II. The Viterbo region. *Journal of Geology. Chicago.* Vol.4, p.826–849. [Ciminite]

WASHINGTON, H.S., 1897A. Italian petrological sketches. III. The Bracciano, Cerveteri and Tolfa regions. *Journal of Geology. Chicago.* Vol.5, p.34–49. [Toscanite]

WASHINGTON, H.S., 1897B. Italian petrological sketches. V. Summary and conclusion. *Journal of Geology. Chicago.* Vol.5, p.349–377. [Santorinite]

WASHINGTON, H.S., 1900. The composition of kulaite. *Journal of Geology. Chicago.* Vol.8, p.610–620. [*Kulaite*]

WASHINGTON, H.S., 1901. The foyaite–ijolite series of Magnet Cove: a chemical study in differentiation. *Journal of Geology. Chicago.* Vol.9, p.607–622. [Arkite, Covite]

WASHINGTON, H.S., 1906. The Roman comagmatic region. *Publications. Carnegie Institution of Washington.* Vol.57, p.1–199. [Tavolatite, Trachivicoite, Vicoite, Viterbite]

WASHINGTON, H.S., 1913. The volcanoes and rocks of Pantelleria. *Journal of Geology. Chicago.* Vol.21, p.683–713. [Gibelite, Khagiarite (Kagiarite), Ponzite]

WASHINGTON, H.S., 1914A. Art. VII – An occurrence of pyroxenite and hornblendite in Bahia, Brazil. *American Journal of Science. New Haven.* Vol.38, 4th Ser., p.79–90. [Bahiaite]

WASHINGTON, H.S., 1914B. The analcite basalts of Sardinia. *Journal of Geology. Chicago.* Vol.22, p.742–753. [Ghizite]

WASHINGTON, H.S., 1920. Art. IV – Italite, a new leucite rock. *American Journal of Science. New Haven.* Vol.50, 4th Ser., p.33–47. [Albanite, Italite, Vesbite]

WASHINGTON, H.S., 1923. Petrology of the Hawaiian Islands. I. Kohala and Mauna Kea, Hawaii. *American Journal of Science. New Haven.* Vol.5, 5th Ser., p.465–502. [*Oligoclasite*]

WASHINGTON, H.S., 1927. The italite locality of Villa Senni. *American Journal of Science. New Haven.* Vol.14, 5th Ser., p.173–198. [Biotitite]

WASHINGTON, H.S. & KEYES, M.G., 1928. Petrology of the Hawaiian Islands: VI. Maui. *American Journal of Science. New Haven.* Vol.15, 5th Ser., p.199–220. [*Oligoclase andesite*]

WASHINGTON, H.S. & LARSEN, E.S., 1913. Magnetite basalt from North Park, Colorado. *Journal of the Washington Academy of Sciences.* Vol.3, p.449–452. [Arapahite]

WATSON, T.L., 1907. Mineral Resources of Virginia. *Bell, Lynchberg, Va.* 618pp. [Nelsonite]

WATSON, T.L., 1912. Art. XLIV – Kragerite, a rutile-bearing rock from Krageroe, Norway. *American Journal of Science. New Haven.* Vol.34, 4th Ser., p.509–514. [Kragerite]

WATSON, T.L. & TABER, S., 1910. The Virginia rutile deposits. In: Hayes, C.W. & Lindgren, W. (Editors) Contributions to Economic Geology (Short papers and preliminary reports) 1909. *Bulletin of the United States Geological Survey. Washington.* No.430, p.200–213. [Gabbro-nelsonite]

WATSON, T.L. & TABER, S., 1913. Geology of the titanium and apatite deposits of Virginia. *Bulletin. Virginia Geological Survey, University of Virginia.* Vol.3A, p.1–308. [Amherstite]

WATTS, W.W., 1925. The geology of South Shropshire. *Proceedings of the Geologists' Association. London.* Vol.36, p.321–363. [Ercalite]

WEED, W.H. & PIRSSON, L.V., 1895A. Highwood Mountains of Montana. *Bulletin of the Geological Society of America. New York.* Vol.6, p.389–422. [Shonkinite]

WEED, W.H. & PIRSSON, L.V., 1895B. Art. LII – Igneous rocks of Yogo Peak, Montana. *American Journal of Science. New Haven.* Vol.50, 3rd Ser., p.467–479. [Yogoite]

WEED, W.H. & PIRSSON, L.V., 1896. Art. XLVI – Missourite, a new leucite rock from the Highwood Mountains of Montana. *American Journal of Science. New Haven.* Vol.2, 4th Ser., p.315–323. [Missourite]

WEINSCHENK, E., 1891. Beiträge zur Petrographie Japans. *Neues Jahrbuch für Mineralogie, Geologie und Paläontologie. Stuttgart. Referate Abt. A.* Bd.7, p.133–151. [Sanukite]

WEINSCHENK, E., 1895. Beiträge zur Petrographie der östlichen Centralalpen speciell des Gross-Venedigerstockes. *Abhandlungen der Mathematisch-Physikalischen Classe der Königlich Bayerischen Akademie der Wissenschaften. München.* Vol.18, p.651–713. [Stubachite]

WEINSCHENK, E., 1899. Zur Kenntnis der Graphitlagerstätten. Chemisch-geologische Studien. *Abhandlungen der Mathematisch-Physikalischen Classe der Königlich Bayerischen Akademie der Wissenschaften. München.* Vol.19, p.509–564. [Bojite]

WELLS, M.K., 1948. In: Cotelo Neiva, J.M., Serpentines and serpentinisation – discussion. *International Geological Congress. Report of the 18th Session. Great Britain, 1948. Part II. Proceedings of Section A. Problems of Geochemistry.* p.88–95. [Serpentinite]

WENTWORTH, C.K., 1935. The terminology of coarse sediments. *Bulletin of the National Research Council. Washington.* No.98, p.225-246. [Block]

WENTWORTH, C.K. & WILLIAMS, H., 1932. The classification and terminology of the pyroclastic rocks. *Bulletin of the National Research Council. Washington.* Vol.89, p.19–53. [Bomb, Coarse (ash) grain, Coarse (ash) tuff, Reticulite]

WERNER, A.G., 1787. Kurze Klassifikationen und Beschreibung der verschiedenen Gebirgsarten. *Dresden.* 28pp. [Klingstein (Clinkstone), Mandelstein, Porphyry]

WESTERMANN, J.H., 1932. The geology of Aruba. *Geographische en Geologische Mededeelingen. Utrecht.* Vol.7, p.1–129.

[Hooibergite]

WHITE, A.J.R., 1979. Sources of granite magma. *Abstracts of papers to be presented at the Annual Meetings of the Geological Society of America and Associated Societies, San Diego, California, November 5-8, 1979.* Vol.11, p.539. [M-type granite]

*WILLEMS, H.W.V., 1934. *Ingénieur in Nederlandsch-Indië. Bandoeng.* Vol.1. [Astridite]

WILLIAMS, G.H., 1886. Art. IV – The peridotites of the "Cortlandt Series" on the Hudson River near Peekskill, N.Y. *American Journal of Science. New Haven.* Vol.31, 3rd Ser., p.26–41. [Cortlandtite]

WILLIAMS, G.H., 1890. The non-feldspathic intrusive rocks of Maryland and the course of their alteration. *American Geologist. Minneapolis.* Vol.6, p.35–49. [Bronzitite, Websterite]

WILLIAMS, H., 1941. Calderas and their origin. *University of California Publications in Geological Sciences. Berkeley.* Vol.25, No.6, p.239–346. [Ash-stone, Aso lava]

WILLIAMS, J.F., 1891. The igneous rocks of Arkansas. *Annual Report of the Geological Survey of Arkansas. Little Rock.* Vol.2, p.1–391. [Fourchite, Pulaskite]

WILLMANN, K., 1919. Die Redwitzite, eine neue Gruppe von granitischen Lamprophyren. *Zeitschrift der Deutschen Geologischen Gesellschaft. Berlin.* Vol.71, p.1–33. [Redwitzite]

WINCHELL, A.N., 1912. Geology of the National mining district, Nevada. *Mining and Science Press, San Francisco.* Vol.105, p.655–659. [Auganite]

WINCHELL, A.N., 1913. Rock classification on three co-ordinates. *Journal of Geology. Chicago.* Vol.21, p.211. [Peralkaline, Rhyodacite]

WINKLER, G.G., 1859. Allgovit (Trapp) in den Allgäuer Alpen Bayerns. *Neues Jahrbuch für Mineralogie, Geognosie, Geologie und Petrefaktenkunde. Stuttgart.* Vol.30, p.641–670. [Allgovite]

WOLF, H., 1866. Trachytsammlungen aus Ungarn. *Verhandlungen der Geologischen Reichsanstalt (Staatanstalt–Landesanstalt). Wien.* Vol.16, p.33–34. [Microtinite]

WOLFF, F. VON, 1899. Beiträge zur Geologie und Petrographie Chile's. *Zeitschrift der Deutschen Geologischen Gesellschaft. Berlin.* Vol.51, p.471–555. [Andennorite]

WOLFF, F. VON, 1914. Der Vulkanismus. *Enke, Stuttgart.* Vol.1, 711pp. [Duckstein]

WOOLLEY, A.R., 1982. A discussion of carbonatite evolution and nomenclature, and the generation of sodic and potassic fenites. *Mineralogical Magazine and Journal of the Mineralogical Society. London.* Vol.46, p.13–17. [Magnesiocarbonatite]

WOOLLEY, A.R., BERGMAN, S.C., EDGAR, A.D., LE BAS, M.J., MITCHELL, R.H., ROCK, N.M.S. & SCOTT SMITH, B.H., 1996. Classification of lamprophyres, lamproites, kimberlites, and the kalsilitic, melilitic and leucitic rocks.. *The Canadian Mineralogist.* Vol.34, p.175-186. [Kalsilitite, Melilite leucitite, Melilite nephelinite, Potassic melilitite, Potassic olivine melilitite]

WOOLLEY, A.R. & KEMPE D.R.C., 1989. Carbonatites: nomenclature, average chemical compositions and element distribution. In: K. Bell (Editor) Carbonatite Genesis and Evolution. *Unwin Hyman, London.* p.1-14. [Calciocarbonatite]

WORTH, R.N., 1884. On trowlesworthite, and certain granitoid rocks near Plymouth. *Transactions of the Royal Geological Society of Cornwall. Penzance.* Vol.10, p.177. [Trowlesworthite]

WYLLIE, B.K.N. & SCOTT, A., 1913. The plutonic rocks of Garabal Hill. *Geological Magazine. London.* Vol.10, Decade 5, p.499–508. [Davainite]

WYLLIE, P.J., 1967. Ultramafic and Related Rocks. *Wiley, New York.* 464pp. [Clinopyroxenite, Hornblende peridotite, Hornblende pyroxenite, Orthopyroxenite, Pyroxene peridotite]

YACHEVSKII, L., 1909. Mestorozhdeniya khrizotila na khrebte Bis-Tag v Minusinskom okruge Eniseiskoi gubernii. *Geologicheskiia Isledovaniia v "Zolotonosnykh Oblastiakh" Sibiri. Sankt Peterburg.* Vol.8, p.31–50. [Bistagite]

YASHINA, R.M., 1957. Shchelochnye porody yugo-vostochnoi Tuvy. *Izvestiia Akademii Nauk SSSR, Seriia Geologicheskaia.* No.5, p.17–36. [Tuvinite]

YODER, H.S. & TILLEY, C.E., 1962. Origin of basalt magmas: an experimental study of natural and synthetic rock systems. *Journal of Petrology. Oxford.* Vol.3, p.342–532. [*Tholeiite*]

YOUNG, G.A., 1906. The geology and petrography of Mount Yamaska. *Annual Report. Geological Survey of Canada. Montreal and Ottawa.* Pt.H, p.1–43. [Yamaskite]

ZAVARITSKII, A.N., 1955. Izverzhennye gornye porody. *Izadatel'stvo Akademii Nauk SSSR, Moskva.* 479pp. [Bereshite, Plagiogranite]

ZAVARITSKII, A.N. & KRIZHANOVSKII, V.I., 1937. Gosudarstvennyi Ilmenskii mineralogicheskii zapovednik. *27th International Geological Congress, Moskva. Uralskaia ekskursiia, iuzhnyi marshrut.* p.5–17. [Sandyite]

ZHIDKOV, A. YA., 1962. Slozhnaya Synnyrskaya intruziya sienitov Severo-Baikalskoi shchelochnoi provintsii. *Geologiia i geofyzika.* Vol.9, p.29–40. [Synnyrite]

ZIRKEL, F., 1866A. Lehrbuch der Petrographie. *Marcus, Bonn.* Vol.1, 607pp. [Ditroite]

ZIRKEL, F., 1866B. Lehrbuch der Petrographie. *Marcus, Bonn.* Vol.2, 635pp. [Corsite, Quartz diorite, *Teschinite*]

ZIRKEL, F., 1870. Untersuchungen über die mikroskopische Zusammensetzung und Structur der Basaltgesteine. *Moncur, Bonn.* 208pp. [Feldspar-basalt, Leucite basalt]

ZIRKEL, F., 1894A. Lehrbuch der Petrographie. *Engelmann, Leipzig.* 2nd Edn, Vol.2, 941pp. [Felsogranophyre, *Leucite phonolite*, Plethorite]

ZIRKEL, F., 1894B. Lehrbuch der Petrographie. *Engelmann, Leipzig.* 2nd Edn, Vol.3, 833pp. [Therolite]

ZLATKIND, Z.G., 1945. Olivinovye turyaity (kovdority) – novye glubinnye melilitovye porody Kolskogo poluostrova. *Sovetskaia Geologiia. Moskva.* Vol.7, p.88–95. [Kovdorite]

Appendix A Lists of participants

This appendix contains the names of all those people who have helped the Subcommission in various ways and consists of two lists.

The **first list** of participants is arranged alphabetically by the country which the participant represented or the country in which they were residing when they corresponded with the Subcommission. The name of each participant is followed in brackets by their affiliation with the Subcommission using the following abbreviations:

M – a Member of the Subcommission acting as a representative of the country

W – a member of one of the Working groups who participated by attending meetings and/or by correspondence

P – a member of the Pyroclastic working group only

C – a Correspondent with the Subcommission who contributed with written submissions in response to Subcommission circulars

G – a Guest at one or more of the meetings.

The **second list** is simply an alphabetical list of participants by name, followed in brackets by their country.

A.1 PARTICIPANTS LISTED BY COUNTRY

ARGENTINA (4)
Andreies, R.R. (C)
Mazzoni, M.M. (C)
Spalletti, L.A. (C)
Teruggi, M.E. (M)

ARMENIA (1)
Shirinian, H.G. (G)

AUSTRALIA (21)
Branch, C.D. (W)
Cundari, A. (W)
Dallwitz, W.B. (M)
Ewart, A. (C)
Flinter, B. (G)
Green, D.H. (W)
Harris, P.G. (C)
Hensel, H.D. (C)
Joyce, E.B. (G)
Le Maitre, R.W. (M)
Mackenzie, D.E. (C)
Matthison, C.J. (C)
Middlemost, E.A.K. (W)

Neall, V. (G)
Rock, N.M.S. (W)
Taylor, S.R. (W)
Vallance, T.G. (W)
Wass, S.Y. (C)
White, A.J.R. (C)
Wilkinson, J.F.G. (M)
Wilson, A.P. (W)

AUSTRIA (4)
Daurer, A. (C)
Exner, C. (C)
Richter, W. (C)
Wieseneder, H. (C)

BELGIUM (4)
Deblond, A. (W)
Duchesne, J.-C. (W)
Michot, J. (W)
Michot, P. (W)

BOLIVIA (1)
Ploskonka, E. (W)

BRAZIL (2)
Coutinho, J.M.V. (W)
Leonardos, Jr., O.H. (W)

BULGARIA (5)
Boyadjiev, St.G. (G)
Georgiev, G.K. (C)
Ivanov, R. (W)
Stefanova, M. (W)
Yanef, Y. (W)

CANADA (25)
Armstrong, R.L. (C)
Ayres, L.D. (M)
Baragar, W.R.A. (W)
Church, B.N. (W)
Currie, K.L. (W)
Davidson, A. (W)
Edgar, A.D. (W)
Frisch, Th. (C)
Gittins, J. (W)
Goodwin, A.M. (M)
Lambert, M.B. (W)
Laurent, R. (C)
Laurin, A.F. (W)
Martignole, J. (W)
McCutcheon, S.R. (C)
Mitchell, R.H. (W.)
Moore, J.M. (C)
Naldrett, A.J. (W)
Robinson, P. (W)
Ruitenberg, A.A. (C)
Scott-Smith, B.H. (W)
Sharma, K.N.M. (W)
Thurston, P.C. (W)
Woodsworth, G.J. (C)
Woussen, G. (C)

CHILE (1)
Vergara, M. (W)

CHINA (4)
Heng Sheng Wang, (M)
Malpas, J. (W)
Sung Shu Ho, (C)
Wang, B. (M)

COLOMBIA (1)
Schaufelberger, P. (C)

CROATIA (2)
Majer, V. (C)
Obradovic, J. (W)

CZECH REPUBLIC (8)
Dudek, A. (M)
Fediuk, F. (C)
Fiala, F. (W)
Kopecky, L. (W)
Neužilová, N. (C)
Palivcova, M. (C)
Suk, M. (C)
Vejnar, Zd. (C)

DENMARK (7)
Berthelsen, A. (C)
Jakobsson, S. (G)
Nielsen, T.F.D. (W)
Noe Nygaard, A. (M)
Sørensen, H. (M)
Wilson, J.R. (C)
Zeck, H.P. (W)

EIRE (1)
Mohr, P. (C)

FINLAND (3)
Härme, M. (C)
Laitakati, I. (C)
Vartiainen, H. (C)

FRANCE (20)
Albarede, F. (W)
Alsac, C. (W)
Bonin, B. (M)
Bordet, P. (G)
Brousse, R. (C)
Chenevoy, M. (C)
Colin, F. (C)
de la Roche, H. (M)
Forestier, F.H. (W)
Giraud, P. (W)
Juteau, Th. (W)

Lameyre, J. (M)
Leterrier, J. (G)
Raguin, E. (C)
Roques, M. (G)
Soler, E. (W)
Tégyey-Dumait, M. (W)
Velde, D. (W)
Vincent, P.M. (W)
Westercamp, D. (W)

GERMANY (45)
Amstutz, G.C. (W)
Arndt, N. (G)
Bambauer, H.U. (C)
Chudoba, K. (C)
Correns, C.W. (W)
Förster, H. (C)
Frechen, J. (C)
Hentschel, H. (W)
Huebscher, H.D. (C)
Jung, D. (W)
Keller, J. (M)
Knetsch, G. (C)
Kramer, W. (W)
Kunz, K. (W)
Lehmann, E. (W)
Lensch, G. (W)
Lorenz, V. (W)
Mädler, J. (C)
Martin, H. (W)
Mathé, G (C)
Matthes, S (C)
Mehnert, K.R. (M)
Mihm, A. (C)
Morteani, G. (C)
Mucke, D. (W)
Müller, P. (W)
Murawski, H. (C)
Okrusch, M. (C)
Pälchen, W. (W)
Paulitsch, P.D. (C)
Pfeiffer, L. (W)
Pichler, H. (W)
Röllig, G. (W)
Rost, F. (W)
Schmincke, H.U. (C)

Seim, R. (C)
Stengelin, R. (W)
Tischendorf, G. (M)
Troll, G. (C)
Vieten, K. (W)
Watznauer, A. (W)
Wedepohl, K.H. (W)
Wimmenauer, W. (M)
Winkler, H.G.F. (W)
Wurm, A. (C)

GUYANA (1)
Singh, S. (W)

HUNGARY (3)
Kubovics, I. (G)
Pantó, G. (M)
Széky-Fux, V. (M)

ICELAND (1)
Thorarinsson, S. (W)

INDIA (8)
Bose, M.K. (W)
Chatterjee, S.C. (M)
Sen, S.K. (W)
Subba Rao, S. (C)
Subbarao, K.V. (W)
Subramaniam, A.P. (W)
Subramaniam, K.V. (G)
Sukheswala, R.N. (M)

ISRAEL (1)
Bentor, Y.K. (W)

ITALY (24)
Balconi, M. (C)
Bellieni, G. (M)
Bertolani, M. (C)
Boriani, A. (G)
Callegari, E. (C)
Cristofolini, R. (C)
D'Amico, Cl. (W)
Dal Piaz, G.V. (C)
De Vecchi, Gp. (G)
Deriu, M. (C)

Galli, M. (C)
Innocenti, F. (C)
Justin-Visentin, E. (W)
Malaroda, R. (C)
Mittempergher, M. (W)
Morbidelli, L. (C)
Piccirillo, E.M. (W)
Rittmann, A. (W)
Sassi, F. (C)
Simboli, G. (W)
Venturelli, A. (C)
Villari, L. (W)
Visentin, E.J. (G)
Zanettin, B (M)

JAPAN (4)
Aoki, K.J. (M)
Aramaki, S. (M)
Katsui, Y. (M)
Yagi, K. (W)

KENYA (1)
Williams, L.A.J. (W)

KUWAIT (1)
Al Mishwat, A.T. (G)

MALAYSIA (1)
Hutchinson, C.S. (C)

NETHERLANDS (6)
de Roever, W.P. (M)
Den Tex, E. (C)
Oen, I.S. (C)
Tobi, A.C. (M)
Uytenbogaardt, W. (C)
van der Plas, L. (C)

NEW ZEALAND (3)
Cole, J.W. (M)
Coombs, D.S. (C)
Neal, V.E. (W)

NORWAY (8)
Barth, T.F.W. (W)
Bryhni, I. (C)

Carstens, H. (W)
Dons, J.A. (C)
Griffin, W.L. (C)
Oftedahl, C. (C)
Strand, T. (C)
Torske, T. (W)

POLAND (2)
Nowakowski, A. (C)
Smulikowski, K. (M)

PORTUGAL (3)
Aires Barros, L. (W)
Mourazmiranda, A. (C)
Schermerhorn, L.J.G. (C)

ROMANIA (9)
Berbeleac, I. (W)
Bercia, J. (G)
Dimitrescu, R. (C)
Giusca, D. (W)
Kräutner, H. (W)
Peltz, S. (W)
Radulescu, D. (M)
Savu, H. (C)
Seclaman, M. (W)

RUSSIAN FEDERATION (41)
Afanass'yev, G.D. (W)
Andreeva, E.D. (W)
Bogatikov, O.A. (C)
Borodaevskaya, M.R. (W)
Borodin, L.S. (W)
Chvorova, L.V. (C)
Daminova, A.M. (W)
Dmitriev, Yu.I. (W)
Efremova, S.V. (M)
Frolova, T.I. (W)
Gladkikh, V.S. (W)
Gon'shakova, V.I. (C)
Gorshkov, G.S. (W)
Gurulev, S.A. (G)
Iakovleva, E.B. (W)
Kononova, V.A. (G)
Kravchenko, S. (G)
L'vov, B.K. (W)

Lazko, E. (W)
Luchitsky, J. (G)
Malejev, E.F. (W)
Markovsky, B.A. (W)
Massaitis, V.L. (W)
Mikhailov, N.P. (M)
Orlova, M.P. (W)
Pertsev, N.N. (W)
Petrova, M.A. (W)
Pjatenko, I.K. (W)
Polunina, L.A. (W)
Rotman, V.K. (W)
Ruminantseva, N.A. (W)
Schtschebakova, M.N. (C)
Sharpenok, L. (G)
Shemyakin, V.M. (W)
Shurkin, K.A. (W)
Sobolev, N. (W)
Sobolev, S.F. (W)
Sobolev, V.S. (G)
Sveshnikova, S.V. (W)
Vorobieva, O.A. (M)
Yashina, R. (G)

SAUDI ARABIA (1)
Wolf, K.H. (W)

SOUTH AFRICA (8)
Dann, J.C. (W)
Evans, D. (W)
Ferguson, J. (M)
Frick, G. (C)
Moore, A.C. (C)
Snyman, C.P. (M)
van Biljon, S. (C)
von Gruenewaldt, G. (M)

SPAIN (7)
Ancochea, E. (W)
Brändle, J.H. (M)
Corretgé, L. (C)
Fuster, J.M. (C)
Muñoz, M. (W)
Pascual, E. (W)
Puga, E. (C)

SURINAM (2)
Dahlberg, E.H. (W)
de Roever, E.W.F. (W)

SWEDEN (5)
Gorbatschev, R. (C)
Koark, H.J. (C)
Lundqvist, T. (W)
Persson, L. (W)
von Eckerman, H. (C)

SWITZERLAND (18)
Ayrton, S. (C)
Bearth, P. (W)
Burckhardt, C.E. (W)
Dietrich, V. (C)
Hänny, R. (C)
Hügi, T. (C)
Kübler, B. (C)
Maggetti, M. (W)
Nickel, E. (C)
Niggli, E. (W)
Persoz, F. (C)
Schmid, R. (M)
Seemann, U. (W)
Streckeisen, A. (M)
Trommsdorff, V. (M)
Vuagnat, M. (W)
Wenk, E. (W)
Woodtli, R. (C)

TAIWAN (2)
Juan, V.C. (C)
Veichow, C.J. (C)

TURKEY (5)
Aslaner, M. (C)
Ataman, G. (C)
Irdar, E. (C)
Kraëff, A. (C)
Kumbazar, I. (C)

UGANDA (1)
Macdonald, R. (W)

UK (46)

Baker, P.E. (C)
Berridge, N.G. (C)
Cheadle, M. (W)
Collins, G.H. (C)
Cox, K.G. (W)
Dearnley, R. (C)
Donaldson, C. (W)
Elliot, R.B. (C)
Ellis, S.E. (W)
Fitch, F.J. (C)
Francis, E.H. (W)
Gibson, S. (W)
Gill, R. (W)
Gillespie, M. (G)
Hall, P. (W)
Harrison, R.K. (W)
Hergt, J. (W)
Hole, M. (W)
Howells, M. (W)
Howie, R.A. (W)
Hughes, D. (W)
Kent, R. (W)
King, B.C. (W)
Lawson, R.I. (W)
Le Bas, M.J. (M)
Macdonald, R. (W)
MacKenzie, W.S. (C)
Menzies, M. (W)
Muir, I.D. (W)
Nesbitt, B. (W)
Nockolds, S.R. (C)
Preston, J (C)
Rea, W.J. (W)
Rollinson, H. (W)
Sabine, P.A. (M)
Stillman, C.J. (C)
Styles, M. (G)
Thompson, R. (W)
Upton, B.J.G. (W)
Wadsworth, W.J. (W)
Weaver, S.D. (W)
Wells, A.K. (W)
Wells, M.K. (W)
Wilson, M. (W)
Woolley, A.R. (M)
Wright, A.E. (W)

USA (80)

Barker, D.S. (W)
Bateman, P.C. (M)
Bergman, S.C. (W)
Best, M.G. (C)
Boone, E.M. (C)
Brooks, E.R. (C)
Chayes, F. (M)
Cheney, J.T. (C)
Chesterman, C.W. (C)
Christiansen, R.L. (C)
Coats, R.R. (W)
Coveney, R.M. (C)
de Waard, D. (W)
Dietrich, R.V. (W)
Duffield, W. (G)
Earton, R.M. (W)
Effimoff, I. (C)
Ekblaw, S.E. (C)
Emerson, D.O. (C)
Ernst, G. (C)
Evans, B.W. (C)
Fisher, R.V. (M)
Forbes, R.B. (C)
Gastil, G. (C)
Grove, T. (W)
Guidotti, C.V. (C)
Harrison, J.E. (C)
Hawkin, J.W. (C)
Hays, J.F. (C)
Heiken, G. (W)
Herz, N. (C)
Hobbs, S.W. (C)
Hollister, L.S. (C)
Honnorez, J. (W)
Hyndman, D.W. (C)
Jackson, E.D. (W)
Krauskopf, K. (C)
Lofgren, G. (M)
Lyons, P.C. (C)
Macdonald, G.A. (W)
Mattson, P.H. (C)
McBirney, A.R. (M)
McCoy, F.W. (W)
McHone, J.G. (C)
Meyer, H.O.A. (W)
Miller, T. (W)

Misch, P. (C)
Miyashiro, A. (C)
Mohr, P.A. (W)
Moore, G.E. (C)
Murray, A.D. (C)
Mutschler, F.E. (C)
O'Connor, J.T. (C)
Peck, D.L. (M)
Peterson, D.W. (W)
Presnall, D. (W)
Putman, G.W. (C)
Rankin, D.W. (C)
Reitan, P.H. (C)
Scott, R.B. (C)
Sheridan, M.F. (W)
Simkin, T. (C)
Snavely, P. (W)
Snyder, G.L. (W)
Taubeneck, W.H. (C)
Thornton, C.P. (M)
Tilling, R.I. (W)
Toulmin III, P. (C)
Tweto, O. (C)

Vidale, R.J. (C)
Vitaliano, C.J. (C)
Walker, G.P.L. (W)
Waters, A.C. (C)
Whitney, J.A. (C)
Whitten, E.H.T. (C)
Wilcox, R.E. (C)
Williams, H. (C)
Wones, D.R. (C)
Wyllie, P.J. (C)
Yoder, H.S.Jr. (W)

VENEZUELA (1)
Urbani, F. (C)

YUGOSLAVIA (1)
Karamata, S. (W)

ZAMBIA (1)
Cooray, P.G. (W)

ZIMBABWE (1)
Wilson, A.H. (W)

A.2 PARTICIPANTS LISTED BY NAME (WITH COUNTRY)

Afanass'yev, G.D. (Russian Federation)
Aires Barros, L. (Portugal)
Al Mishwat, A.T. (Kuwait)
Albarede, F. (France)
Alsac, C. (France)
Amstutz, G.C. (Germany)
Ancochea, E. (Spain)
Andreeva, E.D. (Russian Federation)
Andreies, R.R. (Argentina)
Aoki, K.J. (Japan)
Aramaki, S. (Japan)
Armstrong, R.L. (Canada)
Arndt, N. (Germany)
Aslaner, M. (Turkey)
Ataman, G. (Turkey)
Ayres, L.D. (Canada)
Ayrton, S. (Switzerland)
Baker, P.E. (UK)
Balconi, M. (Italy)
Bambauer, H.U. (Germany)
Baragar, W.R.A. (Canada)
Barker, D.S. (USA)
Barth, T.F.W. (Norway)
Bateman, P.C. (USA)
Bearth, P. (Switzerland)
Bellieni, G. (Italy)
Bentor, Y.K. (Israel)
Berbeleac, I. (Romania)
Bercia, J. (Romania)
Bergman, S.C. (USA)
Berridge, N.G. (UK)
Berthelsen, A. (Denmark)
Bertolani, M. (Italy)
Best, M.G. (USA)
Bogatikov, O.A. (Russian Federation)
Bonin, B. (France)
Boone, E.M. (USA)
Bordet, P. (France)
Boriani, A. (Italy)
Borodaevskaya, M.R. (Russian Federation)
Borodin, L.S. (Russian Federation)
Bose, M.K. (India)
Boyadjiev, St.G. (Bulgaria)
Branch, C.D. (Australia)
Brändle, J.H. (Spain)

Brooks, E.R. (USA)
Brousse, R. (France)
Bryhni, I. (Norway)
Burckhardt, C.E. (Switzerland)
Callegari, E. (Italy)
Carstens, H. (Norway)
Chatterjee, S.C. (India)
Chayes, F. (USA)
Cheadle, M. (UK)
Chenevoy, M. (France)
Cheney, J.T. (USA)
Chesterman, C.W. (USA)
Christiansen, R.L. (USA)
Chudoba, K. (Germany)
Church, B.N. (Canada)
Chvorova, L.V. (Russian Federation)
Coats, R.R. (USA)
Cole, J.W. (New Zealand)
Colin, F. (France)
Collins, G.H. (UK)
Coombs, D.S. (New Zealand)
Cooray, P.G. (Zambia)
Correns, C.W. (Germany)
Corretgé, L. (Spain)
Coutinho, J.M.V. (Brazil)
Coveney, R.M. (USA)
Cox, K.G. (UK)
Cristofolini, R. (Italy)
Cundari, A. (Australia)
Currie, K.L. (Canada)
D'Amico, Cl. (Italy)
Dahlberg, E.H. (Surinam)
Dal Piaz, G.V. (Italy)
Dallwitz, W.B. (Australia)
Daminova, A.M. (Russian Federation)
Dann, J.C. (South Africa)
Daurer, A. (Austria)
Davidson, A. (Canada)
de la Roche, H. (France)
de Roever, E.W.F. (Surinam)
de Roever, W.P. (Netherlands)
De Vecchi, Gp. (Italy)
de Waard, D. (USA)
Dearnley, R. (UK)
Deblond, A. (Belgium)

Den Tex, E. (Netherlands)
Deriu, M. (Italy)
Dietrich, R.V. (USA)
Dietrich, V. (Switzerland)
Dimitrescu, R. (Romania)
Dmitriev, Yu.I. (Russian Federation)
Donaldson, C. (UK)
Dons, J.A. (Norway)
Duchesne, J.-C. (Belgium)
Dudek, A. (Czech Republic)
Duffield, W. (USA)
Earton, R.M. (USA)
Edgar, A.D. (Canada)
Effimoff, I. (USA)
Efremova, S.V. (Russian Federation)
Ekblaw, S.E. (USA)
Elliot, R.B. (UK)
Ellis, S.E. (UK)
Emerson, D.O. (USA)
Ernst, G. (USA)
Evans, B.W. (USA)
Evans, D. (South Africa)
Ewart, A. (Australia)
Exner, C. (Austria)
Fediuk, F. (Czech Republic)
Ferguson, J. (South Africa)
Fiala, F. (Czech Republic)
Fisher, R.V. (USA)
Fitch, F.J. (UK)
Flinter, B. (Australia)
Forbes, R.B. (USA)
Forestier, F.H. (France)
Förster, H. (Germany)
Francis, E.H. (UK)
Frechen, J. (Germany)
Frick, G. (South Africa)
Frisch, Th. (Canada)
Frolova, T.I. (Russian Federation)
Fuster, J.M. (Spain)
Galli, M. (Italy)
Gastil, G. (USA)
Georgiev, G.K. (Bulgaria)
Gibson, S. (UK)
Gill, R. (UK)
Gillespie, M. (UK)
Giraud, P. (France)
Gittins, J. (Canada)
Giusca, D. (Romania)

Gladkikh, V.S. (Russian Federation)
Gon'shakova, V.I. (Russian Federation)
Goodwin, A.M. (Canada)
Gorbatschev, R. (Sweden)
Gorshkov, G.S. (Russian Federation)
Green, D.H. (Australia)
Griffin, W.L. (Norway)
Grove, T. (USA)
Guidotti, C.V. (USA)
Gurulev, S.A. (Russian Federation)
Hall, P. (UK)
Hänny, R. (Switzerland)
Härme, M. (Finland)
Harris, P.G. (Australia)
Harrison, J.E. (USA)
Harrison, R.K. (UK)
Hawkin, J.W. (USA)
Hays, J.F. (USA)
Heiken, G. (USA)
Heng Sheng Wang, (China)
Hensel, H.D. (Australia)
Hentschel, H. (Germany)
Hergt, J. (UK)
Herz, N. (USA)
Hobbs, S.W. (USA)
Hole, M. (UK)
Hollister, L.S. (USA)
Honnorez, J. (USA)
Howells, M. (UK)
Howie, R.A. (UK)
Huebscher, H.D. (Germany)
Hughes, D. (UK)
Hügi, T. (Switzerland)
Hutchinson, C.S. (Malaysia)
Hyndman, D.W. (USA)
Iakovleva, E.B. (Russian Federation)
Innocenti, F. (Italy)
Irdar, E. (Turkey)
Ivanov, R. (Bulgaria)
Jackson, E.D. (USA)
Jakobsson, S. (Denmark)
Joyce, E.B. (Australia)
Juan, V.C. (Taiwan)
Jung, D. (Germany)
Justin-Visentin, E. (Italy)
Juteau, Th. (France)
Karamata, S. (Yugoslavia)
Katsui, Y. (Japan)

Keller, J. (Germany)
Kent, R. (UK)
King, B.C. (UK)
Knetsch, G. (Germany)
Koark, H.J. (Sweden)
Kononova, V.A. (Russian Federation)
Kopecky, L. (Czech Republic)
Kraëff, A. (Turkey)
Kramer, W. (Germany)
Krauskopf, K. (USA)
Kräutner, H. (Romania)
Kravchenko, S. (Russian Federation)
Kübler, B. (Switzerland)
Kubovics, I. (Hungary)
Kumbazar, I. (Turkey)
Kunz, K. (Germany)
L'vov, B.K. (Russian Federation)
Laitakati, I. (Finland)
Lambert, M.B. (Canada)
Lameyre, J. (France)
Laurent, R. (Canada)
Laurin, A.F. (Canada)
Lawson, R.I. (UK)
Lazko, E. (Russian Federation)
Le Bas, M.J. (UK)
Le Maitre, R.W. (Australia)
Lehmann, E. (Germany)
Lensch, G. (Germany)
Leonardos, Jr., O.H. (Brazil)
Leterrier, J. (France)
Lofgren, G. (USA)
Lorenz, V. (Germany)
Luchitsky, J. (Russian Federation)
Lundqvist, T. (Sweden)
Lyons, P.C. (USA)
Macdonald, G.A. (USA)
Macdonald, R. (Uganda)
Macdonald, R. (UK)
Mackenzie, D.E. (Australia)
MacKenzie, W.S. (UK)
Mädler, J. (Germany)
Maggetti, M. (Switzerland)
Majer, V. (Croatia)
Malaroda, R. (Italy)
Malejev, E.F. (Russian Federation)
Malpas, J. (China)
Markovsky, B.A. (Russian Federation)
Martignole, J. (Canada)

Martin, H. (Germany)
Massaitis, V.L. (Russian Federation)
Mathé, G (Germany)
Matthes, S (Germany)
Matthison, C.J. (Australia)
Mattson, P.H. (USA)
Mazzoni, M.M. (Argentina)
McBirney, A.R. (USA)
McCoy, F.W. (USA)
McCutcheon, S.R. (Canada)
McHone, J.G. (USA)
Mehnert, K.R. (Germany)
Menzies, M. (UK)
Meyer, H.O.A. (USA)
Michot, J. (Belgium)
Michot, P. (Belgium)
Middlemost, E.A.K. (Australia)
Mihm, A. (Germany)
Mikhailov, N.P. (Russian Federation)
Miller, T. (USA)
Misch, P. (USA)
Mitchell, R.H. (Canada)
Mittempergher, M. (Italy)
Miyashiro, A. (USA)
Mohr, P. (Eire)
Mohr, P.A. (USA)
Moore, A.C. (South Africa)
Moore, G.E. (USA)
Moore, J.M. (Canada)
Morbidelli, L. (Italy)
Morteani, G. (Germany)
Mourazmiranda, A. (Portugal)
Mucke, D. (Germany)
Muir, I.D. (UK)
Müller, P. (Germany)
Muñoz, M. (Spain)
Murawski, H. (Germany)
Murray, A.D. (USA)
Mutschler, F.E. (USA)
Naldrett, A.J. (Canada)
Neal, V.E. (New Zealand)
Neall, V. (Australia)
Nesbitt, B. (UK)
Neužilová, N. (Czech Republic)
Nickel, E. (Switzerland)
Nielsen, T.F.D. (Denmark)
Niggli, E. (Switzerland)
Nockolds, S.R. (UK)

Noe Nygaard, A. (Denmark)
Nowakowski, A. (Poland)
O'Connor, J.T. (USA)
Obradovic, J. (Croatia)
Oen, I.S. (Netherlands)
Oftedahl, C. (Norway)
Okrusch, M. (Germany)
Orlova, M.P. (Russian Federation)
Pälchen, W. (Germany)
Palivcova, M. (Czech Republic)
Pantó, G. (Hungary)
Pascual, E. (Spain)
Paulitsch, P.D. (Germany)
Peck, D.L. (USA)
Peltz, S. (Romania)
Persoz, F. (Switzerland)
Persson, L. (Sweden)
Pertsev, N.N. (Russian Federation)
Peterson, D.W. (USA)
Petrova, M.A. (Russian Federation)
Pfeiffer, L. (Germany)
Piccirillo, E.M. (Italy)
Pichler, H. (Germany)
Pjatenko, I.K. (Russian Federation)
Ploskonka, E. (Bolivia)
Polunina, L.A. (Russian Federation)
Presnall, D. (USA)
Preston, J (UK)
Puga, E. (Spain)
Putman, G.W. (USA)
Radulescu, D. (Romania)
Raguin, E. (France)
Rankin, D.W. (USA)
Rea, W.J. (UK)
Reitan, P.H. (USA)
Richter, W. (Austria)
Rittmann, A. (Italy)
Robinson, P. (Canada)
Rock, N.M.S. (Australia)
Röllig, G. (Germany)
Rollinson, H. (UK)
Roques, M. (France)
Rost, F. (Germany)
Rotman, V.K. (Russian Federation)
Ruitenberg, A.A. (Canada)
Ruminantseva, N.A. (Russian Federation)
Sabine, P.A. (UK)
Sassi, F. (Italy)

Savu, H. (Romania)
Schaufelberger, P. (Colombia)
Schermerhorn, L.J.G. (Portugal)
Schmid, R. (Switzerland)
Schmincke, H.U. (Germany)
Schtschebakova, M.N. (Russian Federation)
Scott, R.B. (USA)
Scott-Smith, B.H. (Canada)
Seclaman, M. (Romania)
Seemann, U. (Switzerland)
Seim, R. (Germany)
Sen, S.K. (India)
Sharma, K.N.M. (Canada)
Sharpenok, L. (Russian Federation)
Shemyakin, V.M. (Russian Federation)
Sheridan, M.F. (USA)
Shirinian, H.G. (Armenia)
Shurkin, K.A. (Russian Federation)
Simboli, G. (Italy)
Simkin, T. (USA)
Singh, S. (Guyana)
Smulikowski, K. (Poland)
Snavely, P. (USA)
Snyder, G.L. (USA)
Snyman, C.P. (South Africa)
Sobolev, N. (Russian Federation)
Sobolev, S.F. (Russian Federation)
Sobolev, V.S. (Russian Federation)
Soler, E. (France)
Sørensen, H. (Denmark)
Spalletti, L.A. (Argentina)
Stefanova, M. (Bulgaria)
Stengelin, R. (Germany)
Stillman, C.J. (UK)
Strand, T. (Norway)
Streckeisen, A. (Switzerland)
Styles, M. (UK)
Subba Rao, S. (India)
Subbarao, K.V. (India)
Subramaniam, A.P. (India)
Subramaniam, K.V. (India)
Suk, M. (Czech Republic)
Sukheswala, R.N. (India)
Sung Shu Ho, (China)
Sveshnikova, S.V. (Russian Federation)
Széky-Fux, V. (Hungary)
Taubeneck, W.H. (USA)
Taylor, S.R. (Australia)

Tégyey-Dumait, M. (France)

Teruggi, M.E. (Argentina)

Thompson, R. (UK)

Thorarinsson, S. (Iceland)

Thornton, C.P. (USA)

Thurston, P.C. (Canada)

Tilling, R.I. (USA)

Tischendorf, G. (Germany)

Tobi, A.C. (Netherlands)

Torske, T. (Norway)

Toulmin III, P. (USA)

Troll, G. (Germany)

Trommsdorff, V. (Switzerland)

Tweto, O. (USA)

Upton, B.J.G. (UK)

Urbani, F. (Venezuela)

Uytenbogaardt, W. (Netherlands)

Vallance, T.G. (Australia)

van Biljon, S. (South Africa)

van der Plas, L. (Netherlands)

Vartiainen, H. (Finland)

Veichow, C.J. (Taiwan)

Vejnar, Zd. (Czech Republic)

Velde, D. (France)

Venturelli, A. (Italy)

Vergara, M. (Chile)

Vidale, R.J. (USA)

Vieten, K. (Germany)

Villari, L. (Italy)

Vincent, P.M. (France)

Visentin, E.J. (Italy)

Vitaliano, C.J. (USA)

von Eckerman, H. (Sweden)

von Gruenewaldt, G. (South Africa)

Vorobieva, O.A. (Russian Federation)

Vuagnat, M. (Switzerland)

Wadsworth, W.J. (UK)

Walker, G.P.L. (USA)

Wang, B. (China)

Wass, S.Y. (Australia)

Waters, A.C. (USA)

Watznauer, A. (Germany)

Weaver, S.D. (UK)

Wedepohl, K.H. (Germany)

Wells, A.K. (UK)

Wells, M.K. (UK)

Wenk, E. (Switzerland)

Westercamp, D. (France)

White, A.J.R. (Australia)

Whitney, J.A. (USA)

Whitten, E.H.T. (USA)

Wieseneder, H. (Austria)

Wilcox, R.E. (USA)

Wilkinson, J.F.G. (Australia)

Williams, H. (USA)

Williams, L.A.J. (Kenya)

Wilson, A.H. (Zimbabwe)

Wilson, A.P. (Australia)

Wilson, J.R. (Denmark)

Wilson, M. (UK)

Wimmenauer, W. (Germany)

Winkler, H.G.F. (Germany)

Wolf, K.H. (Saudi Arabia)

Wones, D.R. (USA)

Woodsworth, G.J. (Canada)

Woodtli, R. (Switzerland)

Woolley, A.R. (UK)

Woussen, G. (Canada)

Wright, A.E. (UK)

Wurm, A. (Germany)

Wyllie, P.J. (USA)

Yagi, K. (Japan)

Yanef, Y. (Bulgaria)

Yashina, R. (Russian Federation)

Yoder, H.S.Jr. (USA)

Zanettin, B (Italy)

Zeck, H.P. (Denmark)

Appendix B Recommended IUGS names

The 316 rock names and terms that have been defined and recommended for use by the IUGS Subcommission, i.e. all those terms in the Glossary that have been printed in bold capital letters are listed below.

Of these names and terms, 179 are strictly speaking IUGS root names; 103 are subdivisons of these root names, including 33 specific names for the various "foid" root names, e.g. nepheline syenite; and 34 are rock terms.

Acid
Afrikandite (Africandite)
Agglomerate
Agpaite (Agpaitic)
Alaskite
Alkali
Alkali basalt
Alkali feldspar charnockite
Alkali feldspar granite
Alkali feldspar rhyolite
Alkali feldspar syenite
Alkali feldspar trachyte
Alvikite
Analcime basanite
Analcime diorite
Analcime gabbro
Analcime monzodiorite
Analcime monzogabbro
Analcime monzosyenite
Analcime phonolite
Analcime plagisyenite
Analcime syenite
Analcimite
Andesite
Andesitoid
Anorthosite
Ash, Ash grain
Ash tuff
Basalt
Basaltic andesite
Basaltic trachyandesite
Basaltoid
Basanite
Basanitic foidite
Basic

Beforsite
Benmoreite
Block
Bomb
Boninite
Calciocarbonatite
Calcite-carbonatite
Camptonite
Cancrinite diorite
Cancrinite gabbro
Cancrinite monzodiorite
Cancrinite monzogabbro
Cancrinite monzosyenite
Cancrinite plagisyenite
Cancrinite syenite
Carbonatite
Charno-enberbite
Charnockite
Clinopyroxene norite
Clinopyroxenite
Coarse (ash) grain
Coarse (ash) tuff
Colour index (Colour ratio)
Comendite
Comenditic rhyolite
Comenditic trachyte
Crystal tuff
Dacite
Dacitoid
Diabase
Diorite
Dioritoid
Dolerite
Dolomite-carbonatite
Dunite

Dust grain
Dust tuff
Enderbite
Essexite
Fergusite
Ferrocarbonatite
Fine (ash) grain
Fine (ash) tuff
Foid diorite
Foid dioritoid
Foid gabbro
Foid gabbroid
Foid monzodiorite
Foid monzogabbro
Foid monzosyenite
Foid plagisyenite
Foid syenite
Foid syenitoid
Foid-bearing alkali feldspar syenite
Foid-bearing alkali feldspar trachyte
Foid-bearing anorthosite
Foid-bearing diorite
Foid-bearing gabbro
Foid-bearing latite
Foid-bearing monzodiorite
Foid-bearing monzogabbro
Foid-bearing monzonite
Foid-bearing syenite
Foid-bearing trachyte
Foidite
Foiditoid
Foidolite
Gabbro
Gabbroid
Gabbronorite
Granite
Granitoid
Granodiorite
Harzburgite
Haüyne basanite
Haüyne phonolite
Haüynite
Hawaiite
High-K
Hololeucocratic
Holomelanocratic
Hornblende gabbro
Hornblende peridotite

Hornblende pyroxenite
Hornblendite
Hyalo-
Igneous rock
Ijolite
Intermediate
Italite
Jotunite
Kalsilitite
Kamafugite
Kersantite
Kimberlite
Komatiite
Kugdite
Lamproite
Lamprophyre
Lapilli
Lapilli tuff
Lapillistone
Latite
Leucite basanite
Leucite phonolite
Leucite tephrite
Leucitite
Leuco-
Leucocratic
Lherzolite
Limburgite
Liparite
Lithic tuff
Low-K
m-Charnockite, m-Enderbite etc.
Magnesiocarbonatite
Malignite
Mangerite
Medium-K
Meimechite (Meymechite)
Mela-
Melanephelinite
Melanocratic
Melilite leucitite
Melilite nephelinite
Melilitite
Melilitolite
Melteigite
Mesocratic
Miaskite (Miascite)
Minette

Missourite
Monchiquite
Monzodiorite
Monzogabbro
Monzogranite
Monzonite
Mugearite
Natrocarbonatite
Nepheline basanite
Nepheline diorite
Nepheline gabbro
Nepheline monzodiorite
Nepheline monzogabbro
Nepheline monzosyenite
Nepheline plagisyenite
Nepheline syenite
Nepheline tephrite
Nephelinite
Nephelinolite
Norite
Nosean basanite
Noseanite
Obsidian
Okaite
Olivine clinopyroxenite
Olivine gabbro
Olivine gabbronorite
Olivine hornblende pyroxenite
Olivine hornblendite
Olivine melilitite
Olivine norite
Olivine orthopyroxenite
Olivine pyroxene hornblendite
Olivine pyroxenite
Olivine websterite
Olivinite
Opdalite
Orthopyroxene gabbro
Orthopyroxenite
Pantellerite
Pantelleritic rhyolite
Pantelleritic trachyte
Peralkaline
Peralkaline granite
Peralkaline phonolite
Peralkaline rhyolite
Peralkaline trachyte
Peridotite

Pheno-
Phonolite
Phonolitic basanite
Phonolitic foidite
Phonolitic leucitite
Phonolitic nephelinite
Phonolitic tephrite
Phonolitoid
Phonotephrite
Picrite
Picrobasalt
Pitchstone
Plagioclase-bearing hornblende pyroxenite
Plagioclase-bearing hornblendite
Plagioclase-bearing pyroxene hornblendite
Plagioclase-bearing pyroxenite
Plagiogranite
Plutonic
Potassic melilitite
Potassic olivine melilitite
Potassic trachybasalt
Pyroclastic
Pyroclastic breccia
Pyroclastic deposit
Pyroclastic rock
Pyroclasts
Pyroxene hornblende gabbro
Pyroxene hornblende gabbronorite
Pyroxene hornblende norite
Pyroxene hornblende peridotite
Pyroxene hornblendite
Pyroxene peridotite
Pyroxenite
Quartz alkali feldspar syenite
Quartz alkali feldspar trachyte
Quartz anorthosite
Quartz diorite
Quartz gabbro
Quartz latite
Quartz monzodiorite
Quartz monzogabbro
Quartz monzonite
Quartz norite
Quartz syenite
Quartz trachyte
Quartz-rich granitoid
Quartzolite
Rhyolite

Rhyolitoid
Sannaite
Shonkinite
Shoshonite
Silicocarbonatite
Sodalite diorite
Sodalite gabbro
Sodalite monzodiorite
Sodalite monzogabbro
Sodalite monzosyenite
Sodalite plagisyenite
Sodalite syenite
Sodalitite
Sodalitolite
Sövite (Soevite)
Spessartite
Subalkali
Subalkali basalt
Syenite
Syenitoid
Syenodiorite
Syenogabbro
Syenogranite
Tephra
Tephriphonolite
Tephrite
Tephritic foidite

Tephritic leucitite
Tephritic phonolite
Tephritoid
Teschenite (Teschinite)
Theralite
Tholeiitic basalt
Tonalite
Trachyandesite
Trachybasalt
Trachydacite
Trachyte
Trachytoid
Troctolite
Trondhjemite
Tuff
Tuff breccia
Tuffite
Turjaite
Ultrabasic
Ultramafic rock
Uncompahgrite
Urtite
Vitric tuff
Vogesite
Volcanic
Websterite
Wehrlite

Appendix C IUGSTAS software package

By R.W. Le Maitre

In response to several requests, this edition includes a description of a collection of C++ routines for implementing the TAS classification for volcanic rocks (p.33–39). As parts of the TAS classification require the use of the CIPW norm calculation, this code is also included.

The package is only intended as a **basic development kit** with which to write other programs suited to specific needs.

The source code is not included in the book but can be downloaded from the Cambridge University Press website – see section C.5, p.236.

C.1 INTRODUCTION

Wherever possible the code has been written in a style which is largely self-explanatory so that readers who are not experts in programming techniques should be able to follow what is happening and get the package working without too much trouble.

C.1.1 Data input

The package expects to read analyses in **tab-delimited format** in which each item in a row or line is separated from the next by a tab character (entered by the tab key from the keyboard). Each analysis is then one row of a table. Such data can easily be produced from virtually all spreadsheet and word-processing programs, by saving the data in text-only format, and database programs by saving the data in tab-delimited format. The first item of each row must be the specimen name, which can be up to 128 characters long, followed by up to 26 oxide values.

If the first row of the table contains a **header row** (the oxide names) they can be in **any order** but they must be spelt as given in Table C.1 from rows 1 to 26. However whether the text is upper or lower case is unimportant. This feature is useful for reading data from files in which the data is not in the default order required by IUGSTAS.

If an invalid name is found a warning message is shown and the column is ignored. If a valid oxide name occurs more than once, an error message will be reported and the program will abort. Any oxides not specified in the header row will be set to zero.

If the table does not contain a header row (i.e. the first row of a table is an analysis) the package assumes that the oxides will be in the order shown in Table C.1, but not all of them need be given, i.e. they will be oxides 1 to n where n is not greater than 26.

C.1.2 Data output

The results are written to files that can be imported directly back into spreadsheet, database and word-processing programs for further editorial changes. They are of three types:

(1) A simple half-page table containing the analysis and CIPW norm (Table C.3). As the values are spaced with blank characters this table is not ideally suitable for further spreadsheet or database work. This is, therefore, of somewhat limited use except for demonstration purposes.

(2) A tab-delimited format table with a header row and with the analyses in horizontal rows. The order of the items in the row is specimen name, oxides 1 to 29 followed by the normative values 1 to 46 as given in

Table C.1. *List of oxide names and normative values. Normative values preceded by an * were not included in the original CIPW method of calculation. If the input data does not contain a header row the oxides are assumed to be in the order given although not all 26 need be present*

Oxide order		Normative mineral order and composition		
1	SiO_2	1	*Q*	Quartz	SiO_2
2	TiO_2	2	*C*	Corundum	Al_2O_3
3	Al_2O_3	3	*Z*	Zircon	$ZrO_2.SiO_2$
4	Fe_2O_3	4	*or*	Orthoclase	$K_2O.Al_2O_3.6SiO_2$
5	FeO	5	*ab*	Albite	$Na_2O.Al_2O_3.6SiO_2$
6	MnO	6	*an*	Anorthite	$CaO.Al_2O_3.2SiO_2$
7	MgO	7	*ne*	Nepheline	$Na_2O.Al_2O_3.2SiO_2$
8	CaO	8	*lc*	Leucite	$K_2O.Al_2O_3.4SiO_2$
9	Na_2O	9	*kp*	Kaliophilite	$K_2O.Al_2O_3.2SiO_2$
10	K_2O	10	*hl*	Halite	NaCl
11	P_2O_5	11	*nc*	Sodium Carbonate	$Na_2O.CO_2$
12	H_2O+	12	*ac*	Acmite	$Na_2O.Fe_2O_3.4SiO_2$
13	H_2O-	13	*ns*	Sodium metasilicate	$Na_2O.SiO_2$
14	CO_2	14	*ks*	Potassium metasilicate	$K_2O.SiO_2$
15	Other	15	*di*	Diopside	$CaO.(Mg,Fe)O.2SiO_2$
16	ZrO_2	16	*hy*	Hypersthene	$(Mg,Fe)O.SiO_2$
17	Cr_2O_3	17	*wo*	Wollastonite	$CaO.SiO_2$
18	V_2O_3	18	*ol*	Olivine	$2(Mg,Fe)O.SiO_2$
19	NiO	19	*cs*	Dicalcium silicate (larnite)	$2CaO.SiO_2$
20	CoO	20	*cm*	Chromite	$FeO.Cr_2O_3$
21	BaO	21	*hm*	Hematite	Fe_2O_3
22	SrO	22	*mt*	Magnetite	$FeO.Fe_2O_3$
23	Rb_2O	23	*il*	Ilmenite	$FeO.TiO_2$
24	F	24	*tn*	Sphene	$CaO.TiO_2.SiO_2$
25	Cl	25	*pf*	Perovskite	$CaO.TiO_2$
26	S	26	*ru*	Rutile	TiO_2
27	Total	27	*ap*	Apatite	$3(3CaO.P_2O_5).CaF_2$
28	Ox.Eq	28	**hap*	Hydroxyapatite	$3(3CaO.P_2O_5).Ca(OH)_2$
29	TAlk	29	*fr*	Fluorite	CaF_2
		30	*pr*	Pyrite	FeS_2
		31	*cc*	Calcite	$CaO.CO_2$
		32	**mag*	Magnesite	$MgO.CO_2$
		33	**sid*	Siderite	$FeO.CO_2$
		34	**H2O+*		
		35	**H2O-*		
		36	**Other*		
		37	**Fsd*	Final silica deficiency	
		38	**P2O5+*	any excess P_2O_5	
		39	**CO2+*	any excess CO_2	
		40	**Cr2O3+*	any excess Cr_2O_3	
		41	**F+*	any excess F	
		42	**Cl+*	any excess Cl	
		43	**S+*	any excess S	
		44	*Total*	sum of 1 to 43 above	
		45	*MgFe*	$100MgO/(MgO + FeO)$	
		46	**SiUnd*	Silica Undersaturation	

Table C.1. This is ideal for further spreadsheet or database investigation.

(3) A tab-delimited format table with the analyses in vertical columns as shown in Table C.4, p. 232. This format is almost ready for publication and can be finalized with most word-processors. The user has control of the number of columns per page and whether rows with all zero values are to be included or not. For further details see routine WriteAsVertTable() on p. 233.

C.1.3 Error checking

A reasonable amount of error checking is built into the package so that if, for example, you try to write to an output file that has not been previous defined an error message is printed and execution aborts (stops). Similarly, attempts to use a file that is already in use elsewhere will cause execution to abort.

If any of the analytical data contain unexpected characters the row and column in which the error occurs is output and execution is aborted.

C.1.4 Supplied tasks

Certain built-in tasks are supplied as ready-to-use routines. These have all been built into an interactive program in file "Main.cpp" so that you can experiment with them. The built-in tasks are:

TASNamesTest() – this simple task is "hardwired" to read from a set of test analyses supplied in file "TestTAS" which contain examples of analyses from all the TAS root names. The specimen name already attached to each analysis is the TAS root name.

When executed the task determines their TAS names, and outputs to separate files the results with the CIPW norms in the three possible types of table described above. To help you become more familiar with the package the

code of this task is explained in greater detail in the next section.

It is recommended to run this task after you have installed the package to see that everything is working correctly – if it is the name given by the task should be the same as that already present.

TASNames() – this is a more flexible version of the previous task as it requests the following input from the user:

(1) How the analysis is to be recalculated and how CO_2 is to be handled by the CIPW norm calculation. For details of these options see the description of routines Recalculate() and CalcCIPW() on p. 233.

(2) The name of the file from which to read the data.

(3) What types of table are to be output. This is done by asking the user to enter **up to three** characters which must be **S** for a simple half-page table, **H** for a table with the analyses in **h**orizontal rows or **V** for a table with the analyses in **v**ertical columns. The output files use the same name as the input file with the characters "_S", "_H" or "_V" appended to the end, but before any extension. For example input files "Data.txt" and "MoreData" will produce output files such as "Data_H.txt" and "MoreData_S", respectively.

TASNameInteractive() – this is a simple interactive task in which all the data is entered by the user from the keyboard and the result is output to the computer monitor as a simple half-page table. The entered data can then be edited to see what effect it has on the TAS name or the CIPW norm. As in task TASNames(), the user has control over how the analysis is recalculated and CO_2 is allocated during the CIPW norm calculation.

RandomCIPWTest (maxBadTotals) – a fun task to test the CIPW norm calculation. It simple generates random analyses, calculates their CIPW norms and outputs results only if the analysis and normative totals are not the same. Execution stops when the number of bad totals exceeds maxBadTotals or the user aborts execution of the task.

After you have read the next two sections it is recommended that you study the code of the four tasks so that you can modify them to suit you own needs.

C.2 GETTING STARTED WITH C++

The package consists of six text files, which must be compiled and linked by a C++ compiler before you can run the program. They are "IUGSTAS.cpp" and "IUGSTAS.h" which contain the code for the basic low-level routines for doing the hard work. The reader need

only be familiar with a few of these routines which are described in section C.3. Files "Task.cpp" and "Task.h" are a set of higher-level routines built from the routines contained in "IUGSTAS.cpp" for performing the specific tasks described in the previous section. File "Main.cpp" contains the main interactive program which will run any of the tasks taken from "Task.cpp".

A set of analyses in tab-delimited format is also supplied in file "TestTAS" for routine TASNamesTest() to check that the package is working correctly.

As C++ may be unfamiliar to many petrologists, the way in which it can be implemented is briefly described below using the code from some of the tasks supplied with the package.

Table C.2. *Example of the C++ code required in task TASNamesTest() to calculate the TAS name of a series of volcanic rock analyses stored in tab delimited format in a file called "TestTAS". The results are output in each of the three available output formats to separate files named "Table1", "Table2" and "Table3"*

```
void TASNamesTest()
{
    IUGSTAS obj1;                                  //line 1

    obj1.OpenInputFile("TestTAS");                 //line 2
    obj1.OpenOutputFile("Table1", 1);              //line 3
    obj1.OpenOutputFile("Table2", 2);              //line 4
    obj1.OpenOutputFile("Table3", 3);              //line 5

    do {                                           //line 6
        obj1.Init();                               //line 7
        obj1.ReadTabDelimitedRow();                //line 8
        if(obj1.EndOfFile()) break;                //line 9
        obj1.GetTASName(true, kCO2AsCaMgFeCarb);   //line 10
        obj1.WriteTable(1);                        //line 11
        obj1.WriteAsVertTable(2, 7, false);        //line 12
        obj1.WriteAsHorzTable(3);                  //line 13
    } while(true);                                 //line 14

    obj1.CloseInputFile();                         //line 15
    obj1.CloseOutputFile(1);                       //line 16
    obj1.CloseOutputFile(2);                       //line 17
    obj1.CloseOutputFile(3);                       //line 18
}
```

One of the simplest, which can be used as a starting point for many more, is TASNamesTest(), the code of which is shown in Table C.2. This simply reads analyses from a tab-delimited file, determines the TAS names of the analyses, and writes the results in three different formats to separate files. For further details of the routines see section C.3.

Like many computer languages C++ is case sensitive to all names so that the spelling must be **exactly** as shown. Likewise with the various types of brackets.

Creating an Object – the first thing that has to be done in writing any task is to create an object of type IUGSTAS, as without it none of the code or storage required can be accessed. This is done in line 1 with the instruction:

```
IUGSTAS      obj1;
```

which creates an object named obj1. Although it is possible to create an object in other ways, this is the simplest as the object is automatically destroyed when the task is finished. If appropriate to the task in hand you can create as many objects as you like but they must all have different names.

Defining an Input File – next the input file must be defined from which to read the analyses. This can either be from the keyboard (tedious and prone to mistakes) or from a file stored on disk. In the example the analyses are read from the file "TestTAS" in line 2:

```
obj1.OpenInputFile("TestTAS.txt");
```

Note that to use any of the routines in the package the object name followed by a "." character must precede the routine name as shown. As currently written only one input file can be open for each object.

Defining the Output Files – next output files must be defined to store the results. This is done in lines 3, 4 and 5 as shown:

```
obj1.OpenOutputFile("Table1", 1);
obj1.OpenOutputFile("Table2", 2);
obj1.OpenOutputFile("Table3", 3);
```

The package currently supports up to six out-put files. To change this number see p.231 under the description of routine OpenOutputFile().

Reading the Analyses – this is usually done within a do loop (lines 6 and 14) which simply instructs the computer to repeatedly perform all the code between the two lines until instructed to jump out of the loop (in this example the **break** command). Lines 7 to 9 are nearly always together in a task:

```
obj1.Init();
obj1.ReadTabDelimitedRow();
if(obj1.EndOfFile()) break;
```

and initialize the object, read one analysis, and check to see if there is no more data to read respectively. If no more is present, i.e. routine EndOfFile() returns **true**, the break command is executed which causes control to jump out of the do loop to line 15.

Doing some Work – the only calculation performed in this task is in line 10:

```
obj1.GetTASName(kWaterCO2Free,
                kCO2AsNaCarb);
```

where the TAS name and CIPW norm are calculated. See p. 233 for the meaning of the two parameters under the description of routines Recalculate() and CalcCIPW().

Writing the Results – each analysis is then written by lines 11, 12 and 13:

```
obj1.WriteTable(1);
obj1.WriteAsVertTable(2, 7, false);
obj1.WriteAsHorzTable(3);
```

to the three output files in different table formats.

Tidying up – although lines 15 to 18:

```
obj1.CloseInputFile();
obj1.CloseOutputFile(1);
obj1.CloseOutputFile(2);
obj1.CloseOutputFile(3);
```

are not strictly required for a simple task, it is good practice always to close any files that have been opened. This enables several tasks to be executed in succession.

Directly Accessing Values – all the 29 oxide

and 46 normative values shown in Table C.1 can be directly accessed with the following type of syntax:

```
obj1.oxide.named.MgO
obj1.norm.named.ab
```

Note that the names used to access oxide or normative values must be exactly as given in the structures named "SOxideNames" and "SNormNames" in file IUGSTAS.h". For an example of this see task `RandomCIPWTest()` in file "Tasks.cpp" which uses this syntax to directly set the oxide values of an analysis.

As another example you could create a task that would write analyses with normative $Q > 0$ to one file, those with normative $ne > 0$ to another and those with no normative Q or ne to another with the following type of code:

```
if(obj1.norm.named.Q > 0) {
    WriteAsHorzTable(1);
}else if(obj1.norm.named.ne > 0){
    WriteAsHorzTable(2);
}else {
    WriteAsHorzTable(3);
}
```

Executing a Task – to use any of the tasks supplied in "Tasks.cpp" or any that you have created, simply place their names into the program main in file "Main.cpp" as shown below:

```
int main()
{
    MyNewTask();
    return 0;
}
```

If you have closed all the files properly you can also execute any other tasks by adding them to main.

C.3 USEFUL ROUTINES

The following routines are usually the only ones required for most tasks. The others present in "IUGSTAS.cpp" are used internally by these routines. Note, however, that several routines have more than one form which may be useful for experienced programmers to perform special tasks – such routines are preceded by an "*".

C.3.1 Input routines

Init() – initializes all the storage required for the object and **must be called before** each analysis is read and stored within the object. This is usually done by calling routine ReadTabDelimitedRow().

AskForInputFile() – this interactive routine asks the user to enter the name of the file to be opened for input from the keyboard. If the file cannot be opened, because it does not exist or it is already open in another program, the user is asked for another name until a file is opened. If the user enters the return key before entering any text the program quits.

OpenInputFile("aInputFile") – opens for input the file named between the pair of quotes, which must be present as they are part of the argument. This routine is used internally by routine AskForInputFile().

CloseInputFile() – this simply closes the input file. Although not necessary for many tasks, it is useful if you wish to use the same object to read from more than one input file. In other words you can append sets of data sequentially.

ReadTabDelimitedRow() – reads one analysis from the input file and stores it in the object. It automatically checks to see if the tab-delimited data has a header row the first time it is called after the input file has been opened. If any of the analytical values have any invalid characters in them, a warning message is shown together with the row and column in which it occurs. Execution is then aborted.

EndOfFile() – this should be called immediately after a call to ReadTabDelimitedRow() to check that an analysis has been read. If it has the routine returns the value **false**, but if no analysis has been read because the end of file

Table C.3. *Example of a simple half-page table output by routine WriteTable(), using routine CalcCIPW() to calculate the norm of a hypothetical analysis*

```
The specimen name

SiO2    50.00   ZrO2     0.20    Q    1.26   wo   0.00   sid     0.00
TiO2     1.00   Cr2O3    0.10    C    0.00   ol   0.00   H2O+    0.50
Al2O3   12.00   V2O3     0.10    Z    0.30   cs   0.00   H2O-    0.50
Fe2O3    2.00   NiO      0.10    or  12.17   cm   0.15   Other   0.01
FeO      5.00   CoO      0.10    ab  21.69   hm   0.00   Fsd     0.00
MnO      0.05   BaO      0.10    an  15.22   mt   3.05   P2O5+   0.00
MgO     10.00   SrO      0.10    ne   0.00   il   1.90   CO2+    0.00
CaO      9.00   Rb2O     0.10    lc   0.00   tn   0.00   Cr2O3+  0.00
Na2O     3.00   F        1.50    kp   0.00   pf   0.00   F+      0.00
K2O      2.00   Cl       0.50    hl   0.82   ru   0.00   Cl+     0.00
P2O5     0.50   S        0.50    nc   0.00   ap   1.19   S+      0.00
H2O+     0.50   Total   99.96    ac   0.00   hap  0.00   Total  99.09
H2O-     0.50   Ox.Eq   -0.87    ns   0.00   fr   3.02   MgFe   86.47
CO2      1.00                    ks   0.00   pr   0.94
Other    0.01   TAlk     5.00    di   7.70   cc   2.29   SiUnd   0.00
                                 hy  26.39   mag  0.00
```

has been reached it returns **true**.

Note: it is highly recommended that you use this routine to check for the end of file as it also keeps track of how many analyses have been read and written to the output files. It is also responsible for printing a progress message every time a certain number of analyses are read. This number is currently set to 10 but can easily be changed by editing the following line of code in file "IUGSTAS.h":

```
const double reportEveryNth = 10.0;
```

to any other number you prefer.

UserAbort() – this routine is useful to include in certain do loops to detect if an abort sequence has been typed on the keyboard, e.g. Command + period for the Mac OS, Control + c for Windows. It is used in task TASNames() in case the user wishes to abort before reading to the end of the file and in task Random-CIPWTest() to stop execution – the routine EndOfFile() cannot be used in this case as no data is being read from a file. As it is not possible to detect abort sequences in standard C++ this code is machine dependant.

C.3.2 Output routines

AskForOutputTables() – this interactive routine asks the user to enter up to three characters to define the types of tables to be written, i.e. S, H or V for the tables written by routines WriteTable(), WriteAsHorzTable() and WriteAsVertTable() respectively.

OpenOutputFile("anOutputFile", nthFile) – this opens the file specified by the first argument as the nthFile output file. As supplied the value of nthFile must be an integer between 1 and 6. However, if more (or less) are required simply edit the following line of code in file "IUGSTAS.h":

```
const  int  numOutputFiles = 6;
```

to reflect the number you require.

CloseOutputFile(nthFile) – closes the nthFile output file.

CloseAllOutputFiles() – this closes all output files that have been opened.

*WriteTable(nthFile) – this writes to the nthFile output file a single analysis and its normative values as a simple half-page table as

Table C.4. *Example of a vertical table output by routine WriteAsVertTable() which can be imported directly into word-processors or spreadsheet programs as tab-delimited text files almost ready for publication. All the analyses are numbered sequentially. The user controls the number of columns across a page and whether rows with all zero values are to be omitted, as specified in this example*

	1	2	3	4	5	6	7
SiO2	47.37	50.84	53.87	58.70	67.50	73.51	71.35
TiO2	1.69	2.33	1.10	0.88	0.59	0.17	0.61
Al2O3	15.26	14.98	17.59	17.24	16.15	13.97	7.43
Fe2O3	3.60	4.08	3.28	3.31	2.47	0.59	1.96
FeO	6.95	8.32	6.91	4.09	2.33	1.16	6.79
MnO	0.17	0.18	0.43	0.14	0.09	0.05	0.36
MgO	10.85	7.53	3.21	3.37	1.81	0.20	0.00
CaO	9.49	9.59	10.73	6.88	4.38	1.24	0.60
Na2O	3.56	1.75	2.04	3.53	3.85	3.35	6.63
K2O	0.84	0.22	0.86	1.64	0.68	5.64	4.26
P2O5	0.22	0.16	0.00	0.21	0.15	0.12	0.00
Total	100.00	100.00	100.00	100.00	100.00	100.00	100.00
TAlk	4.40	1.96	2.89	5.17	4.53	8.99	10.90
CIPW Norm							
Q	0.00	7.75	10.54	12.48	30.21	29.17	31.94
C	0.00	0.00	0.00	0.00	1.48	0.39	0.00
or	4.99	1.28	5.07	9.70	4.02	33.34	25.19
ab	22.27	14.78	17.24	29.83	32.55	28.34	14.46
an	23.18	32.41	36.31	26.38	20.76	5.35	0.00
ne	4.25	0.00	0.00	0.00	0.00	0.00	0.00
ac	0.00	0.00	0.00	0.00	0.00	0.00	5.67
ns	0.00	0.00	0.00	0.00	0.00	0.00	8.20
di	17.89	11.37	14.05	5.13	0.00	0.00	2.66
hy	0.00	21.68	9.96	9.52	5.95	1.95	10.72
ol	18.48	0.00	0.00	0.00	0.00	0.00	0.00
mt	5.22	5.92	4.75	4.80	3.58	0.86	0.00
il	3.21	4.43	2.09	1.67	1.12	0.32	1.16
hap	0.53	0.39	0.00	0.50	0.36	0.28	0.00
H2O+	-0.01	-0.01	0.00	-0.01	-0.01	-0.01	0.00
Total	100.00	100.00	100.00	100.00	100.00	100.00	100.00
MgFe	82.92	74.57	53.93	75.48	80.34	31.31	0.00
SiUnd	10.93	0.00	0.00	0.00	0.00	0.00	0.00

```
Specimen Names
1) Lava 374 = Basalt (B) [Alkali]
2) Scoria 372 = Basalt (B) [Subalkali]
3) Ash 1956 = Basaltic Andesite (O1) [medium-K]
4) Ash 1873 = Andesite (O2) [medium-K]
5) Dome lava = Dacite (O3) [low-K]
6) Plug 73/43 = Rhyolite (R) [high-K]
7) Obsidian = Rhyolite (R) [Pantelleritic]
```

Note: because the analyses have been recalculated and printed to only two decimal places, some totals may not appear to be the sum of their component values, e.g. TAlk in analyses 2, 3 and 7 are not equal to the sum of $Na_2O + K_2O$ as displayed.

shown in Table C.3.

*WriteAsHorzTable(nthFile) – this writes to the nthFile output file in a tab-delimited format the analyses and normative values in horizontal rows. The table has a header row and the order of the columns is specimen name, oxides 1 to 28 followed by the normative values 1 to 46 as given in Table C.1.

*WriteAsVertTable (nthFile, numCols, printAll) – this writes to the nthFile output file in a tab-delimited format the analyses and normative values in vertical columns as shown in Table C.4. The parameter numCols defines the number of columns per page. As supplied the package has a maximum of 10 columns per page. However, this can be changed by editing the following line of code in file "IUGSTAS.h":

```
const  int  numTableCols = 10;
```
to any suitable number. If the parameter printAll is true all 29 oxide values and 46 normative values are output; if printAll is false any row of the table with all zero values is omitted, as in Table C.4.

WriteTables() – this writes the data to the tables as specified in the call to routine AskForOutputTables().

C.3.3 Calculation routines

*Recalculate(recalcMode) – this recalculates the analysis in a manner which depends upon which of three built-in constants is specified:

kWaterCO2Free – H_2O+, H_2O- and CO_2 are ignored and set to zero before the analysis is recalculated to 100%. This is the procedure recommended by the Subcommission to be adopted before determining the TAS name of an analysis

kWithWaterCO2 – the entire analysis is recalculated to 100%

kDoNotRecalculate – the analysis is not recalculated but is left as it is.

This routine is called internally by routine CalcCIPW().

*CalcCIPW(recalcMode, CO2Usage) – this calculates the CIPW norm of the currently stored analysis after it has been recalculated by routine Recalculate() using the parameter recalMode. The parameter CO2Usage determines how any CO_2 is allocated in the calculation. This can be in one of three ways depending upon which of the three built-in constants is specified:

kCO2AsOthers – CO_2 is added to *Others* and no carbonates are calculated

kCO2AsNaCarb – CO_2 is calculated as Na_2CO_3 (*nc*). If there is an excess of CO_2 this is allocated as for kCO2AsCaMgFeCarb

kCO2AsCaMgFeCarb – CO_2 is calculated first as $CaCO_3$ (*cc*), then if there is CO_2 left over as $MgCO_3$ (*mag*) and then if necessary as $FeCO_3$ (*sid*). Any excess CO_2 (a theoretical possibility) is then reported as *CO2+*.

If the analysis total, corrected for oxygen equivalent if Cl, F or S is present, differs from the normative total by more than 0.0001 the routine returns false, otherwise it returns true.

*GetTASName(recalcMode, CO2Usage) – before each analysis is named the CIPW norm is calculated by routine CalcCIPW() using the two parameters recalcMode and CO2Usage.

Note that any specific changes the user wishes to make to the analyses, such as adjusting the ratio of FeO to Fe_2O_3, must be done before calling this routine.

It then determines the TAS name for the analysis currently stored in the object and appends it to the user supplied specimen name in the following format:

```
lava 5 = Rhyolite (R) [Comenditic]
```
i.e. the user supplied specimen name = the TAS root name (the TAS field number) [any optional additional name] as can be seen in Table C.4.

C.4 THE CIPW NORM CALCULATION

The CIPW norm calculation was originally introduced as a method of classifying igneous rocks (Cross *et. al.*, 1902) and although the classification was never widely used the calculation became extremely popular. An excellent review of this classification method and a comparison with many others classifications can be found in Johannsen (1931, p.83–99). This also contains an excellent description of how to perform the calculation taken almost entirely from Washington (1917). It also contains several worked examples of a wide variety of igneous rock compositions.

The basic principles of the method are simple:
(1) The weight % oxides are converted to molecular proportions by dividing them by their molecular weights.
(2) The molecular proportions are then combined into a set of provisional normative minerals in a **specific order**.
(3) The amount of SiO_2 that is required to satisfy this allocation of provisional normative minerals is then calculated. If enough is present, the excess is reported as quartz (*Q*) and the provisional minerals become permanent. If not enough silica is present the provisional normative minerals are desilicated into related normative minerals, again in a **specific order,** until exactly enough SiO_2 is found to be present.
(4) Finally, the proportions of the normative minerals are then converted to weight % by multiplying them by their molecular weight.

Mathematically this procedure is a linear transformation of the analytical data into a new set of normative minerals (Le Maitre, 1982). However, the linearity is only true while the set of normative minerals remains the same. Once it changes a new linearity appears.

Whereas a **mode** of a rock is the amount of minerals **actually present** in a rock, the **norm** is the amount of minerals that **could theoretically be present**.

C.4.1 Problems

One of the complaints of the original CIPW norm calculation is that the analysis total rarely agrees with the total of the normative minerals. This is due to several causes:
(1) The calculation was done manually, often with the aid of tables such as those found in Johannsen (1931, p.295–308). This introduced considerable rounding-off errors as the molecular proportions were only carried to three decimal places and it was recommended that values less than 0.002 be ignored. Calculating the normative minerals from their molecular proportions then lead to the final values being in fixed increments. For example, orthoclase would appear in increments of 0.556 or 0.001 * mol. wt. of orthoclase (556). Rounded off to two decimal places, this resulted in values such as 2.22, 2.78, 3.34 etc., but nothing in between. Fortunately this method rarely has to be used today.
(2) Although the molecular weights of Mg and Fe-bearing normative minerals were adjusted for the amount of Fe^{2+} replacing Mg, with other substitutions, such as Sr and Ba replacing Ca, no such corrections were made to the Ca-bearing molecules.
(3) Apatite was calculated in an extremely loose manner which could result in some P_2O_5 not being allocated if small amounts of F were present. If no F was present the apatite was made only from CaO and P_2O_5, but when the weight % of apatite was calculated it was treated as if it contained F. The normative total was, therefore, in excess by the amount of F that should have been present.
(4) If S, Cl or F was present in the analysis, the

total was rarely corrected for the oxygen equivalent as it should have been.

C.4.2 IUGSTAS CIPW norm

With the advent of computers all these problems became solvable. Two of the earliest computer programs for performing the CIPW calculation were written in the early 1960s (Kelsey, 1965; Hey & Le Maitre, 1966) which considerable reduced the tedium of the manual calculation. Kelsey gives an excellent description of the procedures involved and produced a better way of desilicating the normative minerals when required.

For the first time it also allowed the norm calculation to be performed in a mathematically correct manner. For example, element substitutions were all taken into account when calculating molecular weights.

This resulted in the analysis total and normative total always being exactly the same. Note, however, that if you add up all the normative minerals (reported to two decimal places) you will often find that this total is not the same as the total displayed by the computer because of rounding-off errors (e.g. see Table C.4).

The main differences between the CIPW norm code presented here and the original version are as follows:

(1) Only the atomic weights of the elements are defined and the values used are not rounded off to the nearest integer. All the required molecular weights and correction factors are calculated from the atomic weights to minimize rounding-off errors. Mathematically it also means that if the atomic weights are given random values and the norm calculation has been programmed correctly, the analysis total will still be the same as the total of the normative minerals.

(2) The molecular weights of the normative

minerals are corrected for all substitutions, i.e. Rb for K; Sr and Ba for Ca; V for Fe^{3+}; Mn, Co and Ni for Fe^{2+}; and finally Fe^{2+}, with any Mn, Co and Ni, for Mg.

(3) Apatite (strictly speaking fluor-apatite) is calculated from the appropriate proportions of CaO, P_2O_5 and F. Any P_2O_5 left over is later calculated as hydroxyapatite (*hap* — not in the original CIPW norm) and the amount of H_2O+ is reduced by the amount that would be required. This can lead to negative values of *H2O+* if not enough was originally present (see examples in Table C.4). Theoretically P_2O_5 may also be in excess at this stage and if it is, it is reported in the norm as *P2O5+*.

(4) The user has much more control over how the CO_2 is allocated as explained under the description of routine CalcCIPW() on p.233.

(5) The logic of the original method did not allow for the excess of certain other (usually minor) molecular proportions which can occasionally occur. These are CO_2, Cr_2O_3, F, Cl and S which are reported in the norm as *CO2+*, *Cr2O3+*, *F+*, *Cl+* and *S+*, respectively.

(6) No other oxides can be in excess. What can happen, however, is that the amount of available SiO_2 may not be sufficient to make all the normative minerals required. If this happens a final silica deficiency (*Fsd*) is reported in the norm.

(7) If the analysis contains S, Cl or F the oxygen equivalent, reported as Ox.Eq in the Tables, is calculated. For an example see Table C.3 in which the sum of the analysis total less the oxygen equivalent is equal to the normative total.

(8) Finally, an extremely useful parameter called *SiUnd* (silica undersaturation) is given. This is the weight % of SiO_2 that would have to be added to the analysis to make quartz (*Q*) just appear in the norm —

remember that mathematically analyses do not have to add up to 100%. In other words *SiUnd* can be thought of as negative normative quartz which makes it an ideal parameter for plotting the status of silica saturation among any series of rocks which transgress over the boundaries of sets of normative minerals, e.g. a series a basaltic rocks which move from *Q*-normative space, through *ol-hy* space, to *ne*-normative space. Note that *SiUnd* is not an essential part of the normative calculation.

The code in IUGSTAS has evolved from the original version written in 1963 in Mercury Autocode through Fortran and Basic to C++. As would be expected the execution speed also increased dramatically from over 5 hours for 1000 calculations on the Ferranti Mercury computer to about 1 second on a Power Macintosh G3.

This code has been used extensively in the CLAIR data system (Le Maitre, 1973, 1976) and was instrumental in the original development of the TAS classification.

C.5 DOWNLOADING IUGSTAS

The IUGSTAS package can be downloaded from the Cambridge University Press website whose URL is given in the footnote below.[†]

Two version of the source code and compiled programs are available; one for the Mac OS and the other for Windows. The only differences are the machine-dependent UserAbort() routine in file "IUGSTAS.cpp" and the key

code for implementing an abort.

C.6 REFERENCES

CROSS, W., IDDINGS, J.P., PIRSSON, L.V. & WASHINGTON, H.S., 1902. A quantitative chemico-mineralogical classification and nomenclature of igneous rocks. *Journal of Geology. Chicago.* Vol.10, p.555–690.

HEY, M.H., LE MAITRE, R.W. AND BUTLER, B.C.M., 1966. A versatile computer program for the recalculation of rock and mineral analyses. *Mineralogical Magazine and Journal of the Mineralogical Society. London.* Vol. 35, p. 788.

JOHANNSEN, A., 1931. A Descriptive Petrography of the Igneous Rocks. *Chicago University Press.* Vol. 1, 267pp.

KELSEY, C.H., 1965. Calculation of the C.I.P.W. norm. *Mineralogical Magazine and Journal of the Mineralogical Society. London.* Vol. 34, p. 276-282.

LE MAITRE, R.W., 1973. Experiences with CLAIR: a computerised library of analysed igneous rocks. *Chemical Geology.* Vol. 12, p. 301-308.

LE MAITRE, R.W., 1976. Chemical variability of some common igneous rocks. *Journal of Petrology. Oxford.* Vol. 17, p. 589-637.

LE MAITRE, R.W., 1982. Numerical Petrology. *Elsevier, Amsterdam.* 281pp.

WASHINGTON, H.S., 1917. Chemical analyses of igneous rocks. *Professional Paper. United States Geological Survey. Washington.* No. 99, p.1–1201.

[†] The Cambridge University Press URL from which to download IUGSTAS is:
http://www.cambridge.org/052166215X